THEORY AND PROBLEMS OF COMBINATORICS

THEORY AND PROBLEMS OF COMBINATORICS

(Common to all P.G. & U.G. Courses:
M.C.A.; M.C.S.; M.Tech.; M.I.T.; M.Sc.; B.C.A.; B.C.S.; B.I.T.; B.E.; B.Tech.; B.Sc.)

(Common to all Indian Universities)

C VASUDEV

Associate Professor
Department of Mathematics
Kammavari Sangha Institute of Technology
Bangalore, Karnataka

Publishing Globally

NEW AGE INTERNATIONAL (P) LIMITED, PUBLISHERS
LONDON • NEW DELHI • NAIROBI
Bangalore • Chennai • Cochin • Guwahati • Hyderabad • Kolkata • Lucknow • Mumbai
Visit us at **www.newagepublishers.com**

Published by New Age International (P) Ltd., Publishers
First Edition: 2005
Reprint: 2020

GLOBAL OFFICES

- **New Delhi** **NEW AGE INTERNATIONAL (P) LIMITED, PUBLISHERS**
  7/30 A, Daryaganj, New Delhi-110002, (INDIA)
  Tel.: (011) 23253771, 23253472, **Telefax:** 23267437, 43551305
  E-mail: contactus@newagepublishers.com • Visit us at www.newagepublishers.com
- **London** **NEW AGE INTERNATIONAL (UK) LTD.**
  27 Old Gloucester Street, London, WC1N 3AX, UK
  E-mail: info@newacademicscience.co.uk • Visit us at www.newacademicscience.co.uk
- **Nairobi** **NEW AGE GOLDEN (EAST AFRICA) LTD.**
  Ground Floor, Westlands Arcade, Chiromo Road (Next to Naivas Supermarket)
  Westlands, Nairobi, KENYA, **Tel.:** 00-254-713848772, 00-254-725700286
  E-mail: kenya@newagepublishers.com

BRANCHES

- **Bangalore** 37/10, 8th Cross (Near Hanuman Temple), Azad Nagar, Chamarajpet, Bangalore- 560 018
  Tel.: (080) 26756823, **Telefax:** 26756820, **E-mail: bangalore@newagepublishers.com**
- **Chennai** 26, Damodaran Street, T. Nagar, Chennai-600 017, **Tel.:** (044) 24353401, **Telefax:** 24351463
  E-mail: chennai@newagepublishers.com
- **Cochin** CC-39/1016, Carrier Station Road, Ernakulam South, Cochin-682 016, **Tel.:** (0484) 2377303, **Telefax:** 4051304
  E-mail: cochin@newagepublishers.com
- **Guwahati** Hemsen Complex, Mohd. Shah Road, Paltan Bazar, Near Starline Hotel, Guwahati-781 008,
  Tel.: (0361) 2513881, **Telefax:** 2543669, **E-mail: guwahati@newagepublishers.com**
- **Hyderabad** 105, 1st Floor, Madhiray Kaveri Tower, 3-2-19, Azam Jahi Road, Near Kumar Theater, Nimboliadda
  Kachiguda, Hyderabad-500 027, **Tel.:** (040) 24652456, **Telefax:** 24652457
  E-mail: hyderabad@newagepublishers.com
- **Kolkata** RDB Chambers (Formerly Lotus Cinema) 106A, 1st Floor, S N Banerjee Road, Kolkata-700 014
  Tel.: (033) 22273773, **Telefax:** 22275247, **E-mail: kolkata@newagepublishers.com**
- **Lucknow** 16-A, Jopling Road, Lucknow-226 001, **Tel.:** (0522) 2209578, 4045297, **Telefax:** 2204098
  E-mail: lucknow@newagepublishers.com
- **Mumbai** 142C, Victor House, Ground Floor, N.M. Joshi Marg, Lower Parel, Mumbai-400 013
  Tel.: (022) 24927869, **Telefax:** 24915415, **E-mail: mumbai@newagepublishers.com**
- **New Delhi** 22, Golden House, Daryaganj, New Delhi-110 002, **Tel.:** (011) 23262368, 23262370, **Telefax:** 43551305
  E-mail: sales@newagepublishers.com

ISBN: 978-81-224-1705-0
C-19-06-12136

Printed in India at A.P. Offset, Delhi.
Typeset at Monu Printer, Delhi.

NEW AGE INTERNATIONAL (P) LIMITED, PUBLISHERS
7/30 A, Daryaganj, New Delhi-110002
Visit us at **www.newagepublishers.com**
(CIN: U74899DL1966PTC004618)

I

Dedicated this book

to

'Maha Kaali Maata'

PREFACE

This text has been carefully designed for flexible use. It is primarily designed to provide an introduction to same fundamental concepts in combinatorial Mathematics for under graduate and post-graduate students.

Each topic is divided into sections of approximately the same length, and each section is divided into subsections that form natural blocks of material for teaching. Instructors can easily pace their lectures using these blocks.

All definitions and theorems in this text are stated extremely carefully so that students will appreciate the precision of language and rigor needed in mathematics. Proofs are motivated and developed slowly, their steps are all carefully justified. Recursive definitions are explained and used extensively.

The writing style in this book is direct and pragmatic. Precise mathematical language is used without excessive formalism and abstraction. Case has been taken to balance the mix of notation and words in mathematical statements.

Over 700 problems are used to illustrate concept, related to different topics, and introduce applications. In most examples, a question is first posed, then its solution is presented with appropriate details. The applications included in this text demonstrate the utility of combinatorial mathematics in the solution of real world problem. This text includes applications to wide-variety of areas, including computer science and engineering.

There are over 1000 exercises in the text with many different types of questions posed. There is an ample supply of straight forward exercises that develop basic skills, a large number of intermediate exercises and many challenging (problem sets) exercise sets. Problem sets are stated clearly and unambiguously, and all are carefully graded for various levels of difficulty. It will be honest on my part to accept that it is not possible to include every thing in one book.

Many people contributed directly or indirectly to the completion of this book. Thanks are due to my friends who were able to convince me that I should write this book.

I am grateful to my students, who always encouraged me and many times thanked me for writing this book.

Special thanks to my teachers, who made me realize that I can indeed write a book on "Combinatorial Mathematics". Some pulled me down, some encouraged me and some gave me constructive suggestions. I am grateful to all of them.

I specially thank my parents, elder brothers, elder sister and maternal uncle, who tolerated me all along while I devoted my time to completing this book.

I express my sincere thanks to M/s New Age International (P) Ltd. Publishers, New Delhi, for their responsible work-done at every level in the publication of the book this high production standards.

Healthy criticism and suggestions to improve the quality and standards of the text are most welcome.

C. Vasudev

Bangalore.

SYLLABUS

1. FUNDAMENTAL PRINCIPLES OF COUNTING

Introduction, Sum rule and product rule, Permutations and Combinations, Permutation with Repetition and Combination with Repetition, The Binomial Theorem, The multinomial Theorem, Pascal's Identity, Vandermonde's Identity, The Catalan Numbers, Stirling Numbers, Bell Numbers, Ramsey Numbers.

2. THE PRINCIPLES OF INCLUSION AND EXCLUSION

Introduction, The Principle of Inclusion and Exclusion, Generalization of the Principle of Inclusion and Exclusion, Derangements, Pigeonhole Principle, Genearlized pigeonhole Principle, Rook polynomials, Arrangements with forbidden positions, Hit Polynomial, The Hatcheck problem.

3. GENERATING FUNCTIONS

Introduction to Generating Functions, Introductory examples, partitions of integers, the summation operator, The Exponential Generating Function, The Binomial Generating Function, The Extended Binomial Theorem, Base-*b* representation, Euler's Theorem, Bernoulli numbers, Dobinski's Equality, Restricted Partitions, Euler's Identities, Calculational Techniques

4. RECURRENCE RELATIONS

Introduction to Recurrence Relations, The First order Linear Recurrence Relation, The problem of Tower of Hanoi and Compound Interest, The Fibonacci Numbers, Codeword Enumeration, Back Tracking Method, Forward chaining Method, Summation Method, The second-order Linear Homogeneous Recurrence Relations with Constant coefficients, the Non-homogeneous Recurrence Relations, Characteristic Equation Method, The Method of Generating Functions.

4. GROUP THEORY

Definition of Group, Subgroup, Finite group, Permutation, Permutation groups and their Cycle Indices, Weight, Cycle Index, Coloring and Equivalent w.r.t. group of Permutation, Pattern, Pattern Inventory, Isomorphic Group, Cycle Group, Abetian group, order of an element, direct Product, Left and Right Cosets, Conjugate, Regular icosahedron Burnside-Frobenius Theorem, Pólya's First Enumeration Theorem, Pólya's Second Enumeration Theorem, Lagrange's Theorem, Characterization Theorem for cycle groups, The Burnside-Frobenius Theorem with Weight, Cauchy's Formula, Cayley's Theorem.

CONTENTS

CHAPTER

1

Fundamental Principles of Counting

INTRODUCTION

Techniques for counting are important in Mathematics and Computer Science, especially in the analysis of algorithms. In this chapter, we present counting techniques, in particular those for addition, multiplication, permutations and combinations, etc. We will introduce the basic techniques of counting in this section.

1.1 THE RULES OF SUM AND PRODUCT

1.1.1 Sum Rule. (The Principle of disjunctive Counting)

If a first task can be done in n_1 ways and a second task in n_2 ways, and if these tasks cannot be done at the same time, then there are $n_1 + n_2$ ways to do one of these tasks.

In other words. If a set X is the union of disjoint non empty subsets $S_1, S_2, ..., S_n$, then $|X| = |S_1| + |S_2| + ... + |S_n|$.

1.1.2 Product Rule. (The principle of Sequential Counting)

Suppose that a procedure can be broken down into a sequence of two tasks. If there are n_1 ways to do the first task and n_2 ways to do the second task after the first task has been done, then there are $n_1 n_2$ ways to do the procedure.

In other words, If $S_1, S_2, ..., S_n$ are non empty sets, then the number of elements in the cartesian product $S_1 \times S_2 \times ...... \times S_n$ is the product $\prod_{i=1}^{n} |S_i|$. That is, $|S_1 \times S_2 \times ...... \times S_n| = \prod_{i=1}^{n} |S_i|$.

Problem 1.1. *A student can choose a computer project from one of three lists. The three lists contain 23, 15 and 19 possible projects, respectively. How many possible projects are there to choose from ?*

Solution. The student can choose a project from the first list in 23 ways, from the second list in 15 ways, and from the third list in 19 ways.

Hence, there are 23 + 15 + 19 = 57 projects to choose from.

Problem 1.2. *Suppose that either a member of the mathematics faculty or a student who is a mathematics major is chosen as a representative to a university committee. How many different choices are there for this representative of there are 37 members of the mathematics faculty and 83 mathematics majors ?*

Solution. The first task, choosing a member of the mathematics faculty, can be done in 37 ways.

The second task, choosing a mathematics major, can be done in 83 days.

From the sum rule it follows that there are 37 + 83 = 120 possible ways to pick this representative.

Problem 1.3. *What is the value of k after the following code has been executed ?*

```
k : = 0
for i1 : = 1 to n1
    k : = k + 1
for i2 : 1 to n2
    k : = k + 1
....................................
....................................
for im : = 1 to nm
    k : = k + 1.
```

Solution. The initial value of k is zero. This block of code is made up of m different loops.

Each time a loop is traversed, 1 is added to k.

Let T_i be the task of traversing the ith loop.

The task T_i can be done in n_i ways, since the ith loop is traversed n_i times.

Since no two of these tasks can be done at the same time, the sum rule shows that the final value of k, which is the number of ways to do one of the tasks T_i, $i = 1, 2, \ldots\ldots, m$, is $n_1 + n_2 + \ldots\ldots + n_m$.

Problem 1.4. *In a version of the computer language BASIC, the name of a variable is a string of one or two alpha numeric characters, where upper case and lower case letters are not distinguished. (An alpha numeric character is either one of the 26 English letters or one of the 10 digits).*

Moreover, a variable name must begin with a letter and must be different from the five string of two characters that are reserved for programming use. How many different variable names are there in this version of BASIC ?

Solution. Let V equal the number of different variable names in this version of BASIC.

Let V_1 be the number of these that are one character long and V_2 be the number of these that are two character long.

Then by the sum rule, $V = V_1 + V_2$.

Note that $V_1 = 26$, since a one-character variable name must be a letter.

Furthermore, by the product rule there are 26 . 36 strings of length two that begin with a letter and end with an alphanumeric character.

However, five of these are excluded, so that

$$V_2 = 26 \,.\, 36 - 5 = 931.$$

Hence, there are $V = V_1 + V_2 = 26 + 931 = 957$ different names for variables in this version of BASIC.

Problem 1.5. *Each user on a computer system has a password, which is six to eight characters long, where each character is an upper case letter or a digit. Each password must conatin at least one digit. How many possible passwords are there ?*

Solution. Let P be the total number of possible passwords, and let P_6, P_7 and P_8 denote the number of possible passwords of length 6, 7, and 8 respectively.

By the sum rule, $P = P_6 + P_7 + P_8$

We will now find P_6, P_7, and P_8. Finding P_6 directly is difficult.

To find P_6 it is easier to find the number of strings of upper case letters and digits that are six characters long, including those with no digits, and subtract from this the number of strings with no digits.

By the product rule, the number of strings of six characters is 36^6 and the number of strings with no digits is 26^6.

Hence, $P_6 = 36^6 - 26^6 = 2,176,782,336 - 308,915,776$

$= 1,867,866,560.$

Similarly, it can be shown that

$$P_7 = 36^7 - 26^7 = 78,364,164,096 - 8,031,810,176$$
$$= 70,332,353,920.$$

and
$$P_8 = 36^8 - 26^8 = 2,821,109,907,456 - 208,827,064,576$$
$$= 2,612,282,842,880.$$

Consequently,

$$P = P_6 + P_7 + P_8$$
$$= 2,684,483,063,360.$$

Problem 1.6. *In how many ways can we draw a heart or a spade from an ordinary deck of playing cards ? A heart or an ace ? An ace or a king ? A card numbered 2 through 10 ? A numbered card or a king ?*

Solution. Since there are 13 hearts and 13 spades we may draw a heart or a spade in

13 + 13 = 26 ways.

We may draw a heart or an ace in 13 + 3 = 16 ways, since there are only 3 aces that are not hearts.

We may draw an ace or a king in 4 + 4 = 8 ways.

These are 9 cards numbered 2 through 10 in each of 4 suits, clubs, diamonds, hearts, or spades. So we may choose a numbered card in 36 ways.

Thus, we may choose a numbered card or a king in 36 + 4 = 40 ways.

Problem 1.7. *How many ways can we get a sum of 4 or of 8 when two distinguishable dice (say one die is red and the other is white) are rolled ? How many ways can we get an even sum ?*

Solution. Let us label the outcome of a 1 on the red die and a 3 on the white die as the ordered pair (1, 3).

Then we see that the out comes (1, 3), (2, 2), and (3, 1) are the only ones whose sum is 4.

Thus, there are 3 ways to obtain the sum.

Likewise, the obtain the sum 8 from the outcomes (2, 6), (3, 5), (4, 4), (5, 3), and (6, 2).

Thus, there are 3 + 5 = 8 outcomes whose sum is 4 or 8.

The number of ways to obtain an even sum is the same as the number of ways to obtain either the sum 2, 4, 6, 8, 10 or 12.

There is 1 way to obtain the sum 2, 3 ways to obtain the sum 4, 5 ways to obtain 6, 5 ways to obtain an 8, 3 ways to obtain a 10, and 1 way to obtain a 12.

Therefore, there are $1 + 3 + 5 + 5 + 3 + 1 = 18$ ways to obtain an even sum.

Problem 1.8. *How many ways can we get a sum of 8 when two indistinguishable dice are rolled ? An even sum ?*

Solution. Had the dice been distinguishable, we should obtain a sum of 8 by the outcomes (2, 6), (3, 5), (4, 4), (5, 3) and (6, 2), but since the dice are similar, the outcomes (2, 6) and (6, 2) and, as well, (3, 5) and (5, 3) cannot be differentiated and thus we obtain the sum of 8 with the roll of two similar dice in only 3 ways. Likewise, we can get an even sum in $1 + 2 + 3 + 3 + 2 + 1 = 12$ ways.

Problem 1.9. *If there are 14 boys and 12 girls in a class, find the number of ways of selecting one student as class representative.*

Solution. Using sum rule, there are $14 + 12 = 26$ ways of selecting one student (either a boy or a girl) as class representative.

Problem 1.10. *If a student is getting admission in 4 different Engineering Colleges and 5 Medical Colleges, find the number of ways of choosing one of the above colleges.*

Solution. Using sum rule, there are $4 + 5 = 9$ ways of choosing one of the colleges.

Problem 1.11. *In how many ways can you get a total of six when rolling two dice ?*

Solution. The event "get a six" is the union of the mutually exclusive subevents.

A_1 : "two 3's"

A_2 : "a 2 and a 4"

A_3 : "a 1 and a 5"

Event A_1 can occur in one way, A_2 can occur in two ways (depending on which die lands 4), and A_3 can occur in two ways, so the numebr of ways to get a six is $1 + 2 + 2 = 5$.

Problem 1.12. *The chairs of an auditorium are to be labeled with a letters and a positive integer not exceeding 100. What is the largest number of chairs that can be labeled differently ?*

Solution. The procedure of labeling a chair consists of two tasks, namely, assigning one of the 26 letters and then assigning one of the 100 possible integers to the seat. The product rule shows that there are $26.100 = 2600$ different ways that a chair can be labeled. Therefore, the largest number of chairs that can be labeled differently is 2600.

Problem 1.13. *There are 32 micro computers in a computer center. Each micro computer has 24 ports. How many different ports to a micro computer in the center are there ?*

Soltuion. The procedure of choosing a port consists of two tasks, first picking a micro computer and then picking a port on this micro computer.

Since there are 32 ways to choose the micro computer and 24 ways to choose the port no matter which micro computer has been selected, the product rule shows that there are $32.24 = 768$ ports.

Problem 1.14. *How many different bit strings are there of length seven ?*

Solution. Each of the seven bits can be chosen in two ways, since each bit is either 0 or 1.

Therefore, the product rule shows there are a total of $2^7 = 128$ different bit strings of length seven.

Problem 1.15. *How many different license plates are available if each contains a sequence of three letters followed by three digits (and no sequences of letters are prohibited, even if they are abscence) ?*

Solution. There are 26 choices for each of the three letters and ten choices for each of the three digits.

Hence, by the product rule there are a total of 26 . 26 . 26 . 10 . 10 . 10 = 17,576,000 possible license plates.

Problem 1.16. *How many functions are there from a set with m elements to one with n elements ? (Counting Functions).*

Solution. A function corresponds to a choice of one of the n elements in the co-domain for each of the m elements in the domain.

Hence, by the product rule there are $n \,.\, n \,......\, n = n^m$ functions from a set with m elements to one with n elements.

For example, there are 5^3 different functions from a set with three elements to a set with 5 elements.

Problem 1.17. *How many one-to-one functions are there from a set with m elements to one with n elements ?*

(Counting one-to-one Functions)

Solution. First note when $m > n$ there are no one-to-one functions from a set with m elements to a set with n elements.

Now let $m \leq n$.

Suppose the elements in the domain are $a_1, a_2, ..., a_m$.

There are n ways to choose the value of the function at a_1.

Since the function is one-to-one, the value of the function at a_2 can be picked in $n - 1$ ways.

(Since the value used for a_1 cannot be used again).

In general, the value of the function at a_k can be choosen in $n - k + 1$ ways.

By the product rule, there are $n(n - 1)(n - 2) ...... (n - m + 1)$ one-to-one functions from a set with m elements to one with n elements.

For example, there are 5.4.3 = 60 one-to-one functions from a set with elements to a set with 5 elements.

Problem 1.18. *The format of telephone numbers in North America is specified by a numbering plan. A telephone number consists of 10 digits, which are split into a 3-digit area code, a 3-digit office code, and a 4-digit station code. Because of signating considerations, these are certain restrictions on some of these digits. To specify the allowable format, let X denote a digit that can take any of the values 0 through 9, let N denote a digit that can take any of the values 2 through 9, and let Y denote a digit that must be a 0 or a 1. Two numbering plans, which will be called the old plan and the new plan, will be discussed. (The old plan, in use in the 1960s, has been replaced by the new plan, but the recent rapid growth in demand for new nubmers will make even this new plan obsolete). As will be shown, the new plan allows the use of more numbers.*

In the old plan, the formats of the area code, office code, and station code are NYX, NNX and XXXX, respectively, so that telephone numbers had the form NYX-NNX-XXXX. In the new plan, the formats of these codes are NXX, NXX and XXXX, respectively, so that telephone numebrs have the form NXX-NXX-XXXX. How many different North American telephone numbers are possible under the old plan and under the new plan ? (The telephone Numbering plane)

Solution. By the product rule, there are 8.2.10 = 160 area codes with format NYX and 8.10.10 = 800 area codes with format NXX.

Similarly, by the product rule, there are 8.8.10 = 640 office codes with formats NNX.

The product rule also shows that there are

$$10.10.10.10 = 10{,}000 \text{ station codes with format XXXX.}$$

Consequently, applying the product rule again, it follows that under the old plan there are

$$160.640.10{,}000 = 1{,}024{,}000{,}000$$

different numebrs available in North America.

Under the new plan there are

$$800.800.10{,}000 = 6{,}400{,}000{,}000$$

different numbers available.

Problem 1.19. *What is the value of k after the following code has been executed ?*

$k := 0$

for $i_1 := 1$ *to* n_1

for $i_2 := 1$ *to* n_2

....

....

for $i_m : 1$ *to* n_m

$k := k + 1.$

Solution. The initial value of k is zero.

Each time the nested loop is traversed, 1 is added to k.

Let T_i be the task of traversing the ith loop.

Then the number of times the loop is traversed is the number of ways to do the tasks $T_1, T_2, \ldots\ldots, T_m$.

The number of ways to carry out the task $T_j, j = 1, 2, \ldots\ldots, m$, is n_j, since the jth loop is traversed once for each integer i_j with $1 \le i_j \le n_j$.

By the product rule, it follows that the nested loop is traversed $n_1 n_2 \ldots\ldots n_m$ times.

Hence, the final value of k is $n_1 n_2 \ldots\ldots n_m$.

Problem 1.20. *Use the product rule to show that the number of different subsets of a finite set S is* $2^{|S|}$.

(Counting subsets of a Finite Set).

Solution. Let S be a finite set. List the elements of S in arbitrary order.

Recall that there is a one-to-one correspondence between subsets of S and bit strings of length | S |.

Namely, a subset of S is associated with the bit string with a 1 in the ith position if the ith element in the list is in the subset, and a 0 in this position otherwise.

By the product rule, there are $2^{|S|}$ bit strings of length | S |.

Hence, $| P(S) | = 2^{|S|}$.

Problem 1.21. *Licence plates in the canadian province of Ontario consist of four letters followed by three of the digits 0 – 9 (not necessarily distinct). How many different licence plates can be made in ontario ?*

Solution. There are 26 ways in which the first letter can be chosen, 26 ways in which the second can be chosen. Similarly, for the third and fourth.

By the multiplication rule, the number of ways in which the three letters can be chosen is

$$26 \times 26 \times 26 \times 26 = 26^4.$$

By the same reasoning there are 10^3 ways in which the final three digits of an ontario licence plate can be selected and, all in all.

$26^4 \times 10^3 = 456{,}976{,}000$ different licence plates which can be manufactured by the government of ontario under its current system.

Problem 1.22. *How many numbers in the range 1000—9999 do not have any repeated digits ?*

Solution. Imagine enumerating all numbers of the desired type in the spirit of Fig. 1.1.

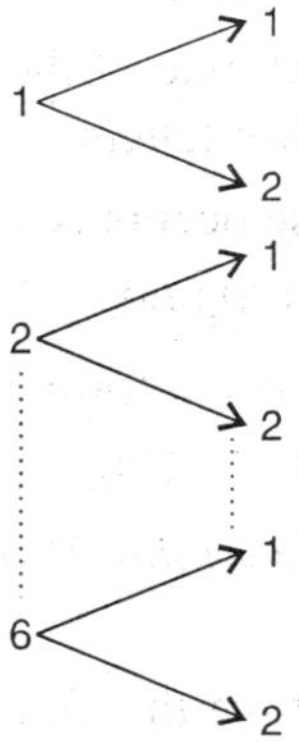

Fig. 1.1.

There are nine choices for the first digit (any of 1—9). Once this has been chosen, there remains still nine choices for the second (the chosen first digit cannot be repeated but 0 can now be used).

There are now eight choices for the third digit and seven for the fourth.

Altogether, there are $9 \times 9 \times 8 \times 7 = 4536$ possible numbers.

Problem 1.23. *How many even numbers in the range 100—999 have no repeated digits ?*

Solution. The question is equivalent to asking for the number of ways in which one can write down an even number in the range 100—999 without repeating digits.

This event can be partitioned into two mutually exclusive cases.

Case 1. The number ends in 0.

In this case, there are nine choices for the first digit (1—9) and then eight for the second (since 0 and the first digit must be excluded).

So there are $9 \times 8 = 72$ numbers of this type.

Case 2. The number does not end in 0.

Now there are four choices for the final digit (2, 4, 6 and 8), then eight choices for the first digit (0 and the last digit are excluded), and eight choices for the second digit (the first and last digits are excluded). There are $4 \times 8 \times 8 = 256$ numbers of this type.

By the addition rule, there are $72 + 256 = 328$ even numbers in the range 100—999 with no repeated digits.

Problem 1.24. *A typesetter (long ago) has before him 26 trays, one for each letter of the alphabet. Each tray contains ten copies of the same letter. In how many ways can he form a three letter "word" which requires atmost two different letters ? By "word", we mean any sequence of three letters—x pt, for example—not necessarily a real word from the dictionary. Two "ways" are different unless they use the identical pieces of type.*

Solution. The event "atmost two different letters" is comprised of two mutually exclusive cases :

Case 1. The first two letters are the same. Here, the third letter can be arbitrary, that is, any of the 258 letters which remain after the first two are set can be used.

So the number of ways in which this case can occur is

$$260 \times 9 \times (260 - 2) = 603{,}720.$$

Case 2. The first two letters are different.

In this case, the third letter must match one of the first two, so it must be one of the 18 letters remaining in the two trays used for the first two letters.

The number of ways in which this case occurs is

$$260 \times 250 \times 18 = 1{,}170{,}000.$$

By the addition rule, the number of ways to form a word using at most two different letters is

$$603{,}720 \times 1{,}170{,}000 = 1{,}773{,}720.$$

Theorem 1.1. *A set of cardinality n contains 2^n subsets (including the empty set and the entire set itself).*

Proof. There are several ways to prove this fundamental result. We present one here which uses the ideas of this section.

Given n objects $a_1, a_2, \ldots\ldots, a_n$, each subset corresponds to a sequence of choices. Is a_1 in the subset. Is a_2 in the subset. Finally, is a_n in the subset.

There are two answers to the first question, two for the second, and so on.

In all, there are

$$\underbrace{2 \times 2 \times \ldots\ldots \times 2}_{n \text{ factors}} = 2^n$$

Ways in which all n choices can be made.

Thus, these are 2^n subsets.

Problem 1.25. *If 2 distinguishable dice are rolled, in how many ways can they fall ? If 5 distinguishable dice are rolled, how many possible outcomes are there ? How many if 100 distinguishable dice are tossed ?*

Solution. The first dice can fall (event E_1) in 6 ways and the second can fall (event E_2) in 6 ways.

Thus, there are $6 \, . \, 6 = 6^2 = 36$ outcomes when 2 dice are rolled.

Also, the third, fourth, and fifth die each have 6 possible outcomes so there are $6.6.6.6.6 = 6^5$ possible outcomes when all 5 dice are tossed.

Likewise there are 6^{100} possible outcomes when 100 dice are tossed.

Problem 1.26. *Suppose that the licence plates of a certain state require 3 English letters followed by 4 digits.*

(*a*) *How many different plates can be manufactured if repetition of letters and digits are allowed ?*

(*b*) *How many plates are possible if only the letters can be repeated ?*

(*c*) *How many are possible if only the digits can be repeated ?*

(*d*) *How many are possible if no repetitions are allowed at all ?*

Solution. (*a*) $26^3 \, . \, 10^4$ since there are 26 possibilities for each of the 3 letters and 10 possibilities for each of 4 digits.

(*b*) $26^3 \, .10.9.8.7$

(*c*) $26.25.24.10^4$

(*d*) 26.25.24.10.9.8.7.

Problem 1.27. (*a*) *How many 3-digit numbers can be formed using the digits 1, 3, 4, 5, 6, 8 and 9 ?*

(*b*) *How many can be formed if no digit can be repeated ?*

Solution. There are 7^3 such 3-digit numbers in

(*a*) Since each of the 3-digits can be filled with 7 possibilities. Likewise, the answer to question.

(*b*) is 7.6.5 since these are 7 possibilities for the hundreds digit but once one digit is used it is not available for the tens digit (since no digit can be repeated in this problem).

Thus, there are only 6 possibilities for the tens digit, and then for the same reason there are only 5 possibilities for the units digit.

Problem 1.28. *How many different licence plates are there that involve 1, 2 or 3 letters followed by 4 digits ?*

Solution. We can form plates with 1 letter followed by 4 digits in 26.10^4 ways, plates with 2 letters followed by 4 digits in $26^2 \, . \, 10^4$ ways, and plates with 3 letters followed by 4 digits in $26^3 \, . \, 10^4$ ways.

These separate events are mutually exclusive so we can apply the sum rule to conclude that there are $26 \, . \, 10^4 + 26^2 \, . \, 10^4 + 26^3 \, . \, 10^4 = (26 + 26^2 + 26^3) \, 10^4$ plates with 1, 2 or 3 letters followed by 4-digits.

Problem 1.29. *How many different plates are there that involve 1, 2 or 3 letters followed by 1, 2, 3 or 4 digits ?*

Solution. We see that there are $(26 + 26^2 + 26^3) \, 10$ ways to form plates of 1, 2 or 3 letters followed by 1 digit. $(26 + 26^2 + 26^3) \, 10^2$ plates of 1, 2 or 3 letters followed by 2 digits,

$(26 + 26^2 + 26^3)10^3$ plates of 1, 2 or 3 letters followed by 3 digits, and $(26 + 26^2 + 26^3)10^4$ plates of 1, 2 or 3 letters followed by 4 digits.

Thus, we can apply the sum rule to conclude that there are

$$(26 + 26^2 + 26^3)10 + (26 + 26^2 + 26^3)10^2 + (26 + 26^2 + 26^3)10^3 + (26 + 26^2 + 26^3)10^4$$
$$= (26 + 26^2 + 26^3)(10 + 10^2 + 10^3 + 10^4)$$

ways to form plates of 1, 2 or 3 letters followed by 1, 2, 3, or 4 digits.

Problem 1.30. *(a) How many 2 digit or 3-digit numebrs can be formed using the digits 1, 3, 4, 5, 6, 8 and 9 if no repetition is allowed ?*

(b) How many numbers can be formed using the digits 1, 3, 4, 5, 6, 8 and 9 if no repitition are allowed ?

Solution. (*a*) There are 7.6.5, three-digit numbers possible. Likewise, we can apply the product rule to see that these are 7.6 possible 2-digit numbers.

Hence, these are 7.6 + 7.6.5 possible two-digit or three-digit numbers.

(*b*) The number of digits are not specified in this problem so we can form one-digit numbers, two-digit numbers, or three digit numbers, etc.

But since no repetitions are allowed and we have only the 7 integers to work with, the maximum number of digits would have to be 7.

Applying the product rule, we see that we may form 7 one-digit numebrs, 7.6 = 42 two digit numbers 7.6.5 three digit numbers, 7.6.5.4 four digit numbers, 7.6.5.4.3 five digit nubmers, 7.6.5.4.3.2 six-digit numbers, and 7.6.5.4.3.2.1 seven-digit numbers.

The events of forming one-digit numbers, two digit numbers, three digit numbers, etc., are mutually exclusive events so we apply the sum rule to see that there are 7 + 7.6 + 7.6.5 + 7.6.5.4 + 7.6.5.4.3 + 7.6.5.4.3.2 + 7.6 + 7.6.5.4.3.2.1 different numbers we can form under the restrictions of this problem.

Problem 1.31. *How many three-digit numbers are there which are even and have no repeated digits ? (Here we are using all digits 0 through 9).*

Solution. For a number to be even it must end in 0, 2, 4, 6 or 8. There are two cases to consider.

First, suppose that the number ends in 0 ; then there are 9 possibilities for the first digit and 8 possibilities for the second since no digit can be repeated. Hence there are 9.8 three-digit numbers that end in 0. Now suppose the number does not end in 0.

Then there are 4 choices for the last digit (2, 4, 6 or 8).

When this digit is specified, then there are only 8 possibilities for the first digit, since the number cannot begin with 0. Finally, there are 8 choices for the second digit and therefore there are 8.8.4 numbers that do not end in 0. Accordingly since these two cases are mutually exclusive, the sum rule gives 9.8 + 8.8.4 even three-digit numbers with no repeated digits.

Problem 1.32. *Suppose that we draw a card from a deck of 52 cards and replace it before the next draw. In how many ways can 10 cards be drawn so that the tenth card is a repetition of a previous draw ?*

Solution. First we count the number of ways, we can draw 10 cards so that the 10th card is not a repetition.

First choose what the 10th card will be. This can be done in 52 ways.

If the first 9 draws are different from this, then each of the 9 draws can be chosen from 51 cards.

Thus there are 51^9 ways to draw the first 9 cards different from the 10th card.

Hence, there are $(51^9)(52)$ ways to choose 10 cards with the 10th card different from any of the previous 9 draws.

Hence, there are $52^{10} - (51^9)(52)$ ways to draw 10 cards where the 10th is a repetition since there are 52^{10} ways to draw 10 cards with replacements.

Problem 1.33. *In how many ways can 10 people be seated in a row so that a certain pair of them are not next to each other ?*

Solution. There are 10 ! ways of seating all 10 people. Thus, by indirect counting, we need only count the number of ways of seating the 10 people where the certain pair of people (say, A and B) are seated next to each other. If we treat the pair AB as one entity, then there are 9 total entities to arrange in 9 ! ways.

But A and B can be seated next to each other in 2 different orders, namely AB and BA.

Thus, there are (2)(9 !) ways of seating all 10 people where A and B are next to each other.

The answer to our problem then is 10 ! – (2)(9 !).

Problem 1.34. *In how many ways can one select two books from different subjcets from among six distinct computer science books, three distinct mathematics books, and two distinct chemistry books ?*

Solution. Using product rule one can select two books from different subjects as follows :

(*i*) one from computer science and one from mathematics in 6.3 = 18 ways.

(*ii*) one from computer science and one from chemistry in 6.2 = 12 ways.

(*iii*) one from mathematics and one from chemistry in 3.2 = 6 ways.

Since these sets of selections are pairwise disjoint one can use the sum rule to get the required number of ways which is 18 + 12 + 6 = 36.

Problem 1.35. *For a set of six true or false questions, find the number of ways of answering all questions.*

Solution. The number of ways of answering the first question is 2.

The second question can also be answered in 2 ways and similarly for other 4 questions.

Hence, the total number of ways of answering all the questions is $2^6 = 64$.

Problem 1.36. *Three persons enter into car, where there are 5 seats. In how many ways can they take up their seats ?*

Solution. The first person has a choice of 5 seats and can sit in any one of those 5 seats.

So, there are 5 ways of occupying the first seat. The second person has a choice of 4 seats.

Similarly, the third person has a choice of 3 seats. Hence, the requried number of ways in which all the three persons can seat is $5 \times 4 \times 3 = 60$.

Problem 1.37. *There are four roads from city X to Y and five roads from city Y to Z, find*

(*i*) *how many ways is it possible to travel from city X to city Z via city Y.*

(*ii*) *how different round trip routes are there from city X to Y to Z to Y and back to X.*

Solution. (*i*) In going from city X to Y, any of the 4 roads may be taken.

In going from city Y to Z, any of the 5 roads may be taken.

So by the product rule, there are 5.4 = 20 ways to travel from city X to Z via city Y.

(*ii*) A round trip journey can be performed in the following four ways.

(1) From city X to Y

(2) From city Y to Z

(3) From city Z to Y

(4) From city Y to X.

(1) Can be performed 4 ways, 5 ways to perform 2, 5 ways to perform 3 and 4 ways to perfor:n 4. By product rule, there are 4.5.5.4 = 400 round trip routes.

Problem 1.38. *How many bit strings of length eight either start with a 1 bit or end with the two bits 00 ?*

Solution. The first task, constructing a bit string of length eight beginning with a 1 bit, can be done in $2^7 = 128$ ways.

This follows by the product rule, since the first bit can be chosen in only one way and each of the other seven bits can be chosen in two ways.

The second task, constructing a bit string of length eight ending with the two bits 00, can be done in $2^6 = 64$ ways.

This follows by the product rule, since each of the first six bits can be chosen in two ways and the last two bits can be chosen in only one way.

Both tasks, constructing a bit string of length eight that begins with a 1 and ends with 00, can be done in $2^5 = 32$ ways.

This follows by the product rule, since the first bit can be chosen in only one way, each of the second through the sixth bits can be chosen in two ways, and the last wto bits can be chosen in one way.

Consequently, the number of bit strings of length eight that begin with a 1 or end with a 00, which equals the number of ways to do either the first task or the second task, equals $128 + 64 - 32 = 160$.

1.2 PERMUTATIONS

Suppose that, we have three letters *a*, *b* and *c*. Then, all possible arrangements of any two letters out of these three letters can be enumerated as : *ab, ac, bc, ba, ca* and *cb*. If we make an arrangement of all three letters out of these three, then we have *abc, acb, bac, bca, cab* and *cba* as possible arrangements.

Each of the distinct order of arrangements of a given set of distinct objects, taking some or all of them at a time (with or without repetition), is called a **permutation** of the objects. The total number of permutations of n distinct objects, taken n at a time $(r \leq n)$, is equal to the total nubmer of ways of placing n objects in r boxes. This is denoted as $P(n, r)$ or ${}^nP_r$. This number is equal to $n(n-1) \ldots\ldots (n-r+1)$.

Notation ${}^n\mathbf{P}_r$:

We know that ${}^nP_r$ is the number of permutation of n distinct objects taken r at a time, and this is equal to $n(n-1)(n-2) \ldots\ldots (n-r+1)$.

$$\text{Therefore, } {}^nP_r = \frac{n(n-1)(n-2)\ldots\ldots(n-r+1) * (n-r)!}{(n-r)!} = \frac{n!}{(n-r)!}$$

Therefore $^nP_r = \dfrac{n!}{(n-r)!}$

$$\left(^1P_1 = \frac{1!}{(1-1)!} = \frac{1}{0!} = 1 \quad \because \; 0! = 1 \right)$$

The number of permutations of n distinct objects, taken n at a time, is given by $^nP_n = n!$.

- Permutation of objects when all are not distinct, in this case, the number of permutations is given by

$$x = \frac{n!}{p!q!r!}$$

- The number of distinguishable permutations that can be formed from a collection of n objects, taken all n at a time, in which the first object appears k_1 times, the second object k_2 times, and so on, is given by

$$\frac{n!}{k_1!k_2!k_3!\ldots\ldots k_r!}$$

where $k_1 + k_2 + \ldots\ldots + k_r = n$.

- The number of permutations of n distinct objects, taken r at a time, when repetitions are allowed, is given by n^r.
- The number of ways of arranging objects under some restriction = the number of arrangements of the same number of object without restriction – the number of arrangements of the same number of arrangements with the opposite restriction.
- Generating function for permutation :

The coefficient of $\dfrac{x^r}{r!}$ in a polynomial P(x) is ^nP_r and ^nP_r is the number of permutations of n objects taken r at a time.

Theorem 1.2. *The number of r-permutations of a set with n distinct elements is*

$$P(n, r) = n(n-1)(n-2) \ldots\ldots (n-r+1) = {}^nP_r = \frac{n!}{(n-r)!}$$

Proof. The first element of the permutation can be chosen in n ways, since there are n elements in the set. There are $n - 1$ ways to choose the second element of the permutation, since there are $n - 1$ elements left in the set after using the element picked for the first position. Similarly, there are $n - 2$ ways to choose the third element and so on, until there are exactly $n - (r - 1) = n - r + 1$ ways to choose the rth element.

Consequently, by the product rule, there are

$$n(n-1)(n-2) \ldots\ldots (n-r+1)$$

r-permutations of the set.

It follows that

$$P(n, r) = n(n-1)(n-2) \ldots\ldots (n-r+1) = \frac{n!}{(n-r)!}.$$

Example 1. Let A be {1, 2, 3, 4}. Then the sequences 124, 421, 341 and 243 are same permutations of A taken 3 at a time. The sequences 12, 43, 31, 24, and 21 are examples of different permutations of A taken two at a time.

The total number of permutations of A taken three at a time is $4P_3$ or 4.3.2 or 24.

The total number of permutations of A taken two at a time is $4P_2$ or 4.3 or 12.

Note. When $r = n$, we are counting the distinct arrangements of the elements of A, with $|A| = n$, into sequences of length n. Such a sequence is simply called a permutation of A.

The number of permutations of A is thus ${}^nP_n$ or

$n \,.\, (n-1) \,.\, (n-2) \ldots\ldots 2.1$, if $n \geq 1$. This number is also written n ! and is read n factorial.

Both ${}^nP_r$ and n ! are built in functions on many calculators.

Example 2. Let A be $\{a, b, c\}$. Then the possible permutations of A are the sequences *abc, acb, bac, bca, cab and cba.*

For convenience, we define 0 ! to be 1. Then for every $n \geq 0$ the number of permutations of n objects is n !.

If $n \geq 1$ and $1 \leq r \leq n$.

Example 3. Let A consist of all 52 cards in an ordinary deck of playing cards. Suppose that this deck is shuffled and a hand of five cards is dealt. A list of cards in this hand, in the order in which they were dealt, is a permutation of A taken five at a time. Examples would include AH, 3D, 5C, 2H, JS, 2H, 3H, 5H, QH, KD ; JH, JD, JS, 4H, 4C ; and 3D, 2H, AH, JS, 5C.

Note that the first hand and last hands are the same, but they represent different permutations since they were dealt in a different order.

The number permutations of A taken five at a time is

$${}^{52}P_5 = \frac{52!}{47!} \text{ or } 52.51.50.49.48 \text{ or } 311,875,200.$$

This is the number of five-card hands that can be dealt if we consider the order in which they were dealt.

Exercise 4. *The number of distinguishable "words" that can be formed from the letters of MISSISSIPPI is* $\dfrac{11!}{1!\,4!\,4!\,2!}$ *or 34, 650.*

Theorem 1.3. *Suppose that two tasks T_1 and T_2 are to be performed in sequence. If T_1 can be performed in n_1 ways, and for each of these ways T_2 can be performed in n_2 ways, then the sequence T_1T_2 can be performed in n_1n_2 ways. (Multiplication principle of counting).*

Proof. Each choice of a method of performing T_1 will result in a different way of performing the task sequence. There are n_1 such methods, and for each of these we may choose n_2 ways of performing T_2.

Thus, in all, there will be $n_1 n_2$ ways of performing the sequence T_1T_2. See Fig. 1.2 for the case where n_1 is 3 and n_2 is 4.

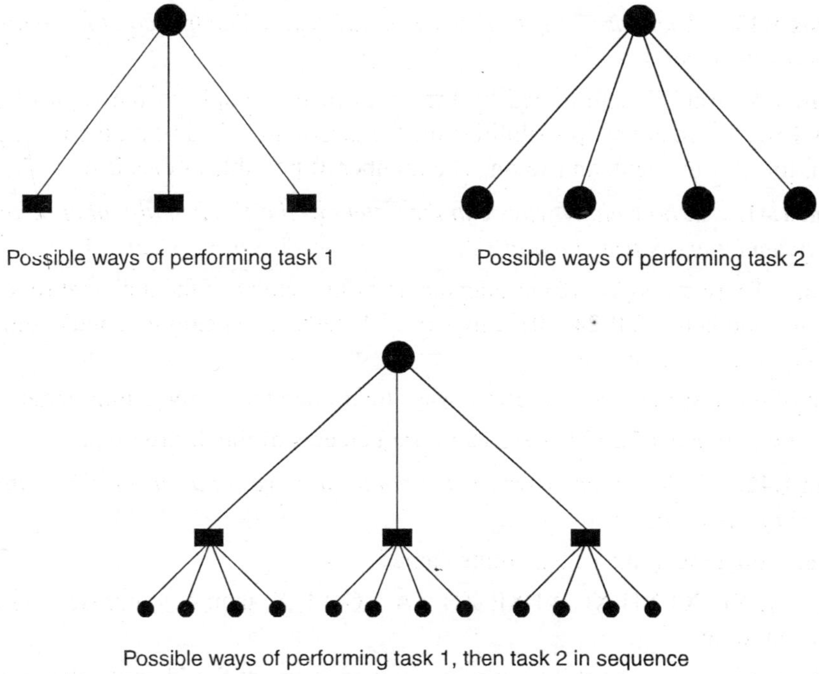

Fig. 1.2.

Theorem 1.4. *Suppose that tasks T_1, T_2, ... T_k are to be performed in sequence. If T_1 can be performed in n_1 ways, and for each of these ways T_2 can be performed in n_2 ways, and for each of these n_1n_2 ways of performing T_1T_2 in sequence, T_3 can be performed in n_3 ways, and so on, then the sequence T_1T_2 ...... T_k can be performed in exactly n_1n_2 ......n_k ways.*

(Extended Multiplication principle of counting).

Theorem 1.5. *Given natural numbers r and n with $r \le n$, the number of ways to place r marbles of different colours into n numbered boxes, at most one marble to a box, is P(n, r).*

Notice that $\underbrace{n(n-1)(n-2)\ldots\ldots(n-r+1)}_{P(n,\,r)}\ \underbrace{(n-r)(n-r-1)\ldots\ldots(3)(2)(1)}_{(n-r)!} = n\,!$

Thus, $P(n, r) = \dfrac{n\,!}{(n-r)\,!}$ a formula which holds also for $r = 0$ and $r = n$ because $P(n, 0) = 1$ and $0\,! = 1$.

Theorem 1.6. *The number of permutations of n symbols is n !. The number of r-permutations of n symbols is P(n, r).*

Problem 1.39. *How many pairs of dance partners can be selected from a group of 12 women and 20 men ?*

Solution. The first woman can be paired with any of 20 men, the second woman with any of the remaining 19 men, the third with any of the remaining 18, and so on.

These are 20.19.18......9 = P(20, 12) possible couples.

Problems 1.40. *There are 7 ! = 5040 ways in which seven people can form a line. In how many ways can seven people form a circle ?*

Solution. A circle is determined by the order of the people to the right of any one of the individuals, say Eric. There are six possibilities for the person in Eric's right, then five possibilities for the next person, four for the next, and so on. The number of possible circles is 6 ! = 720.

Problem 1.41. *In how many ways can the letters of the English alphabet be arranged so that there are exactly ten letters between a and z ?*

Solution. There are P(24, 10) arrangements of the letters of the alphabet (excluding *a* and *z*) taken ten at a time, and hence 2.P(24, 10) strings of 12 letters, each beginning and ending with an *a* and *az*(either letter coming first in a string).

For each of these strings, there are 15 ! ways to arrange the 14 remaining letters and the string.

So there are altogether 2.P(24, 10) . 15 ! arrangements of the desired type.

Problem 1.42. *In how many ways can ten adults and five children stand in a line so that no two children are next to each other ?*

Solution. Imagine a line of ten adults named

A, B, ..., *j*, XD XJ XH XC XI XE XB XA XG XF X, then X's representing the 11 possible locations for the children.

For each such line, the first child can be positioned in any of the 11 spots, the second child in any of the remaining 10, and so on.

Hence, the children can be positioned in 11.10.9.8.7 = P(11, 5) ways.

For each such positioning, there are 10 ! ways of ordering the adults A, ..., J, so, by the multiplication rule, the number of lines of adults and children is 10 ! P(11, 5).

Problem 1.43. *A label identifier, for a computer system, consists of one letter followed by three digits. If repetitions are allowed, how many distinct label identifiers are possible ?*

Solution. There are 26 possibilities for the beginning letter and there are 10 possibilities for each of three digits.

Thus, by the extended multiplication principle, there are 26 × 10 × 10 × 10 or 26,000 possible label identifiers.

Problem 1.44. *Let A be a set with n elements. How many subsets does A have ?*

Solution. We know that, each subset of A is determined by its characteristic function, and if A has n elements, this function may be described as an array of 0's and 1's having length n.

The first element of the array can be filled in two ways (with a 0 or a 1), and this is true for all succeeding elements as well.

Thus, by the extended multiplication principle, there are

$$\underbrace{2\,.\,2\,......\,2}_{n \text{ factors}} = 2^n$$

ways of filling the array, and therefore 2^n subsets of A.

Problem 1.45. *How many different sequences, each of length r, can be formed using elements from A if*

(*a*) *elements in the sequence may be repeated ?*

(*b*) *all elements in the sequence must be distinct ?*

Solution. First we note that any sequence of length r can be formed by filling r boxes in order from left to right with elements of A.

In case (*a*), we may use copies of elements of A.

			...		
box 1	box 2	box 3		box $r-1$	box r

Let T_1 be the task "fill box 1", let T_2 be the task "fill box 2", and so on. Then combined task $T_1 T_2 \ldots T_r$ represents the formation of the sequence.

Case (*a*). T_1 can be accomplished in n ways, since we may copy any element of A for the first position of the sequence. The same is true for each of the tasks T_2, T_2, ... T_r.

Then by the extended multiplication principle, the number of sequences that can be formed is

$$\underbrace{n \,.\, n \,\ldots\ldots\, n}_{r \text{ factors}} = n^r$$

Now we consider case (*b*), here also T_1 can be performed in n ways, since any element of A can be chosen for the first position. Which ever element is chosen, only $(n-1)$ elements remain, so that T_2 can be performed in $(n-1)$ ways, and so on, until finally T_r can be performed in

$$n-(r-1) \quad \text{or} \quad (n-r+1) \text{ ways.}$$

Thus, by the extended principle of multiplication, a sequence of r distinct elements from A can be formed in $n(n-1)(n-2) \ldots\ldots (n-r+1)$ ways.

Problem 46. *How many three-letter "words" can be formed from letters in the set {a, b, y, z}. If repeated letters are allowed ?*

Solution. Here n is 4 and r is 3, so the number of such words is 4^3 or 64.

Problem 47. *How many "words" of three distinct letters can be formed from the letters of the word MAST ?*

Solution. The number is ${}^4P_3 = \dfrac{4\,!}{(4-3)\,!}$ or $\dfrac{4\,!}{1\,!}$ or 24.

Problem 48. *How many distinguishable permutations of the letters in the word BANANA are there ?*

Solution. We begin by tagging the A's and N's in order to distinguish between them temporarily.

For the letters B, A_1, N_1, A_2, N_2, A_3, there are 6 ! or 720 permutations. Some of these permutations are identical except for the order in which the N's appear.

For example, $A_1A_2A_3BN_1N_2$ and $A_1A_2A_3BN_2N_1$.

In fact, the 720 permutations can be listed in pairs whose members differ only in the order of the two N's. This means that if the tags are dropped from the N's only $\frac{720}{2}$ or 360 distinguishable permutations remain.

Reasoning in a similar way we see that these can be gruoped in groups of 3 ! or 6 that differ only in the order of the three A's.

For example, one group of 6 consists of $BNNA_1A_2A_3$, $BNNA_1A_3A_2$, $BNNA_2A_1A_3$, $BNNA_2A_3A_1$, $BNNA_3A_1A_2$, $BNNA_3A_2A_1$.

Dropping the tags would change these 6 into the single permutation BNNAAA.

Thus, there are $\frac{360}{6}$ or 60 distinguishable permutations of the letters of BANANA.

Problem 49. *A man, a woman, a boy, a girl, a dog, and a cat are walking down a long and winding road one after the other.*

(a) In how many ways can this happen ?

(b) In how many ways can this happen if the dog comes first ?

(c) In how many ways can this happen if the dog immediately follows the boy ?

(d) In how many ways can this happen if the dog (and only the dog) is between the man and the boy ?

Solution. (*a*) There are 6 ! = 720 ways for six creatures to form a line.

(*b*) If the dog comes first, the others can form = 5 ! lines behind.

(*c*) If the dog immediately follows the boy, then the dog-boy pair should be thought of as a single object to be put into a line with four others.

There are 5 ! = 120 such lines.

(*d*) If the man, dog, and boy appear in this order, then thinking of man-dog-boy as a single object to be put into a line with three others, we see that there are 4 ! possible lines. Similarly, there are 4 ! lines in which the boy, dog, and man appear in this order.

So, by the addition rule, there are 4 ! + 4 ! = 48 lines in which the dog (and only the dog) is between the man and the boy.

Problem 1.50. *In how many ways can ten adults and five children stand in a circle so that no two children are next to each other ?*

Solution. Arrange the adults into a circle in one of 9 ! ways. There are then 10 locations for the first child, 9 for the second, 8 for the third, 7 for the fourth, and 6 for the fifth. The answer is 9 ! (10.9.8.7.6) = 9 ! P(10, 5).

Problem 1.51. *Find (a) P(5, 3), P(4, 4) and P(7, 2)*

(b) $\frac{20!}{17!}, \frac{100!}{98!}$ *and P(7, 0).*

Solution. (*a*) P(5, 3) = 5.4.3 = 60,

P(4, 4) = 4.3.2.1 = 24,

P(7, 2) = 7.6 = 42.

(*b*) $\dfrac{20!}{17!} = 20.19.18 = 6840,$

$$\frac{100!}{98!} = 100.99 = 9900$$

$$P(7, 0) = \frac{7!}{7!} = 1.$$

Problem 1.52. *A gentleman has 6 friends to invite. In how many ways can he send invitation cards to them, if he has three servants to carry the cards ?*

Solution. A card can be send to any one friend by any one of the three servants.

Let us take the tasks of sending cards to six friends as

$$T_1, T_2, T_3, T_4, T_5 \text{ and } T_6.$$

Each of the tasks can be completed in three distinct ways according to the number of servants to carry the cards.

Thus, by the multiplication principle of counting the tasks $T_1T_2T_3T_4T_5T_6$ can be performed in $3 \times 3 \times 3 \times 3 \times 3 \times 3 = 729$ ways.

Problem 1.53. *How many numbers of three digits can be formed with the digits 1, 2, 3, 4 and 5 if the digits in the same number are not repeated ? How many such numbers are possible between 100 and 10,000 ?*

Solution. Here we have to find the number of permutations of 5 distinct objects (digits) taken 3 at a time.

This is given by ${}^5P_3 = 5 \times 4 \times 3 = 60.$

The numbers between 100 and 10,000 are numbers of three digits and of four digits.

The total number of three digits numbers, formed with the given digits, is calculated above and it is equal to 60.

Similarly, the total number of four digits numbers, formed with 1, 2, 3, 4 and 5, is given by 0.

$${}^5P_4 = 5 \times 4 \times 3 \times 2 = 120.$$

Thus, the required number is $60 + 120 = 180$.

Problem 1.54. *A telegraph has 5 arms and each arm is capable of 4 distinct positions, including the position of rest. What is the total number of signals that can be made ?*

Solution. There are five arms say T_1, T_2, T_3, T_4 and T_5.

Each arm can be in any one of the four positions.

For each of the position of arm T_1, there are four possible positions for the second arm T_2, for each of the possible positions for T_1T_2, there are four possible positions for the third arm T_3 and so on.

Thus, by the multiplication principle of counting the total possible positions for $T_1T_2T_3T_4T_5$ is $4 \times 4 \times 4 \times 4 \times 4 = 1024$.

Since each distinct position is a distinct signal, so total number of possible signals is 1024 including the signal (meaningless) corresponding to the situation when all arms are in rest position.

Therefore, the total number of signals that can be generated is $1024 - 1 = 1023$.

Problem 1.55. *Find the number of positive integers greater than a million that can be formed with the digits 2, 3, 0, 3, 4, 2 and 3.*

Solution. The numbers greater than a million must be of 7 digits.

In the given set of digits, 2 appear twice, 3 appear thrice and all others are distinct.

Thus, the total number of seven digit numbers that can be formed with given digits is $\frac{7!}{2!3!} = 420.$

The set of these 420 positive integers, include some numbers which begin with 0.

Clearly, these nubmers are less than a million and they must not be counted in our answer.

The number of such numbers is given by the permutations of 6 non-zero digits and is equal to $\frac{6!}{2!3!} = 60.$

Therefore, the number of positive integers greater than a million that can be formed with given digits is equal to $420 - 60 = 360$.

Problem 1.56. *How many distinguishable permutations of the letters in the word BANANA are there ?*

Solution. The word "BANANA" has 6 letters. All the letters are not distinct. Let us use sub-script to distinguish them temporarily.

Let the letters be B, A_1, N_1, A_2, N_2, A_3.

Thus, the number of permutations are $6 ! = 720$.

Some of the permutations are identical like $A_1A_2A_3BN_1N_2$ and $A_1A_2A_3BN_2N_1$ except the order in which the N's appear.

This means that if we drop the subscripts, the total number of permutations will be $\frac{6!}{2!} = 360.$

Similarly, if we drop subscript with A's then total number of distinguishable permutations are $\frac{360}{3!} = 60.$

Problem 1.57. *There are 10 stalls for animals in an exhibition. Three animals ; lion, pussycat and horse are to be exhibited. Animals of each kind are not less than 10 in number. What is the possible number of ways of arranging the exhibition.*

Solution. There are three types of animals and 10 stalls.

One stall can be filled by any of the three animals. Once the first stall is filled, the second stall can be filled again in three ways by placing any of the three animals in it.

We have to fill 10 such stalls and number of each animal is greater than equal to 10, so we have $3^{10} = 59049$ ways to fill the stalls.

Thus, we can arrange the exhibition in 59049 ways.

Problem 1.58. *In how many ways can 10 different examination papers be scheduled so that*

(i) the best and the worst always come together

(ii) the best and the worst never come together ?

Solution. (*i*) Let us consider the best and the worst paper together and consider them as one object.

We have, now, 9 objects (papers).

These 9 objects can be arranged in 9 ! ways.

And in each of these 9 ! arrangements, the best and the worst papers can be arranged in 2 ! ways.

Therefore, the numebr of ways in which the 10 papers can be scheduled in this situation

$$= 2\,! * 9\,! = 725760.$$

(*ii*) Without any restriction, the 10 papers can be scheduled in 10 ! ways.

We have just calculated in part (*i*) that total number of ways in which the 10 papers can be scheduled so that the best and the worst always come together = 725760.

Therefore, the number of ways of scheduling 10 papers so that the best and the worst never come together = 10 ! – 725760 = 3628800 – 725760 = 2903040.

Problem 1.59. *In how many different ways can 5 men and 5 women sit around a table, if*

(*i*) *there is no restriction*

(*ii*) *no two women sit together ?*

Solution. The problem is related to circular permutation of 10 objects (5 men and 5 women). If there is no restriction then the number of permutations is (10 – 1) ! = 9 ! = 362880.

Notice here the difference in arrangement between clockwise and anticlockwise.

In the second case, there is a restriction that no two women are allowed so sit side by side.

To meet this restriction each woman should occupy a sit between two men.

The number of ways five men can sit around a table is 4 ! = 24.

Once these five men have sat on alternate chairs, the five women can occupy the 5 empty chairs in 5 ! ways.

Thus, total number of ways, in this case, will be 24 * 5 ! = 24 * 120 = 2880.

Problem 1.60. *Find the sum of all the four-digit numbers that can be formed with the digits 3, 2, 3 and 4.*

Solution. One thing is worth noticing here that a four-digit number so formed does not conatin a repeated digit except the digit 3.

This is implied from the question, because if it were not so, 3 should not have been repeated in the list of the digits.

To find the sum of the four-digit numbers formed with 3, 2, 3 and 4, we have to calculate the sum of digits at unit place in all such numbers.

The sum of the digits at ten, hundred and thousand place will be the same, only theirs place value will change.

The number of four-digit numbers in which 2 appears at unit place is determined by the number of permutations of digits 3, 3 and 4 to fill ten, hundred and thousand place. And this is $\frac{3\,!}{2\,!} = 3$.

Similarly, the number of four-digit numbers in which 3 appears at unit place is 3 ! = 6.

The number of four-digit numbers in which 4 appears at unit place is $\frac{3!}{2!} = 3$.

Therefore, sum of the digits in the unit place of all the numbers = $3 \times 2 + 6 \times 3 + 3 \times 4 = 36$.

As stated above, the sum of the digits in all such numbers at ten, hundred and thousand places is 36 each.

Thus, the sum of all such numbers

$$= 36 \times 1000 + 36 \times 100 + 36 \times 10 + 36$$
$$= 39996.$$

Problem 1.61. *We are asked to make slips for all numbers up to five-digit. Since the digits 0, 1, 6, 8 and 9 can be read as 0, 1, 9, 8 and 6 when they are read upside down, there are pairs of numbers that can share same slip if the slips are read upside down or right sideup (e.g. 89166 and 99168).*

Find the number of slips required for all five-digit numbers.

Solution. We have 10 digits. We have to make all five digit numbers.

The total such numbers is equal to 10^5.

Here we have to make slips for these many numbers. The numbers made of digits 0, 1, 6, 8 and 9 can be read upside down or right side up.

And, there are 5^5 many such five-digit numbers (all those five-digit numbers made of digits 0, 1, 6, 8 and 9).

Out of these 5^5 many numbers, however, there are some numbers that read the same either upside down or right side up.

For example, 91816, and there are 3×5^2 such numbers (center place filled with 0, 1 or 8).

Thus, there are $5^5 - 3 \times 5^2$ numbers that can be read upside down or right side up.

And, for these numbers we need only

$$\frac{1}{2}(5^5 - 3 * 5^2) \text{ number of slips.}$$

Therefore, number of slips required to be made is $10^5 - \frac{1}{2}(5^5 - 3 * 5^2)$.

Problem 1.62. *How many binary sequences of r-bits long have even numebr of 1's ?*

Solution. There will be 2^r possible binary sequences of r-bits long.

This can be verified by the permutation of objects when repetitions are allowed.

There are r places and two objects (0 and 1).

The first place can be filled in 2 ways, for each of these, the second place can be filled in 2 ways, so we have 2×2 ways to fill first two places. Extending the sequence upto the rth steps, we have 2^r possible arrangements and hence 2^r possible binary sequences.

We can now make pairs of binary sequences in such a way that two sequences differ only at r^{th} place.

There will 2^{r-1} such pairs, and in each pair one sequence will have even number of 1's.

Thus, number of binary seqeunces of r bits long having even number of 1's = 2^{r-1}.

Problem 1.63. *How many different words can be made from letters of the word 'committee'.*

Solution. The word committee contains letters c, o and i once and m, t and e twice.

When a word is formed from these letters, a letter may appear at the most the number of time it appear in the word committee or not at all.

So generating function for c, o and i is given by $(1 + x)$ each, whereas, for c, m and e is given by $\left(1 + x + \frac{x^2}{2!}\right)$ each.

Thus, generating function for the problem is then given by

$$(1 + x)^3 \left(1 + x + \frac{x^2}{2!}\right)^3 = (1 + 3x + 3x^2 + x^3) \left(1 + x + \frac{x^2}{2!}\right)^3$$

If words are to be formed taking all the letters at once, then the numbers of such words is given by the coefficient of $\frac{x^9}{9!}$ and this is equal to $\frac{9!}{2!2!2!}$.

Problem 1.64. *A fair six-sided die is tossed four times and the numbers shown are recorded in a sequence. How many different sequences are there ?*

Solution. Let us assume, here, that each toss is an object and the number appearing on the face of a die is the number of times it occurs. Thus each object occurs at least once and at the most 6 times.

Also the order of appearing of different 1's (conceptually) to make it 2 or 3 or 4 or 5 or 6 is fixed and is one and only one way.

The generating function for first toss is given as $x + x^2 + x^3 + \ldots\ldots + x^6$.

The sum of all the coefficients, here, is 6 and number of possible sequences of numbers, one face of a die in one toss, is 6 only, die is tossed four times,

Therefore, generating function for the problem is

$$\{x + x^2 + x^3 + \ldots\ldots + x^6\}^4.$$

The number of sequence is the sum of coefficients of all the terms in this generating function. This is equal to 6^4.

Problem 1.65. *Find the generating function, also called enumerator, for permutations of n objects with the following specified conditions :*

(*a*) *each object occurs at the most twice*

(*b*) *each object occurs at least twice*

(*c*) *each object occurs at least once and at the most k times.*

Solution. (*a*) Each object occurs at the most twice implies that an object may occur 0, 1 or 2 times. The exponential generating fnction for an object under this condition is given as

$$1 + x + \frac{x^2}{2!}.$$

There are n objects, so the generating function for the problem is written as

$$\left(1+x+\frac{x^2}{2!}\right)^n.$$

(*b*) In this case, an objcet occurs at least twice. This implies that an object may appear 2 or more times. The exponential generating function for an object under this condition is $\frac{x^2}{2!}+\frac{x^3}{3!}+\frac{x^4}{4!}+......$

Therefore, for the problem dealing with n objects, the generating function can be written as

$$\left(\frac{x^2}{2!}+\frac{x^3}{3!}+\frac{x^4}{4!}+......\right)^n.$$

(*c*) Here, each object occurs at least once and at the most k times.

That is to say that an object may occur 1 or 2 or 3 or ...... or k times.

The exponential generating function for an object under this condition is

$$x+\frac{x^2}{2!}+\frac{x^3}{3!}+......+\frac{x^k}{k!}$$

Since the problem for n objects, the generating function for the problem is written as

$$\left(x+\frac{x^2}{2!}+\frac{x^3}{3!}+......+\frac{x^k}{k!}\right)^n.$$

Problem 1.66. *How many ways are there to select a first prize winner, a second prize winner, and a third-prize winner from 100 different people who have entered a contest ?*

Solution. Because it matters which person wins which prize, the number of ways to pick the three prize winners is the number of ordered selections of three elements from a set of 100 elements, that is, the number of 3-permutations iof a set of 100 elements.

Consequently, the answer is

$$P(100, 3) = 100.99.98 = 970{,}200.$$

Problem 1.67. *Suppose that a saleswoman has to visit eight different cities. She must begin her trip in a specified city, but she can visit the other seven cities in any order she wishes. How many possible orders can the saleswoman use when visiting these cities ?*

Solution. The number of possible paths between the cities is the number of permutations of seven elements, since the first city is determined, but the remaining seven can be ordered arbitrarily.

Consequently, there are $7! = 7.6.5.4.3.2.1 = 5040$ ways for the saleswoman to choose her tour.

If, for instance, the saleswoman wishes to find the path between the cities with minimum distance, and she computes the total distance for each possible path, she must consider a total of 5040 paths.

Problem 1.68. *Suppose that there are eight runners in a race. The winner receives a gold medal the second-place finisher receives a silver medal, and the third-place finisher receives a bronze medal. How many different ways are there to award these medals, if all possible outcomes of the race can occur and there are no ties ?*

Solution. The number of different ways to award the medals is the number of 3-permutations of a set with eight elements.

Hence, there are P(8, 3) = 8.7.6 = 336 possible ways to award the medals.

Problem 1.69. *How many permutations of the letters ABCDEFGH contain the string ABC ?*

Solution. Because the letters ABC must occur as a block, we can find the answer by finding the number of permutations of six objects, namely, the block ABC and the individual letters D, E, F, G and H.

Because these six objects can occurs in any order, there are 6 ! = 720 permutations of the letters ABCDEFGH in which ABC occurs as a block.

1.3 COMBINATIONS

Let A be a set with | A | = n, and let $1 \le r \le n$. Then the number of combinations of the elements of A, taken r at a time that is the number of r-element subsets of A is

$$\frac{n!}{r!(n-r)!}.$$

The number of combinations of A, taken r at a time, does not depend on A, but only on n and r. This number is often written nC_r and is called the number of combinations of n objects taken r at a time.

We have $\;{}_nC_r = \dfrac{n!}{r!(n-r)!}$.

• Suppose k selections are to be made from n items without regard to order and that repeats are allowed, assuming at least k copies of each of the n items. The number of ways these selections can be made is $(n + k - 1)\,C_k$.

Problem 1.70. *Show that* ${}_nC_r = {}_nC_{n-r}$.

Solution. We have $^{n}C_r = \dfrac{n!}{r!(n-r)!} = \dfrac{n!}{(n-(n-r))!(n-r)!} = {}_nC_{n-r}$.

Problem 1.71. *Compute the number of distinct five-card hands than can be dealt from a deck of 52 cards.*

Solution. This number is ${}_{52}C_5$ because the order in which the cards were dealt is irrelevant.

$${}_{52}C_5 = \frac{52!}{5!47!} \quad \text{or} \quad 2{,}598{,}960.$$

Problem 1.72. *In how many ways can a prize winner choose three CDs from the top ten list if repeats are allowed ?*

Solution. Here n is 10 and k is 3.

There are $(10 + 3 - 1)C_3$ or ${}_{12}C_3$ ways to make the selections.

The prize winner can make the selection in 220 ways.

Problem 1.73. *How many different seven-person committees can be formed each containing three women from an available set of 20 women and four men from an available set of 30 men ?*

Solution. In this case a committee can be formed by performing the following two tasks in succession :

Task 1 : Choose three women from the set of 20 women.

Task 2 : Choose four men from the set of 30 men.

Here order does not matter in the individual choices, so we are merely counting the number of possible subsets.

Thus task 1 can be performed in ${}_{20}C_3$ or 1140 ways and task 2 can be performed in ${}_{30}C_4$ or 27, 405 ways.

By the multiplication principle, there are (1140) (27405) or 31,241,700 different committees.

Theorem 1.7. *Let n and r be integers with $n \geq 0$ and $0 \leq r \leq n$. The number of ways to choose r objects from n is* $\binom{n}{r}$.

Proof. If $r = 0$, the result is true, because there is just one way to choose 0 objects (do nothing !), while $\binom{n}{0} = \dfrac{n!}{0!(n-0)!} = 1$, because $0! = 1$.

Thus, we may assume that $r \geq 1$ and hence $n \geq 1$.

Let N be the number we are seeking, that is, there are N ways to choose r objects from the n given objects.

Notice that for each way of choosing r objects, there are $r!$ ways to order them.

By the multiplication rule, the number of r-permutations of n objects (which we know is $P(n, r)$) is the number of ways to choose r objects multiplied by $r!$, the number of ways to order the r objects.

$$P(n, r) = N \times r!$$

Therefore, $N = \dfrac{P(n, r)}{r!} = \dfrac{n!}{r!(n-r)!} = \binom{n}{r}$.

Corollary. The number of r-combinations of n objects is $\binom{n}{r}$.

Problem 1.74. *Wanda is going to toss a coin eight times. In how many ways can she get five heads and three tails ?*

Solution. Wanda might get a string of five heads followed by three tails (denote this possibility HHHHHTTT), or a string of three tails followed by five heads, TTTHHHHH, or the sequence HTHHTHTH, and so on.

The number of such sequences is the number of ways of selecting five occasions (from the eight) on which the heads should arise or, equivalently, the number of ways of selecting the three occasions on which tails should come up.

The answer is $\binom{8}{5} = \binom{8}{3} = 56$.

Problem 1.75. *Explain why* $\binom{n}{r} = \binom{n}{n-r}$.

Solution. Suppose we have n white marbles and we wish to point r of them black.

Choosing the r marbles is equivalent to choosing the $n-r$ marbles which are to remain white.

Thus, each choice of r marbles from n corresponds to a choice of the remaining $n-r$, so the numbers of choices are the same.

There are $\binom{n}{r}$ and $\binom{n}{n-r}$, respectively.

Problem 1.76. *Mr. Hiscock has ten children but his car holds only five people (including driver). When he goes to the circus, in how many ways can he select four children to accompany him ?*

Solution. The question involves choosing, not order.

$$\text{There are } \binom{10}{4} = \frac{10!}{4!6!} = 210 \text{ different ways.}$$

Theorem 1.8. *C(n, r) = C(n, n – r).*

Proof. (Algebraic) :

We have that $C(n, n-r) = \dfrac{n!}{[n-(n-r)]!(n-r)!}$

Simplifying, we find that

$$C(n, n-r) = \frac{n!}{r!(n-r)!},$$ which in turn is equal to $C(n, r)$, as was to be shown.

Theorem 1.9. *$C(n, k) = C(n-1, k) + C(n-1, k-1)$ for $n > k > 0$.*

Proof. (Combinatorial)

Let A be a set of cardinality n and let k be an integer such that $0 < k < n$.

There are $C(n, k)$ different subsets of A of cardinality k.

Let y be an element of A. Every subset of A either includes y or does not.

There are $C(n-1, k)$ different subsets of A of cardinality k that do not include y.

(We form such a subset from the $(n-1)$ elements of A that are not equal to y).

There are $C(n-1, k-1)$ ways to choose a subset of A of cardinality k that includes y.

(To form such a subset, we must choose $(k-1)$ elements in addition to y from the $(n-1)$ elements in A that are not equal to y).

Adding, we find that there are $C(n-1, k) + C(n-1, k-1)$ ways in which to choose a subset of cardinality k from A.

Thus, $C(n, k) = C(n-1, k-1) + C(n-1, k)$.

Algebraic Proof. By definition, we have $C(n-1, k) + C(n-1, k-1)$

$$= \frac{(n-1)!}{[(n-1)-k]!\,k!} + \frac{(n-1)!}{[(n-1)-(k-1)]!\,(k-1)!}$$

Simplifying, finding a common denominator, and adding, we find that the sum is equal to

$$\frac{(n-k)\,(n-1)!+(n-1)!\,k}{(n-k)!\,k!}.$$

Using the distributive law, we find that the latter sum is in turn equal to $\frac{n!}{(n-k)!\,k!}$, as was to be proved.

- In general, combinations of n items taking r at a time where r lies between a and b where $1 \le a$ and $b \le n$ is given by the sum
$$^nC_a + {}^nC_{a+1} + {}^nC_{a+2} + \ldots\ldots + {}^nC_b.$$
- The number of ways of distributing $p + q$ different objects between two distinguishable groups in such a ways that one group gets P objects and other gets q objects is given by $\frac{(p+q)!}{p!\,q!}$.
- If the two groups are indistinguishable in the above case then, the number of ways of distribution is given by
$$\frac{(p+q)!}{2!\,p!\,q!}.$$
- In general, if n different objects are to be distributed among m distinguishable groups containing $P_1, P_2, P_3, \ldots\ldots P_m$ objects, where $P_1 + P_2 + \ldots\ldots + P_m = n$.
The number of ways in which this takes can be completed is given by
$$\frac{n!}{P_1!\,P_2!\,P_3!\ldots\ldots P_m!}.$$
- In the previous result if the m groups are indistinguishable, the number of ways of distribution is given by $\frac{n!}{m!\,P_1!\,P_2!\,P_3!\ldots\ldots P_m!}$.

Problem 1.77. *How many committees of five people can be chosen from 20 men and 12 women.*
(*a*) *if exactly three men must be on each committee ?*
(*b*) *if at least four women must be on each committee ?*

Solution. (*a*) We must choose three men from 20 and then two women from 12.

The answer is $\binom{20}{3}\binom{12}{2} = 1140(66) = 75{,}240$.

(*b*) We calculate the case of four women and five women separately and add the results (using the addition rule).

The answer is $\binom{12}{4}\binom{20}{1} + \binom{12}{5}\binom{20}{0} = 495(20) + 792 = 10{,}692.$

Problem 1.78. *In how many ways can 20 students out of a class of 32 be chosen to attend class on a late Thursday afternoon (and take notes for the others) if*

(*a*) *Paul refuses to go to class ?*

(*b*) *Michelle insists on going ?*

(*c*) *Jim and Michelle insist on going ?*

(*d*) *either Jim or Michelle (or both) go to class ?*

(*e*) *just one of Jim and Michelle attend ?*

(*f*) *Paul and Michelle refuse to attend class together ?*

Solution. (*a*) The answer is $\binom{31}{20} = 84{,}672{,}315$, since, in effect, it is necessary to select 20 students from the 31 students excluding Paul.

(*b*) Now the number of possibilities is $\binom{31}{19} = 141{,}120{,}525$ since 19 students must be chosen from 31.

(*c*) The answer is $\binom{30}{18} = 86{,}493{,}225$, it being necessary to choose the remaining 18 students from a group of 30.

(*d*) Let J be the set of classes of 20 which contain Jim and M the set of classes of 20 which contain Michelle. The question asks for $|J \cup M|$.

Using the principle of Inclusion-Exclusion, we obtain,

$$|J \cup M| = |J| + |M| - |J \cap M| = \binom{31}{19} + \binom{31}{19} - \binom{30}{18} = 195{,}747{,}825.$$

(*e*) Using the formula $|J \oplus M| = |J| + |M| - 2|J \cap M|$

We obtain, $\binom{31}{19} + \binom{31}{19} - 2\binom{30}{18} = 109{,}254{,}600.$

(*f*) The number of classes containing Paul and Michelle is $\binom{30}{18}$. So the number which do not contain both is

$$\binom{32}{20} - \binom{30}{18} = 139{,}299{,}615.$$

Problem 1.79. *A group of 30 people have been trained as astronauts to go on the first mission to Mars. How many ways are there to select a crew of six people to go on this mission (assuming that all crew members have the same job) ?*

Solution. The number of ways to select a crew of six from the pool of 30 people is the number of 6-combinations of a set with 30 elements, because the order in which these people are chosen does not matter.

The number of such combinations is

$$C(30, 6) = \frac{30!}{6!24!} = \frac{30.29.28.27.26.25}{6.5.4.3.2.1} = 593{,}775.$$

Problem 1.80. *How many ways are there to select a committee to develop a discrete mathematics course at a school if the committee is to consists of three faculty members from the mathematics department and four from the computer science department, if there are nine faculty members of the mathematics department and 11 of the computer science department ?*

Solution. By the product rule, the answer is the product of the number of 3-combinations of a set with nine elements and the number of 4-combinations of a set with 11 elements.

The number of ways of select the committee is

$$C(9, 3) . C(11, 4) = \frac{9!}{3!6!} \cdot \frac{11!}{4!7!}$$

$$= 84.330 = 27{,}720.$$

Problem 1.81. *How many ways are there to select five players from a 10-members tennis team to make a trip to a match at another school ?*

Solution. The answer is given by the number of 5-combinations of a set with ten elements.

The number of such combinations is

$$C(10, 5) = \frac{10!}{5!5!} = 252.$$

Problem 1.82. *How many bit strings of length n contain exactly r 1 s ?*

Solution. The positions of r 1s in a bit string of length n form an r-combination of the set

$$\{1, 2, 3, ..., n\}$$

Hence, there are $C(n, r)$ bit strings of length n that contain exactly r 1s.

Problem 1.83. *A person has 8 children of them he takes 3 at a time to a circus. He does not take the same three children twice to the circus. How many times he will have to go to circus to ensure that every three children have seen the circus together ? In this case find the number of times a particular child has visited the circus.*

Solution. Here we have to find the numebr of combinations of 8 children taken 3 at a time.

Note that order of selection of child is not important in this case.

This selection can be made in ${}^8C_3$ ways.

We can make ${}^8C_3 = 56$ distinct groups of three children, and for each such group, the person will have to go to circus once.

Therefore, the person will have to visit circus 56 times.

A particular child goes to circus with every possible pair of two children out of remaining 7 children.

Number of such possible pair is ${}^7C_2 = 21$.

Therefore, a particular child goes to circus 21 times.

Problem 1.84. *In a party of 30 people, each shakes hand with others. How many handshakes took place in the party ?*

Solution. In a normal case, a handshake involves two persons. This case is of counting 2-elements subsets of a set containing 30 elements.

And this count is ${}^{30}C_2 = 30 \times \frac{29}{2} = 435$.

Problem 1.85. *From 8 men and 4 women and team of 5 is to be formed. In how many ways can this be done so as to include at least one woman ?*

Solution. This is a case of restricted combination.

Total number of persons = 8 men + 4 women = 12 persons.

A team of 5 has to be made, and this can be made in ${}^{12}C_5$ ways.

This count includes the case of teams containing all five men (*i.e.*, no women in the team) which is equal to ${}^8C_5$.

Thus, number of ways in which the specified team can be selected = ${}^{12}C_5 - {}^8C_5$

$$= \frac{12!}{5!7!} - \frac{8!}{5!3!}$$

$$= \frac{12 \times 11 \times 10 \times 9 \times 8}{5 \times 4 \times 3 \times 2} - \frac{8 \times 7 \times 6}{3 \times 2}$$

$$= 792 - 56 = 736.$$

Problem 1.86. *There are 10 points in a 2-D plane. Four of these are co-linear. Find the number of different straight lines that can be drawn by joining these points.*

Solution. Any two points are always co-linear. So, a line can be drawn between any two points.

If there are three non-co-linear points (a single line cannot be drawn joining all these three points).

We can draw $3 = {}^3C_2$ distinct lines.

A triangle is an example of this. Thus, we can draw ${}^{10}C_2$ distinct lines joining 10 points. Out of these 10 points, 4 are co-linear. So ${}^4C_2$ lines will be same and we consider them as one line.

Therefore, actual number of lines that can be drawn = ${}^{10}C_2 - {}^4C_2 + 1 = 45 - 6 + 1 = 40$.

Problem 1.87. *What is the number of diagonals that can be drawn in a polygon of n sides ?*

Solution. A polygon having n sides has n vertices.

A diagonal is a line between two points, which are not adjacent to each other.

The total number of line that can be drawn in a polygon of n vertices = ${}^nC_2$.

There are already n sides (lines) which are not diagonal,

the remaining lines will be diagonals.

Therefore, the number of diagonals that can be drawn in a polygon of side n is equal to

$$^nC_2 - n = \frac{n!}{2!(n-2)!} - n = \frac{n(n-1)}{2} - n$$

$$= \frac{n^2 - n - 2n}{2} = \frac{n(n-3)}{2}.$$

Problem 1.88. *If three dice are rolled, and we make a set of numbers shown on the three dice. How many different sets are possible.*

Solution. Rolling three dice is equivalent to selecting three numbers from the list of six numbers 1, 2, 3, 4, 5 and 6 with repetitions allowed. Because sequence 111, 121 etc. are possible.

Thus the different possible combinations is

$$^{6+3-1}C_3 = {}^8C_3 = 56.$$

Problem 1.89. *A bookstore allows the recipient of a gift coupon to choose 6 books from the combined list of 10 best-selling fiction books and 10 best-selling non-fiction books. In how many different ways can the selection of 6 books be made ?*

Solution. The number of different types of books is 10 + 10 = 20.

A gift coupon recipient may select any 6 books, possible 6 copies of a single book.

This is a case of selection of 6 objects from 20 objects with repetitions allowed.

The number of ways the selection can be made is

$$^{20+6-1}C_6 = {}^{25}C_6 = 177100.$$

Problem 1.90. *In an election the number of candidates is one more than the number of vacancies. If a voter can vote in 30 different ways, find the number of candidates.*

Solution. Let the number of candidates be x. An elector may vote for any one or, any two or, any three up to maximum of any $x - 1$ candidates from total of 4, because number of vacancies is $x - 1$ only.

Therefore, number of ways in which an elector can cast his vote is

$^xC_1 + {}^xC_2 + {}^xC_3 + \ldots\ldots + {}^xC_{x-1}$ and this value is given to be 30.

Thus, $^xC_1 + {}^xC_2 + {}^xC_3 + \ldots\ldots + {}^xC_{x-1} = 30.$

or $\quad ^xC_0 + {}^xC_1 + {}^xC_2 + {}^xC_3 + \ldots\ldots + {}^xC_{x-1} + {}^xC_x - {}^xC_0 - {}^xC_x = 30.$

$\Rightarrow \quad 2^x - 2 = 30$

$\Rightarrow \quad x = 5.$

Therefore, the number of candidates is 5.

Problem 1.91. *In an election, there are four candidates contesting for three vacancies, an elector can vote for any number of candidates not exceeding the number of vacancies. In how many ways can one cast his votes ?*

Solution. An elector may vote for any one or, any two, or any three candidates out of total 4.

Therefore, an elector may vote in ${}^4C_1 + {}^4C_2 + {}^4C_3$

$= 4 + 6 + 4 = 14$ different possible ways.

Problem 1.92. *Find the total number of selections of at least one red ball from 4 red balls and 3 green balls, if*

(*a*) *the balls of the same color are different*

(*b*) *the balls of the same color are identical.*

Solution. (*a*) From 4 different red balls and 3 different green balls, we have to find number of selections taking at least one red ball and any number of (including 0) 3 green balls.

The total number of ways of selecting at least one red ball from 4 different red balls

$$= {}^4C_1 + {}^4C_2 + {}^4C_3 + {}^4C_4 = 15.$$

Corresponding to each of these selections, the number of ways of selecting green balls

$$= {}^3C_0 + {}^3C_1 + {}^3C_2 + {}^3C_3 = 8.$$

Therefore, total number of different ways of selection

$$= 15 \times 8 = 120.$$

Problem 1.93. *In an examination a candidate has to pass in each of the 5 papers. How many different combinations of papers are there so that a student may fail ?*

Solution. For a student to pass the examination, he/she will have to pass in each of the five papers.

To fail, a student may fail in any one or, in any two or so on including in all the five papers.

Thus, a student may fail in as many as

$${}^5C_1 + {}^5C_2 + {}^5C_3 + {}^5C_4 + {}^5C_5 = 2^5 - 1 = 31 \text{ ways.}$$

Problem 1.94. *In how many ways can a pack of 52 cards be equally divided into four gruops ? If the cards are to be distributed equally among four players, then find the number of ways of this distribution.*

Solution. First part :

When 52 cards are distributed equally among four groups, each group contains 13 cards.

Since groups are indistinguishable, the number of ways of distribution is given by

$$\frac{52!}{4!13!13!13!}$$

Second part :

Here four groups (players) are distinguishable, thus number of ways of distribution is given by

$$\frac{52!}{13!13!13!13!}.$$

Problem 1.95. *A library has 5 black books, 4 red books and 3 yellow books, all with different titles. How many distinguishable ways can a student take home 6 books, 2 of each color ?*

Solution. In this case, the books of a particular color are distinguishable by their title.

Each book can be either selected or not, giving the possible number of selection for each book as 0 or 1.

This can be written as, in polynomial, form, as $1 + x$.

Thus, the generating function for 5 black books is $(1 + x)^5$.

Similarly, the generating function for 4 red books is $(1 + y)^4$ and for 3 yellow books is $(1 + z)^3$.

The count of number of ways 6 books, 2 of each color, can be selected, we take coefficient of $x^2y^2z^2$ in the generating of function

$$f(x) = (1 + x)^5(1 + y)^4(1 + z)^3.$$

The coefficient of $x^2y^2z^2$ is ${}^5C_2\,{}^4C_2\,{}^3C_2$

$$= 10 \times 6 \times 3 = 180.$$

Problem 1.96. *A library has 5 indistinguishable black books, 4 indistinguishable red books and 3 indistinguishable yellow books. In how many distinguishable ways can a student take home (a) 6 books ? (b) 6 books taking atleast 1 of each color ? (c) 6 books taking 2 of each color ?*

Solution. Here all books of a particular color are indistinguishable. A black (or red or yellow) book can be selected or, not selected.

If selected then maximum number of black books that can be selected is 5,

i.e., possible ways of selections for black books are 0, 1, 2, 3, 4 or 5.

Similarly, the possible ways of selections for red books are 0, 1, 2, 3 or 4 and for yellow books are 0, 1, 2 or 3.

If we take three variables x, y and z for the numbers of black, red and yellow books selected by a student, respectively then the number of solutions to the equations $x + y + z = 6$ where $0 \le x \le 5$, $0 \le y \le 4$ and $0 \le z \le 3$ is the required number of ways in which a student can take home 6 books.

The generating function for x is $(1 + x + x^2 + x^3 + x^4 + x^5)$, for y is $(1 + y + y^2 + y^3 + y^4)$ and for z is $(1 + z + z^2 + z^3)$.

If we replace y and z by x, we get the generating function for the above problem as

$$f(x) = (1 + x + x^2 + x^3 + x^4 + x^5)(1 + x + x^2 + x^3 + x^4)(1 + x + x^2 + x^3)$$

The coefficient of x^6 in $f(x)$ is the required number, and this number is 18.

In the second part, at least one book of each colour has to be selected, so generating function $f(x)$ is given as

$$f(x) = (x + x^2 + x^3 + x^4 + x^5)(x + x^2 + x^3 + x^4)(x + x^2 + x^3)$$

The coefficient of x^6 in $f(x)$ is the required number, and this number is 9.

In the third part, two books of each color are to be selected.

So generating function $f(x)$ is given as

$$f(x) = x^2 * x^2 * x^2.$$

The coefficient of x^6 in $f(x)$ is the required number, and this number is 1.

Problem 1.97. *In how many ways can one choose n pieces of fruit, assuming there are an infinitely large number of apples, bananas, oranges and pears, and he(she) wants an even number of apples, an odd number of bananas, not more than 4 oranges and atleast two pears ?*

Solution. The generating function for the selection of apple can be written as $(1 + x^2 + x^4 + x^6 + \ldots\ldots)$, for the selection of bananas can be written as $(x + x^3 + x^5 + x^7 + \ldots\ldots)$, for the selection of oranges can be written as $(1 + x + x^2 + x^3 + x^4)$ and for pears it can be written as $(x^2 + x^3 + x^4 + \ldots\ldots)$.

Therefore, the generating function for this problem is

$$f(x) = (1 + x^2 + x^4 + x^6 + \ldots\ldots)(x + x^3 + x^5 + x^7 + \ldots\ldots)(1 + x + x^2 + x^3 + x^4)(x^2 + x^3 + x^4 + \ldots\ldots)$$

Problem 1.98. *In how many ways 2n + 1 items can be distributed among 3 persons so that the sum of the number of items received by any two persons should exceed the number of items received by the other ?*

Solution. Let us take three variables x, y and z for the number of items received by the three persons. Then number of ways $2n + 1$ items can be distributed among three persons is equal to the number of positive integer solutions to the equation

$$x + y + z = 2n + 1 \text{ where } 1 \le x, y, z \le n.$$

To ensure that the sum of the two variables must exceed the third one, it is important to define the range of values for each variable as above.

Therefore, the generating function for the problem is written as

$$f(x) = (x + x^2 + x^3 + \ldots\ldots + x^n)(x + x^2 + x^3 + \ldots\ldots + x^n)(x + x^2 + x^3 + \ldots\ldots + x^n)$$

$$= (x + x^2 + x^3 + \ldots\ldots + x^n)^3$$

$$= \left\{\frac{x(1 - x^n)}{1 - x}\right\}^3$$

$$= x^3(1 - 3x^n + 3x^{2n} - x^{3n})(1 - x)^{-3}.$$

In this function $f(x)$, the coefficient of x^{2n+1} is the count we are looking for.

And, this coefficient is equal to the coefficient of x^{2n-2} in $(1 - 3x^n + 3x^{2n} - x^{3n})(1 - x)^{-3}$.

This value is ${}^{3+2n-2-1}C_{2n-2} - 3 \times {}^{3+n-2-1}C_{n-2}$

or,
$$ {}^{2n}C_{2n-2} - 3 \times {}^{n}C_{n-2} = \frac{2n(2n-1)}{2!} - 3\frac{n(n-1)}{2!}$$

$$= \frac{n}{2}(4n - 2 - 3n + 3)$$

$$= \frac{n(n+1)}{2}.$$

Problem 1.99. *A valid password consists of seven symbols. Symbols are chosen from digits and Roman capital alphabets. The first symbol of the password must be a Roman capital alphabet. How many different passwords are possible ?*

Solution. Number of symbols = 26 + 10 = 36

The first place of seven characters password can be chosen in 26 ways.

The remaining 6 places can be filled in 36 ways each *i.e.*, the second symbol for a password can be chosen in 36 ways, and for each of this, the third place can be filled in 36 ways and so on.

Thus, the number of different possible passwords

$$= 26 \times (36)^6 = 56596340736.$$

1.4 PERMUTATIONS AND COMBINATIONS WITH REPETITIONS

Factorials :

Frequently it is useful to have a simple notation for products such as

$$4.3.2.1, \; 6.5.4.3.2.1 \quad \text{or} \quad 7.6.5.4.$$

For each positive integer we define $n\,! = n(n-1)(n-2)\;......\;3.2.1$

$= $ the product of all integers from 1 to n.

Also define $0\,! = 1$, note that $1\,! = 1$

Thus, $\quad 4! = 4.3.2.1,\; 6\,! = 6.5.4.3.2.1$

and $$7.6.5.4 = \frac{7\,!}{3\,!} = \frac{7.6.5.4.3.2.1}{3.2.1}$$

We read n ! as "n factorial".

It is true that $4\,! = 24$ and $6\,! = 720$ but frequently we leave our answers in factorial form rather than evaluating the factorials.

Nevertheless, the relation $n\,! = n[(n-1)\,!]$ enables us to compute the values of n ! for small n fairly quickly.

For example :

$0\,! = 1,$	$1\,! = 1,$	$2\,! = 2$
$3\,! = 6,$	$4\,! = 24,$	$5\,! = 120$
$6\,! = 720,$	$7\,! = 5040,$	$8\,! = 40320$
$9\,! = 362880,$	$10\,! = 3628800,$	$11\,! = 39916800.$

Theorem 1.10. *Enumerating r-permutations without repetitions*

$$P(n, r) = n(n-1)\;......\;(n-r+1) = \frac{n!}{(n-r)!}\,.$$

Proof. Since there are n distinct objects, the first position of an r-permutation may be filled in n ways.

This done, the second position can be filled in $n - 1$ ways since no repetitions are allowed and there are $n - 1$ objects left to choose from.

The third can be filled in $n - 2$ ways and so on until the rth position is filled in $n - r + 1$ ways . (See figure below).

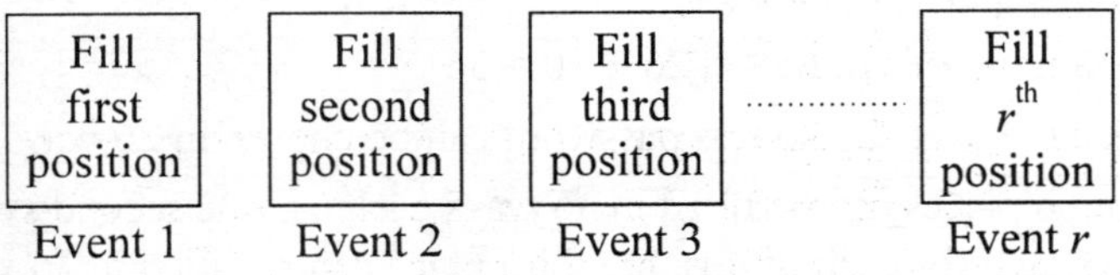

Fig. 1.3.

By applying the product rule, we conclude that

$$P(n, r) = n(n-1)(n-2)\ldots\ldots (n-r+1)$$

From the definition of factorials, it follows that

$$P(n, r) = \frac{n!}{(n-r)!}.$$

When $r = n$, this formula becomes

$$P(n, n) = \frac{n!}{0!} = n!.$$

Corollary

There are n ! permutations of n distinct objects.

Example 1. *There are* 3 ! = 6 permutations of $\{a, b, c\}$. There are 4 ! = 24 permutations of $\{a, b, c, d\}$. The number of 2-permutations of $\{a, b, c, d, e\}$ is $P(5, 2) = \frac{5!}{(5.2)!} = 5.4 = 20.$

The number of 5-letters words using the letters a, b, c, d and e at most once is P(5, 5) = 120.

Example 2. *There are P(10, 4) = 5040, 4-digit numbers that contain no repeated digits since each such number is just an arrangement of four of the digits 0, 1, 2, 3, ..., 9*

(leading zeros are allowed)

There are P(26, 3), 3-letters words formed from the English alphabet with no repeated letters.

Thus, there are P(26, 3) P(10, 4) license places formed by 3-distinct letters followed by 4-distinct digits.

Theorem 1.11. *Enumerating r-permutations with unlimited repetitions*

$$U(n, r) = n^r.$$

Proof. Each of the r positions can be filled in n ways and so by the product rule, $U(n, r) = n^r$.

Theorem 1.12. (Enumerating n-Permutations with constrained repetitions)

$$P(n\,;\, q_1, \ldots\ldots, q_t) = \frac{n!}{q_1!\, q_2! \ldots\ldots q_t!}$$

$$= C(n, q_1)\, C(n-q_1, q_2)\, C(n-q_1-q_2, q_3) \ldots\ldots C(n-q_1-q_2 \ldots\ldots - q_{t-1}, q_t).$$

Proof. Let $x = P(n\,;\, q_1, q_2, \ldots\ldots, q_t)$

If the q_1 a_1's were all different there would be $(q_1!)\,x$ permutations since each old permutation would give rise of q_1 ! new permutations corresponding to the number of ways of arranging the q_1 distinct objects in a row. If the q_2 a_2's were all replaced by distinct objects, then by similar reasoning there would be $(q_2!)(q_1!\,x)$ permutations.

If we repeat this procedure until all the objects are distinct. We will have $(q_t!) \ldots\ldots (q_2!)(q_1!)x$ permutations.

However, we know that there are n ! permutations of n distinct objects.

Equating these two quantities and solving for x gives the first equality of the theorem.

The second equality is obtained as follows :

First choose the q_1 positions for the a_1's ; then from the remaining $n - q_1$ positions, choose q_2 positions for the a_2's and so on.

Note that at the last we will have left $n - q_1 - q_2 \ldots\ldots - q_{t-1} = q_t$

Positions to fill with the q_t a_t's, so

$$C(n - q_1 - q_2 \ldots\ldots - q_{t-1}, q_t) = C(q_t, q_t).$$

The last equality of the theorem follows because both numbers represent the same number of permutations.

Example :

The number of arrangements of letters in the word TALLAHASSEE is

$$P(11 ; 3, 2, 2, 2, 1, 1) = \frac{11!}{3!2!2!2!1!1!}$$

Since this equals the nubmer of permutations of {3.A, 2.E, 2.L, 2.S, 1.H, 1.T}.

The number of arrangements of these letters that begin with T and end with E is $\frac{9!}{3!1!2!2!1!}$.

Theorem 1.13. Enumerating r-combinations wtih unlimited repetitions

$V(n, r)$ = the number of r-combinations of n distinct objects wtih unlimited repetitions

= the number of nonnegative integral solutions to $x_1 + x_2 + \ldots\ldots + x_n = r$

= the number of ways of distributing r similar balls into n numbered boxes

= the number of binary numbers with $n - 1$ one's and zeros.

$= C(n - 1 + r, r) = C(n - 1 + r, n - 1)$

$$= \frac{(n + r - 1)!}{[r!(n - 1)!]}.$$

Remark. The number of r-combinations of $\{\infty . a_1, \infty . a_2, \ldots\ldots, \infty . a_n\}$ is the same as the number of r-combinations of $\{r . a_1, r . a_2, \ldots\ldots, r . a_n\}$.

Theorem 1.14. The number of integarl solutions of $x_1 + x_2 + \ldots\ldots + x_n = r$ where each $x_i > 0$

= the number of ways of distributing r similar balls into n numbered boxes with at least one ball in each box

$= C(n - 1 + (r - n), r - n) = C(r - 1, r - n)$

$= C(r - 1, n - 1)$.

Likewise, suppose that $r_1, r_2, \ldots\ldots, r_n$ are integers. Then the number of integral solutions of $x_1 + x_2 + \ldots\ldots + x_n = r$, where $x_1 \geq r_1, x_2 \geq r_2, \ldots\ldots$, and $x_n \geq r_n$.

= the number of ways of distributing r similar balls into n numbered boxes where there are atleast r_1 balls in the first box, atleast r_2 balls in the second box, ..., and atleast r_n balls in the nth box.

$= C(n - 1 + r - r_1 - r_2 - \ldots\ldots - r_n, r - r_1 - r_2 - \ldots\ldots - r_n)$

$= C(n - 1 + r - r_1 - r_2 - \ldots\ldots - r_n, n - 1)$.

Problem 1.100. *The results of 50 football games (win, lose or, tie) are to be predicted. How many different forecasts can contain exactly 28 correct results ?*

Solution. Choose 28 correct results C(50, 28) ways.

Each of the remaining 22 games has 2 wrong forecasts. Thus, there are C(50, 28). 2^{22} forecasts with exactly 28 correct predictions.

Problem 1.101. *A telegraph can transmit two different signals : a dot and a dash. What length of these symbols is needed to encode the 26 letters of the English alphabet and the ten digits 0, 1, ..., 9 ?*

Solution. Since there are two choices for each character, the number of different sequences of length k is 2^k.

The number of non trivial sequences of length n or less is

$$2 + 2^2 + 2^3 + \ldots\ldots + 2^n = 2^{n+1} - 2.$$

If $n = 4$ this total is 30, which is enough to encode the letters of the English alphabet, but not enough to also encode the digits.

To encode the digits we need to allow sequences of length upto 5 for then there are possibly $2^{5+1} - 2 = 62$ total sequences.

Problem 1.102. *How many 10-digit binary numbers are there with exactly six 1's ?*

Solution. The key to this problem is that we can specify a binary number by choosing the subset of 6 positions where the 1's go (or the subset of 4 positions for the 0's)

Thus, there are C(10, 6) = C(10, 4) = 210 such binary numbers.

Problem 1.103. *There are 21 consonants and 5 vowels in the English alphabet. Consider only 8-letter words with 3 different vowels and 5 different consonants.*

(*a*) *How many such words can be formed ?*

(*b*) *How many such words contain the letter a ?*

(*c*) *How many contain the letters a and b ?*

(*d*) *How many contain the letters b and c ?*

(*e*) *How many contain the letters a, b and c ?*

(*f*) *How many begin with a and end with b ?*

(*g*) *How many begin with b and end with c ?*

Solution. (*a*) C(5, 3) C(21, 5) 8 !

(Choose the vowels, choose the consonants, and then arrange the 8 letters.)

(*b*) C(4, 2) C(21, 5) 8 !

(*c*) C(4, 2) C(20, 4) 8 !

(*d*) C(5, 3) C(19, 3) 8 !

(*e*) C(4, 2) C(19, 3) 8 !

(*f*) C(4, 2) C(20, 4) 6 !

(*g*) C(5, 3) C(19, 3) 6 !.

Problem 1.104. *There are 30 females and 35 males in the junior class while there are 25 females and 20 male in the senior class.In how many ways can a committee of 10 be chosen so that there exactly 5 females and 3 juniors on the committee ?*

Solution. Let us draw a chart illustrating the possible male-female and junior senior constitution of the committee.

Juniors		*Seniors*		
Female	*Male*	*Female*	*Male*	*Number of ways of selecting*
0	3	5	2	C(30, 0) C(35, 3) C(25, 5) C(20, 2)
1	2	4	3	C(30, 1) C(35, 2) C(25, 4) C(20, 3)
2	1	3	4	C(30, 2) C(35, 1) C(25, 3) C(20, 4)
3	0	2	5	C(30, 3) C(35, 0) C(25, 2) C(20, 5)

Thus, the total number of ways is the sum of the terms in the last column:

C(30, 0) C(35, 3) C(25, 5) C(20, 2) + C(30, 1) C(35, 2) C(25, 4) C(20, 3) + C(30, 2) C(35, 1) C(25, 3) C(20, 4) + C(30, 3) C(35, 0) C(25, 2) C(20, 5).

Problem 1.105. *There are 25 true or false questions on an examination. How many different ways can a student do the examination if he or she can also choose to leave the answer blank ?*

Solution. 3^{25}.

Problem 1.106. (*a*) *How many different outcomes are possible by tossing 10 similar coins ?*

(*b*) *How many different outcomes are possible from tossing 10 similar dice ?*

(*c*) *How many ways can 20 similar books be placed on 5 different shelves ?*

(*d*) *Out of a large supply of pennies, nickels, dimes, and quarter, in how many ways can 10 coins be selected ?*

(*e*) *How many ways are there to fill a box with a dozen dough nuts chosen from 8 different varieties of dough nuts ?*

Solution. (*a*) This is the same as placing 10 similar balls into two boxes labeled "heads" and "tails".

$$C(2 - 1 + 10, 10) = C(11, 10) = 11.$$

(*b*) This is the same as placing 10 similar balls into 6 numbered boxes.

Therefore there are C(15, 10) = 3,003 possibilities.

(*c*) C(5 – 1 + 20, 20) = C(24, 20).

(*d*) C(4 – 1 + 10, 10) = C(13, 10) since this is equivalent to placing 10 similar balls in 4 numbered boxes labeled "pennies", "nickels", "dimes", and "quarters".

(*e*) First, we observe that relative positions in the box are immaterial so that order does not count.

Therefore, this is a combination problem

Secondly, a box might consist of a dozen of one variety of doughnut, so that we see that this problem allows unlimited repetitions.

The answer then is C(8 – 1 + 12, 12) = C(19, 12).

Problem 1.107. *In how many ways can the letters of the word attention be rearranged ?*

Solution. The word attention has nine letters, three of one kind, two of another, and four other different letters.

The number of rearrangements of this word is $\dfrac{9!}{3!2!} = 30{,}240$.

Problem 1.108. *Suppose there are ten players to be assigned to three teams. The Xtreme, the Maniax, and the Enforcers. The Xtreme and the Maniax are to receive four players each and the Enforcers are to receive two. In how many ways can this be done ?*

Solution. The assignment of players is accomplished by choosing four players from ten for the Xtreme, then choosing four players from the remaining six for the Maniax, and assigning the remaining two players to the Enforcers. The number of possible teams is $\binom{10}{4} \times \binom{6}{4} = 210(15) = 3150$.

Problem 1.109. *In how many ways can 14 men be divided into six named teams, two with three players and four with two ? In how many ways can 14 men be divided into two unnamed teams of three and four teams of two ?*

Solution. If the teams are named, the answer is

$$\binom{14}{3} \times \binom{11}{3} \times \binom{8}{2} \times \binom{6}{2} \times \binom{4}{2} = 364(165)(28)(15)(6)$$

$$= 151,\ 351,\ 200.$$

This number is $2!\,4! = 48$ times the number of divisions into unnamed teams of sizes 3, 3, 2, 2, 2 and 2, because for each division into unnamed teams, there are two ways to name the teams of three and $4!$ ways to name the teams of two.

The answer in the case of unnamed teams is

$$\frac{151,351,200}{48} = 3,\ 153,\ 150.$$

Problem 1.110. *In how many ways can the letters of the word REARRANGE be rearranged ?*

Solution. There are $\dfrac{9!}{3!2!2!} = 15{,}120$ rearrangements of the letters fo the word REARRANGE.

Problem 1.111. (*a*) *In how many ways can the letters of the word easy be rearranged ?*

(*b*) *In how many ways can the letters of the word ease be rearranged ?*

Solution. (*a*) The question just asks for the number of permutations of four different letters.

The answer is $4! = 24$.

(*b*) This is a slightly different Problem because of the repeated *e*'s, when the first and last letters of easy are interchanged, we get two different arrangements of the four letters, but when the first and last letters of ease are interchanged we get the same word.

To see how to count the ways in which the four letters of ease can be arranged, we imagine the list of all these arrangements.

$s \quad e \quad a \quad e$

$a \quad e \quad e \quad s$

$e \quad e \quad s \quad a$

$\vdots$

Pretending for a moment that the two *e*'s are different (say one is a capital E), then each "word" in this list will produce two different arrangements of the letters easE (See Fig. (1.3(*a*)))

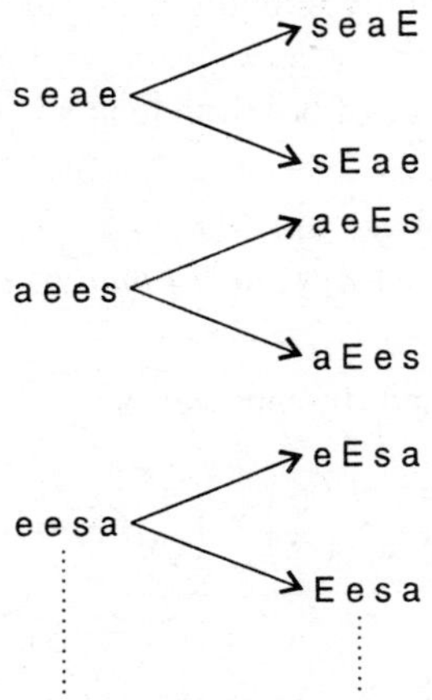

Fig. 1.3(*a*).

The list on the right contains the 4 ! = 24 arrangements of the four letters *e a s* E, so the list on the left contains half as many.

There are $\frac{4\,!}{2\,!} = 12$ ways in which the letters of ease can be arranged.

Theorem 1.15. *The number of ways to put r identical marbles into n boxes is* $\binom{n+r-1}{r}$.

Problem 1.112. *Doughnuts come in 30 different varieties and catherine wants to buy a dozen. How many choices does she have ?*

Solution. Imagine that the 30 varieties are in $n = 30$ boxes labeled "chocolate white", "Boston creme", "peanut crunch", and so on.

Catherine can indicate her choice by dropping $r = 12$ (identical) marbles into the boxes.

So there are $\binom{30+12-1}{12}$ = 7,898,654,920 possibilities.

Problem 1.113. *David wants to buy 30 doughnuts and finds just 12 varieties available. In how many ways can he make his selection ?*

Solution. We imagine the 12 varieties in boxes.

Each of David's possible decisions can be indicated by dropping 30 identical marbles into these boxes.

There are $\binom{12+30-1}{30} = 3{,}159{,}461{,}968$ possibilities ($r = 30$, $n = 12$).

Problem 1.114. *Consider the set {a, b, c, d}. In how many ways can we select two of these letters (repetition is not allowed) when (i) order matters (ii) order does not matter.*

Solution. (*i*) If order matters but repetition is not allowed , $n = 4$ and $r = 2$ and hence the number of ways of selecting two letters from four letters is

$$P(4, 2) = \frac{4!}{(4-2)!} = 12 \text{ and } 12 \text{ possibilities are}$$

ab *ba* *ac* *da*

ac *bc* *cb* *db*

ad *bd* *cd* *dc*

(*ii*) If order does not matter and repetition is not allowed then $C(4, 2) = \frac{4!}{2!(4-2)!} = 6$ and 6 possibilities

are *ab* *bc* *cd*

ac *bd* *ad*

Problem 1.115. *A man 7 relatives, 4 of them are ladies and 3 gentlemen, his wife has 7 relatives and 3 of them are ladies and 4 gentlemen. In how many ways can they invite a dinner party of 3 ladies and 3 gentlemen so that there are 3 of man's relative and 3 of wife's relatives ?*

Solution. They can invite in four possible ways :

(*i*) 3 ladies from husband's side and three from wife's side.

Number of ways = C(4, 3) × C(4, 3) = 16.

(*ii*) 3 gents from husband's side and 3 ladies from wife's side.

Number of ways = C(3, 3) × C(3, 3) = 1.

(*iii*) 2 ladies and 1 gent from husband's side and one lady and 2 gents from wife's side.

Number of ways in this case

$$= \{C(4, 2) \times C(3, 1)\} \times \{C(3, 1) \times C(4, 2)\}$$
$$= 324.$$

(*iv*) One lady and 2 gents from husband's side and 2 ladies and 1 gent from wife's side.

Number of ways in this case

$$= \{C(4, 1) \times C(3, 2)\} \times \{C(3, 2) \times C(4, 1)\}$$
$$= 144$$

The total number of ways = 16 + 1 + 324 + 144 = 485.

Problem 1.116. *In how many ways can the letters of the word MONDAY be arranged ? How many of them begin with M and end with Y ? How many of them do not begin with M but end with ?*

Solution. The word MONDAY consists of six letters which can be arranged in P(6, 6) = 6 ! = 720 ways.

If M occupies the first place and Y the last place, then there are 4 letters (O, N, D, A) left to be arranged in 4 places in between M and Y.

This can be done in 4 ! = 24 ways.

If M does not occupy the first place but Y occupies the last place, the first place can be occupied in 4 ways by any one of the letters O, N, D, A.

For the second place, again 4 letters are available including M.

The third, fourth, and fifth places can be filled by 3, 2, 1 ways.

Hence, by product rule, the required number of arrangements are $4 \times 4 \times 3 \times 2 \times 1 = 96$.

Problem 1.117. *In how many ways can a cricket team of eleven be chosen out of batch of 15 players ? How many of them will*

(a) include a particular player

(b) exclude a particular player ?

Solution. The number of ways of selecting 11 players out of 15 is C(15, 11) = 1365

(*a*) The number of ways is which a particular player is included is C(14, 10) = 1001

(*b*) The number of ways in which a particular player is excluded is C(14, 11) = 364.

Problem 1.118. *A computer password consists of a letter of the alphabet followed by 3 or 4 digits. Find (a) the total number of password that can be formed, and (b) the number of passwords in which no digit repeats.*

Solution. (*a*) Since there are 26 alphabets and 10 digits and the digits can be repeated, by product rule the number of 4-character password is 26.10.10.10 = 26000.

Similarly the number of 5-character password is

$$26.10.10.10.10 = 260000$$

Hence the total number of password is

$$26000 + 260000 = 286000.$$

(*b*) Since the digits are not repeated, the first digit after alphabet can be taken from any one out of 10, the second digit from remaining 9 digits and so on. Thus the number of 4-character password is

26.10.9.8 = 18720 and the number of 5-character password is

26.10.9.8.7 = 131040 by the product rule.

Hence, the total number of passwords is 149760.

Problem 1.119. *A bit is either 0 or 1 : a byte is a sequence of 8 bits.*

Find (a) the number of bytes that can be formed

(b) the number of bytes that begin with 11 and end with 11

(c) the number of bytes that begin with 11 and do not end with 11 and

(d) the number of bytes that begin with 11 or end with 11.

Solution. (*a*) Since the bits 0 or 1 can repeat, the eight positions can be filled up either by 0 or 1 in 2^8 ways.

Hence the number of bytes that can be formed is 256.

(*b*) Keeping two positions at the beginning by 11 and the two positions the end by 11, there are four open positions which can be filled up in $2^4 = 16$ ways.

Hence the required number is 16.

(*c*) Keeping two positions at the beginning by 11, the remaining six open positions can be filled up by $2^6 = 46$ ways.

Hence the required number is $64 - 16 = 48$.

(*d*) 64 bytes begin with 11, likewise, 64 bytes end with 11. In the sum of these numbers, $64 + 64 = 128$, each byte that both begins and ends with 11 is counted twice.

Hence the required number is $128 - 16 = 112$ bytes.

Problem 1.120. *Consider the set {a, b, c, d}. In how many ways can we select two of these letters when repetition is allowed.*

Solution. If order matters and repetition is allowed, there are $2^4 = 16$ possible selections and they are

$$\begin{array}{cccc} aa & ba & ca & da \\ ab & bb & cb & db \\ ac & bc & cc & dc \\ ad & bd & cd & dd \end{array}$$

If order does not matter but repetitions are allowed, there are C(4 + 2 – 1, 2) = C(5, 2) = 10 possibilities and these possibilities are

$$\begin{array}{cccccc} aa & bb & cc & dd & & \\ ab & bc & cd & ad & ac & bd \end{array}$$

Problem 1.121. *How many solutions are there of $x + y + z = 17$ subject to the constraints $x \geq 1$, $y \geq 2$ and $z \geq 3$.*

Solution. Put $x = 1 + u, y = 2 + v$, and $z = 3 + w$.

The given equation becomes $u + v + w = 11$ and we seek in non negative integers *u, v, w*.

The number of solutions is therefore

$$C(11 + 3 - 1, 11) = C(13, 11) = C(13, 2) = 78.$$

Problem 1.122. *How many solutions does the equation $x + y + z = 17$ have, where x, y, z are non negative integers ?*

Solution. Each solution of the given equation is equivalent to selecting 17 items from the set {*x, y, z*}, repetitions allowed.

Hence, the required number of solutions

$$= C(17 + 3 - 1, 17) = C(19, 2) = 171.$$

Problem 1.123. *Find the number of ways in whcih 7 different beads can be arranged to form a necklace.*

Solution. Fixing the position of one bead, the remaining beads can be arranged in 6 ! ways.

But this is a ring permutation, so the required number of arrangements is $\begin{pmatrix}1\\2\end{pmatrix}(6\,!) = 360$.

Problem 1.124. *In how many ways can 7 persons form a ring ? In how way can 7 gentlemen and 7 ladies sit down at a round table, no two ladies being together ?*

Solution. Seven persons can be seated along a circle in 6 ! ways.

First, let all the gentlemen be seated along the round table in 6 ! ways. Between any two men let a women be seated.

Hence all the seven ladies can be seated in 7 intermediate places in 7 ! ways.

Therefore, 7 gentlemen and 7 ladies can be seated along a round table in 6 ! × 7 ! ways.

Problem 1.125. *Find the number of unordered samples of size five (repetition allowed) from the set {a, b, c, d, e, f}.*

(*a*) *No further restrictions*

(*b*) *a occurs at least twice*

(*c*) *a occurs exactly twice.*

Solution. (*a*) Here $n = 6$, $r = 5$

So, the required member C(6 + 5 – 1, 5) = C(10, 5)

$$= \frac{10.9.8.7.6}{5.4.3.2} = 252.$$

(*b*) Since *a* occurs at least twice, we have to find the number of unordered samples of size 3 (repetitions allowed) from 6-element set.

So, the required number is C(6 + 3 – 1, 3) = C(8, 3)

$$= \frac{8.7.6}{3.2} = 56.$$

The form of the samples are *a a x y z, a a a y z* etc.

(*c*) Since *a* occurs exactly twice, we have to find the number of unordered samples of size 3(repetition allowed) from 5-element set {*b, c, d, e, f*}.

So, the required number is

$$C(5 + 3 - 1, 3) = C(7, 3) = \frac{7.6.5}{3.2} = 35.$$

Problem 1.126. *In how many ways can 12 balloons be distribuetd at a Birth day party among 10 children ?*

Solution. This is an unordered selection with repetition, of 12 objects from 10 types.

Hence the number of selection is C(10 + 12 – 1, 12) = C(21, 12).

If we want to ensure that every child gets at least one balloon, we must give a balloon to each child, then distribute the remaining two balloons which can be done is C(10 + 2 – 1, 2) = C(11, 2) = 55 ways.

Problem 1.127. *In how many ways can the letters of the English alphabet be arranged so that there are exactly 5 letters between the letters a and b ?*

Solution. There are P(24, 5) ways to arrange the 5 letters between *a* and *b*, 2 ways to place *a* and *b*, and then 20 ! ways to arrange any 7-letter word treated as one unit along with the remaining 19 letters. The total is P(24, 5)(20 !)(2).

Problem 1.128. *In how many ways can 7 women and 3 men be arranged in a row if the 3 men must always stand next to each other ?*

Solution. There are 3 ! ways of arranging the 3 men.

Since the 3 men always stand next to each other, we treat them as a single entity, which we denote by X. Then if W_1, W_2, ..., W_7 represents the women, we next are interested in the number of ways of arranging $\{X, W_1, W_2, W_3, ..., W_7\}$.

There are 8 ! permutations of these 8 objects.

Hence there are (3 !)(8 !) permutations altogether.

Problem 1.129. *How many 6-digit numbers without repetition of digits are there such that the digits are all non zero and 1 and 2 do not appear consecutively in either order ?*

Solution. We are asked to count cetain 6-permutations of the 9 integers 1, 2, ..., 9.

In the following table we separate these 6-permutations into 4 disjoint classes and count the number of permutations in each class.

Class	*Number of permutations in the class*
(*i*) Neither 1 nor 2 appears as a digit	7 !
(*ii*) 1, but not 2, appears as a digit	6P(7, 5)
(*iii*) 2, but not 1, appears	6P(7, 5)
(*iv*) Both 1 and 2 appear	(2)(7)(4) P(6, 3) + (4)(7)(6)(3) P(5, 2)
Total	7 ! + (2)(6) P(7, 5) + (56) P(6, 3) + (504) P(5, 2)

Let us explain how to count the elements in class (*iv*).

1. The hundred thousands digit is 1 (and thus the ten thousands digit is not 2). The second digit can be chosen in 7 ways. Choose the position for 2 in 4 ways ; then fill the other 3 positions P(6, 3) ways.

   Hence, there are (7) 4P(6, 3) numbers in this category.
2. The units digit is 1 (and hence the tens digit is not 2). Likewise, there are (7) 4P(6, 3) numbers in this category.
3. The ingeter 1 appears in a position different from the hundered thousands digit and the units digit. Hence, 2 cannot appear immediately to the left or to the right of 1. Since 1 can be any one of the digits from the tens digit up to the ten thousands digit, 1 can be placed in 4 ways. The digit immediately to the left of 1 can be filled in 7 ways, while the digit immediately to the rigth of 1 can be filled in 6 ways. The integer 2 can be placed in any of the remaining positions in 3 ways and then the other 2 digits are a 2-permutation of the remaining 5 inegers.

Hence, there are (4)(7)(6)(3) P(5, 2) numbers in this category.

Thus, there are (2)(7)(4) P(6, 3) + (4)(7)(6)(3) P(5, 2) numbers in class (*iv*).

Then, by the sum rule, there are

P(7, 6) + (2)(6) P(7, 5) + (56) P(6, 3) + (504) P(5, 2) elements in all four classes.

Problem 1.130. *In how many ways can 5 children arrange themselves in a ring ?*

Solution. Here, the 5 children are not assigned to particular places but are only arranged relative to one another.

Thus, the arrangements (see figure below) are considered the same if the children are in the same order clockwise.

Hence, the position of child C_1 is immaterial and it is only the position of the 4 other children relative to C_1 that counts.

Therefore, keeping C_1 fixed in position, there are 4 ! arrangements of the remaining children.

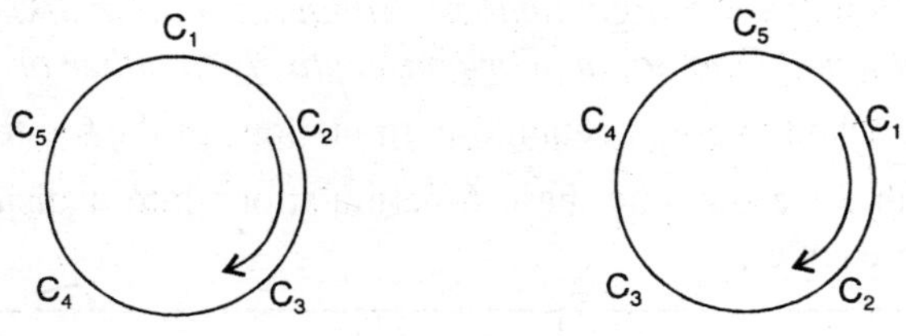

Fig. 1.4.

Problem 1.131. *In how many ways can a hand of 5 cards be selected from a deck of 52 cards ?*

Solution. Each hand is essentially a 5-combination of 52 cards.

Thus there are $C(52, 5) = \dfrac{52!}{5!47!} = \dfrac{52.51.50.49.48}{5.4.3.2.1}$

$= 52.51.10.49.2 = 2,598,960$ such hands.

Problem 1.132. (*a*) *How many 5-card hands consist only of hearts ?*

(*b*) *How many 5-card hands consist of cards from a single suit ?*

(*c*) *How many 5-card hands have 2 clubs and 3 hearts ?*

(*d*) *How many 5-card hands have 2 cards of one suit and 3-cards of a different suit ?*

(*e*) *How many 5-card hands contain 2 aces and 3 kings ?*

(*f*) *How many 5-card hands contain exactly 2 of one kind and 3 of another kind ?*

Solution. (*a*) Since there are 13 hearts to choose from, each such hand is a 5-combination of 13 objects.

Thus, there is a total of

$$C(13, 5) = \frac{13!}{5!8!} = \frac{13.12.11.10.9}{5.4.3.2.1} = 13.11.9 = 1,287.$$

(*b*) For each of the 4 suits, spades, hearts, diamonds, or clubs, there are C(13, 5) 5-card hands. Hence, there are a total of 4C(13, 5) such hands.

(*c*) C(13, 2), C(13, 3).

(*d*) For a fixed choice of 2 suits there are 2C(13, 2), C(13, 3) ways to choose 2 from one of the suits and 3 from the other. We can choose the 2 suits in C(4, 2) ways.

Thus, there are 2C(13, 2) C(13, 3) C(4, 2) such 5-card hands. Recall that two of a kind means 2 aces, 2 kings, 2 queens etc. Similarly, 3 tens are called three of a kind.

Thus, there are 13 kinds in a deck of 52 cards.

(*e*) C(4, 2) C(4, 3)

(*f*) Choose the first kind 13 ways, choose 2 of the first kind C(4, 2) ways, choose the second kind 12 ways and choose 3 of the second kind in C(4, 3) ways.

Hence there are (13)C(4, 2) (12)C(4, 3), 5-card hands with 2 of one kind and 3 of another kind.

Problem 1.133. (*a*) *In how many ways can a committee of 5 be chosen from 9 people ?*

(*b*) *How many committees of 5 or more can be chosen from 9 people ?*

(*c*) *In how many ways can a committee of 5 teachers and 4 students be chosen from 9 teachers and 15 students ?*

(*d*) *In how many ways can the committee in (C) be formed if teacher A refuses to serve if student B is on the committee ?*

Solution. (*a*) C(9, 5) ways.

(*b*) C(9, 5) + C(9, 6) + C(9, 7) + C(9, 8) + C(9, 9)

(*c*) The teachers can be selected in C(9, 5) ways while the students can be chosen in C(15, 4) ways so that the committee can be formed in C(9, 5) C(15, 4) ways.

(*d*) We answer this question by counting indirectly. First we count the number of committees where both A and B are on the committee. Thus, there are only 8 teachers remaining from which 4 teachers are to be chosen.

Likewise, there are only 14 students remaining from which 3 more students are to be chosen.

There are C(8, 4), C(14, 3) committees containing both A and B, and hence there are

C(9, 5) C(15, 4) – C(8, 4) C(14, 3) committees that do not have both A and B on the committee.

Problem 1.134. *How many strings of length n can be formed from the English alphabet ?*

Solution. By the product rule, since there are 26 letters, and since each letter can be used repeadtly, we see that there are 26^n strings of length n.

Problem 1.135. *How many ways are there to select four pieces fo fruit from a bowl containing apples, oranges, and pears if the order in which the pieces are selected does not matter, only the type of fruit and not the individual piece matters, and there are at least four pieces of each type of fruit in the bowl ?*

Solution. To solve this problem we list all the ways possible to select the fruit. There are 15 ways :

4 apples	4 oranges	4 pears
3 apples, 1 orange	3 apples, 1 pear	3 oranges, 1 apple
3 oranges, 1 pear	3 pears, 1 apple	3 pears, 1 orange

2 apples, 2 oranges 2 apples, 2 pears 2 oranges, 2 pears

2 apples, 1 orange, 1 pear 2 oranges, 1 apple, 1 pear 2 pears, 1 apple, 1 orange

The solution is the number of 4-combinations with repetition allowed from a three element set, {apple, orange, pear}.

Problem 1.136. *Suppose that a cookie shop has four different kinds of cookies. How many different ways can six cookies be chosen ? Assume that only the type of cookie, and not the individual cookies or the order in which they are chosen, matters.*

Solution. The number of ways to choose six cookies is the number of 6-combinations of a set with four elements.

We have, there are $C(n + r - 1, r)$ r-combinations from a set with n elements when repetition of elements is allowed.

This equals $C(4 + 6 - 1, 6) = C(9, 6)$

Since $C(9, 6) = C(9, 3) = \dfrac{9.8.7}{1.2.3} = 84.$

There are 84 different ways to choose the six cookies.

Problem 1.137. *What is the value of k after the following pseudocode has been executed ?*

```
    k : = 0
for i_1 = 1 to n
for i_2 = 1 to i_1
    ..................

    ..................
for i_m = 1 to i_(m-1)
    k : = k + 1
```

Solution. Note that the initial value of k is 0 and that 1 is added to k each time the nested loop is traversed with a sequence of integers $i_1, i_2, ..., i_m$ such that

$$1 \le i_m \le i_{m-1} \le ...... \le i_1 \le n.$$

The number of such sequences of integers is the number of ways to choose m integers from {1, 2, ..., n}, with repetition allowed.

Note that once such a sequence has been selected, if we order the integers in the sequence in non decreasing order, this uniquely defines an assignment of $i_m, i_{m-1}, ..., i_1$.

Conversely, every such assignment corresponds to a unique unordered set.

Hence, it follows that $K = C(n + m - 1, m)$ after this code has been executed.

Problem 1.138. *How many solutions does the equation $x_1 + x_2 + x_3 = 11$ have, where x_1, x_2 and x_3 are non negative integers ?*

Solution. To count the number of solutions, we note that a solution corresponding to a way of selecting 11 items from a set with three elements, so that x_1 items of type one, x_2 items of type two, and x_3 items of type three are chosen.

Hence, the number of solutions is equal to the number of 11-combinations with repetition allowed from a set with three elements.

We have, there are $C(n + r - 1, r)$ r-combination from a set with n elements when repetition of elements is allowed, it follows that there are

$$C(3 + 11 - 1, 11) = C(13, 11) = C(13, 2) = \frac{13.12}{1.2} = 78 \text{ solutions.}$$

The number of solutions of this equation can also be found when the variables are subject to constraints.

For instance, we can find the number of solutions where the variables are integers with $x_1 \geq 1$, $x_2 \geq 2$ and $x_3 \geq 3$.

A solution to the equation subject to these constraints corresponds to a selection of 11 items with x_1 items of type one, x_2 items of type two, and x_3 items of type three where, in addition, there is atleast one item of type one, two items of type two, and three items of type three. So, choose one item of type one, two of type two, and three of type three. Then select five additional items this can be done in

$$C(3 + 5 - 1, 5 = C(7, 5) = C(7, 2) = \frac{7.6}{1.2} = 21 \text{ ways.}$$

Thus, there are 21 solutions of the equation subject to the given constraints.

Problem 1.139. *How many ways are there to place ten indistinguishable balls into eight distinguishable bins ?*

Solution. The number of ways to place ten indistinguishable balls into eight bins equals the number of 10 combinations from a set with eight elements when repetition is allowed. Consequently, there are

$$C(8 + 10 - 1, 10) = C(17, 10) = \frac{17!}{10!7!} = 19{,}448.$$

Problem 1.140. *How many different strings can be made by reordering the letters of the word SUCCESS ?*

Solution. Because some of the letters of SUCCESS are the same, the answer is not given by the number of permutations of seven letters.

This word contains three Ss, two Cs, one U, and one E. To determine the number of different strings that can be made by reordering the letters, first note that the three Ss can be placed among the seven positions in C(7, 3) different ways, leaving four positions free.

Then the two Cs can be placed in C(4, 2) ways, leaving two free positions. The U can be placed in C(2, 1) ways, leaving just one position free.

Hence E can be placed in C(1, 1) way.

Consequently, from the product rule, the number of different strings that can be made is

$$C(7, 3)\, C(4, 2)\, C(2, 1)\, C(1, 1)$$

$$= \frac{7!}{3!4!} \cdot \frac{4!}{2!2!} \cdot \frac{2!}{1!1!} \cdot \frac{1!}{1!0!}$$

$$= \frac{7!}{3!2!1!1!} = 420.$$

Problem 1.141. *How many ways are there to distribute hands of 5 cards to each of four players from the standard deck of 52 cards ?*

Solution. We will use the product rule to solve this problem.

Note that, the first player can be dealt 5 cards in C(52, 5) ways.

The second player can be dealt 5 cards in C(47, 5) ways, since only 47 cards are left.

The thrid player can be dealt 5 cards in C(42, 5) ways.

Finally, the fourth player can be dealt 5 cards in C(37, 5) ways.

Hence, the total number of ways to deal four players 5 cards each is

$$C(52, 5)\, C(47, 5)\, C(42, 5)\, C(37, 5)$$

$$= \frac{52!}{47!5!} \cdot \frac{47!}{42!5!} \cdot \frac{42!}{37!5!} \cdot \frac{37!}{32!5!}$$

$$= \frac{52!}{5!5!5!5!32!}.$$

Problem 1.142. *What is the next largest permutation in lexicographic order after 362541 ?*

Solution. The last pair of inetgers a_j and a_{j+1} where $a_j < a_{j+1}$ is $a_3 = 2$ and $a_4 = 5$. The least integer to the right of 2 that is greater than 2 in the permutation is $a_5 = 4$.

Hence, 4 is placed in the third position.

Then the integers 2, 5 and 1 are placed in order in the last three positions, giving 125 as the last three positions of the permutation.

Hence, the next permutation is 364125.

Problem 1.143. *Generate the permutations of the integers 1, 2, 3 in Lexicographic order.*

Solution. Begin with 123. The next permutation is obtained by interchanging 3 and 2 to obtain 132.

Next, since 3 > 2 and 1 < 3, permute the three integers in 132.

Put the smaller of 3 and 2 in the first position, and then put 1 and 3 in increasing order in positions 2 and 3 to obtain 213.

This is followed by 231, obtained by interchanging 1 and 3, since 1 < 3.

The next largest permutation has 3 in the first position followed by 1 and 2 in increasing order, namely, 312.

Finally, interchange 1 and 2 to obtain the last permutation 321.

Problem 1.144. *Find the next largest bit string after 1000100111.*

Solution. The first bit from the right that is not a 1 is the fourth bit from the right.

Change this bit to *a* 1 and change all the following bits to 0s.

This produces the next largest bit string, 1000101000.

Problem 1.145. *Find the next largest 4-combination of the set {1, 2, 3, 4, 5, 6} after {1, 2, 5, 6}.*

Solution. The last term among the terms a_i with $a_1 = 1$, $a_2 = 2$, $a_3 = 5$ and $a_4 = 6$ such that $a_i \neq 6 - 4 + i$ is $a_2 = 2$.

To obtain the next largest 4-combination, increment a_2 by 1 to obtain $a_2 = 3$.

Then set $a_3 = 3 + 1 = 4$ and $a_4 = 3 + 2 = 5$.

Hence the next largest 4-combination is $\{1, 3, 4, 5\}$.

Problem 1.146. *Suppose that a valid computer password consists of seven characters, the first of which is a letter chosen from the set {A, B, C, D, E, F, G} and the remaining six characters are letters chosen from the English alphabet or a digit. How many different passwords are possible ?*

Solution. A password can be constructed by performing the tasks T_1 and T_2 in sequence.

Task 1 : Choose a starting letter from the set given

Task 2 : Choose a sequence of letter and digits. Repeats are called.

Task T_1 can be performed in ${}_7C_1$ or 7 ways.

Since there are 26 letters and 10 digits that can be chosen for each of the remaining six characters, and since repeats are allowed, task T_2 can be performed in 36^6 or 2, 176, 782, 336 ways.

By the multiplication principle, there are 7.2176782336 or 15, 237, 476, 352 different passwords.

Problem 1.147. *Show that* ${}_{2n}C_2 = 2 \,.\, {}_nC_2 + n^2$.

Solution. We have ${}_nC_r = \dfrac{n!}{r!(n-r)!}$

$$2 \,.\, {}_nC_2 + n^2 = 2 \,.\, \frac{n!}{(n-2)!2!} + n^2$$

$$= \frac{2.n! + 2!(n-2)!n^2}{(n-2)!}$$

$$= n(2n-1) \,.\, \frac{(2n-2)!}{(2n-2)!} \,.\, \frac{2}{2}$$

$$= \frac{(2n)!}{2!(2n-2)!} = {}_{2n}C_2.$$

1.5 THE BINOMIAL THEOREM

Theorem 1.16. *Let n be a positive integer. Then for all x and y,*

$$(x + y)^n = C(n, 0)x^n + C(n, 1)x^{n-1}y + C(n, 2)x^{n-2}y^2 + \ldots\ldots + C(n, r)x^{n-r}y^r + \ldots\ldots + C(n, n)y^n.$$

$$= \binom{n}{0}x^n + \binom{n}{1}x^{n-1}y + \binom{n}{2}x^{n-2}y^2 + \ldots\ldots \binom{n}{r}x^{n-r}y^r + \ldots\ldots + \binom{n}{n}y^n.$$

$$= \sum_{r=0}^{n} C(n, r)\, x^{n-r}y^r = \sum_{j=0}^{n} \binom{n}{r} x^{n-r}y^r.$$

The binomial coefficients $C(n, r) = \binom{n}{r}$ receive their name from their appearance in the expansion of powers of a binomial.

Proof. First Proof. Write $(x + y)^n$ as a product of n factors $(x + y)(x + y) \ldots\ldots (x + y)$.

Then expand this product until no parentheses remain. We can do this in many ways. One way is to select an x or a y from each factor, multiply and arrange these into a term of the from $x^{n-r}y^r$ for $r = 0, 1, \ldots\ldots, n$.

The collection of all terms with the same exponents on x and y will determine the coefficients in the expansion of $(x + y)^n$.

Thus for any given r we need only determine the number of terms of the form $x^{n-r}y^r$ obtained as described. But such a term is obtained by selecting y from r of the factors and then x from the remaining $n - r$ factors.

The number of such terms is therefore the number of ways of choosing r of the n factors from which to choose the y.

Since this can be done in $C(n, r)$ ways the coefficient of $x^{n-r}y^r$ is $C(n, r)$ as stated in the theorem.

Second proof. This proof is by mathematical induction on n.

If $n = 1$, the formula becomes $(x + y)^1 = C(1, 0)x + C(1, 1)y = x + y$, and this is clearly true.

We now assume the formula is true for a positive integer n and prove that it is true when n is replaced by $n + 1$.

We write $(x + y)^{n+1} = (x + y)(x + y)^n$ and by the inductive hypothesis this becomes

$$(x + y)\left(\sum_{r=0}^{n} C(n, r)x^{n-r}\,y^r\right) = x\left(\sum_{r=0}^{n} C(n, r)x^{n-r}\,y^r\right) + y\left(\sum_{r=0}^{n} C(n, r)x^{n-r}\,y^r\right)$$

$$= C(n, 0)x^{n+1} + \sum_{r=0}^{n} C(n, r)x^{n-r+1}\,y^r + \sum_{r=0}^{n-1} C(n, r)x^{n-r}\,y^{r+1} + C(n, n)y^{n+1}$$

If we set $r = k - 1$, in the third term above, then as r runs from 0 to $n - 1$, k runs from 1 to n, $\sum_{r=0}^{n-1} C(n, r)x^{n-r}\,y^{r+1}$ becomes $\sum_{k=1}^{n} C(n, k-1)x^{n+i-k}\,y^k$.

Now the letter of the dummy variable is immaterial so now replace k by r and the third term becomes $\sum_{r=1}^{n} C(n, r-1)x^{n+1-r}\,y^r$ and $(x + y)^{n+1} = x^{n+1} + \sum_{r=1}^{n} [C(n, r) + C(n, r-1)]\,x^{n+1-r}y^r + y^{n+1}$.

But using Pascal's identify, we then have

$$(x + y)^{n+1} = x^{n+1} + \sum_{r=1}^{n} C(n+1, r)x^{n+1-r}\;y^r + y^{n+1}$$

or $$(x+y)^{n+1} = \sum_{r=0}^{n+1} C(n+1, r)x^{n+1-r} \,.\, y^r$$

Thus, the formula is true for $n + 1$ and the theorem is proved by mathematical induction.

Note. The binomial theorem can be written in several other equivalent forms

$$(x+y)^n = \sum_{r=0}^{n} C(n, n-r)x^{n-r}y^r$$

$$= \sum_{r=0}^{n} C(n, r)x^r y^{n-r}$$

$$= \sum_{r=0}^{n} C(n, n-r)x^r y^{n-r}.$$

Note. Let n be a positive integer. Then for all x,

$$(x+1)^n = \sum_{r=0}^{n} C(n, r)x^{n-r} = \sum_{r=0}^{n} C(n, n-r)x^{n-r} = \sum_{r=0}^{n} C(n, r)x^r$$

Replacing x by $-x$ we have

$$(1-x)^n = \sum_{r=0}^{n} C(n, r)(-x)^r = \sum_{r=0}^{n} C(n, r)(-1)^r \, x^r.$$

Third Proof. A combinatorial proof of the theorem will be given. The terms in the product when it is expanded are of the form $x^{n-r} y^r$ for $i = 0, 1, 2, \ldots\ldots n$.

To count the number of terms of the form $x^{n-r} y^r$.

Note that to obtain such a term it is necessary to choose $n - r$ xs from the n sums (so that the other r terms in the product are ys)

Therefore, the coefficient of $x^{n-r} y^r$ is $\binom{n}{n-r}$, which is equal to $\binom{n}{r}$.

This proves the theorem.

Corollary 1. *Let n be a non negative integer. Then*

$$\sum_{k=0}^{\infty} \binom{n}{k} = 2^n.$$

Proof. Using the Binomial theorem with $x = 1$ and $y = 1$, we see that

$$2^n = (1+1)^n = \sum_{k=0}^{n} \binom{n}{k} 1^k \, 1^{n-k} = \sum_{k=0}^{n} \binom{n}{k}.$$

This is the desired result.

Combinatorial Proof. A set with n elements has a total of 2^n different subsets. Each subset has zero elements, one element, two elements, ... or n elements in it.

There are $\binom{n}{0}$ subsets with zero elements, $\binom{n}{1}$ subsets with one element, $\binom{n}{2}$ subsets with two elements, ..., and $\binom{n}{n}$ subsets with n elements.

Therefore, $\sum_{k=0}^{n}\binom{n}{k}$ counts the total number of subsets of a set with n elements.

This shows that $\sum_{k=0}^{n}\binom{n}{k} = 2^n$.

Corollary 2. *Let n be a positive integer. Then*

$$\sum_{k=0}^{n}(-1)^k\binom{n}{k} = 0.$$

Proof. By the Binomial theorem it follows that

$$0 = 0^n = [(-1)+1]^n = \sum_{k=0}^{n}\binom{n}{k}(-1)^k\,1^{n-k} = \sum_{k=0}^{n}\binom{n}{k}(-1)^k.$$

This proves the corollary.

Remark. Corollary 2 implies that

$$\binom{n}{0}+\binom{n}{2}+\binom{n}{4}+\ldots\ldots = \binom{n}{1}+\binom{n}{3}+\binom{n}{5}+\ldots\ldots$$

Corollary 3. *Let n be a non-negative integer. Then*

$$\sum_{k=0}^{n}2^k\binom{n}{k} = 3^n.$$

Proof. We recognize that the left-hand side of this formula is the expansion of $(1+2)^n$ provided by the Binomial theorem.

Therefore, by Binomial theoerm, we see that

$$(1+2)^n = \sum_{k=0}^{n}\binom{n}{k}1^{n-k}\,2^k = \sum_{k=0}^{n}\binom{n}{k}2^k.$$

Hence $\sum_{k=0}^{n}2^k\binom{n}{k} = 3^n$.

1.5.1 Pascal's Identity

Theorem 1.17. *Let n and k be positive integers with n ≥ k. Then*

$$\binom{n+1}{k} = \binom{n}{k-1} + \binom{n}{k}.$$

Proof. Suppose that T is a set containing $n + 1$ elements. Let a be an element in T, and let S = T – $\{a\}$.

Note that there are $\binom{n+1}{k}$ subsets of T containing k elements.

However, a subset of T with k elements either contains a together with $k - 1$ elements of S, or contains k elements of S and does not contain a.

Since there are $\binom{n}{k-1}$ subsets of $k - 1$ elements fo S, there are $\binom{n}{k-1}$ subsets of k elements of T that contain a.

And there are $\binom{n}{k}$ subsets of k elements of T that do not cotnain a, since there are $\binom{n}{k}$ subsets of k elements of S.

Consequently, $\binom{n+1}{k} = \binom{n}{k-1} + \binom{n}{k}$.

Remark. A combinatorial proof of Pascal's identity has been given. It is also possible to prove this identity by algebraic manipulation from the formula for $\binom{n}{r}$.

Remark. Pascal's identity, together with the initial conditions $\binom{n}{0} = \binom{n}{n} = 1$ for all integers n, can be used to recursively define binomial coefficients. This recursive definition is useful in the computation of binomial.

Coefficients since only the addition of integers is needed to use this recursive definition.

Pascal's identity is the basis for a geometric arrangement of the binomial coefficients in a triangle, as shown in Figure below.

The nth row in the triangle consists of the binomial coefficients $\binom{n}{k}$, $k = 0, 1, \ldots\ldots, n$.

This triangle is known as Pascal's triangle.

Pascal's identity shows that when two adjacent binomial coefficients in this triangle are added, the binomial coefficient in the next row between these two coefficients is produed.

$$\begin{array}{c}
\binom{0}{0}\\
\binom{1}{0}\ \binom{1}{1}\\
\binom{2}{0}\ \binom{2}{1}\ \binom{2}{2}\\
\binom{3}{0}\ \binom{3}{1}\ \binom{3}{2}\ \binom{3}{3}\\
\binom{4}{0}\ \binom{4}{1}\ \binom{4}{2}\ \binom{4}{3}\ \binom{4}{4}\\
\binom{5}{0}\ \binom{5}{1}\ \binom{5}{2}\ \binom{5}{3}\ \binom{5}{4}\ \binom{5}{5}\\
\binom{6}{0}\ \binom{6}{1}\ \binom{6}{2}\ \binom{6}{3}\ \binom{6}{4}\ \binom{6}{5}\ \binom{6}{6}\\
\binom{7}{0}\ \binom{7}{1}\ \binom{7}{2}\ \binom{7}{3}\ \binom{7}{4}\ \binom{7}{5}\ \binom{7}{6}\ \binom{7}{7}\\
\binom{8}{0}\ \binom{8}{1}\ \binom{8}{2}\ \binom{8}{3}\ \binom{8}{4}\ \binom{8}{5}\ \binom{8}{6}\ \binom{8}{7}\ \binom{8}{8}\\
\cdots\cdots\cdots\\
\cdots\cdots\cdots
\end{array}$$

(*a*)

$$\begin{array}{c}
1\\
1\quad 1\\
1\quad 2\quad 1\\
1\quad 3\quad 3\quad 1\\
1\quad 4\quad 6\quad 4\quad 1\\
1\quad 5\quad 10\quad 10\quad 5\quad 1\\
1\quad 6\quad 15\quad 20\quad 15\quad 6\quad 1\\
1\quad 7\quad 21\quad 35\quad 35\quad 21\quad 7\quad 1\\
1\quad 8\quad 28\quad 56\quad 70\quad 56\quad 28\quad 8\quad 1\\
\cdots\cdots\cdots\\
\cdots\cdots\cdots
\end{array}$$

(*b*)

Fig. 1.5. Pascal's Triangle

By Pascal's identity $\binom{6}{4} + \binom{6}{5} = \binom{7}{5}$.

1.5.2. Vander Monde's Identity

Theorem 1.18. *Let m, n and r be non-negative integers with r not exceeding either m or n.* Then

$$\binom{m+n}{r} = \sum_{k=0}^{r} \binom{m}{r-k}\binom{n}{k}.$$

Proof. Suppose that there are m items in one set and n items in a second set.

Then the total number of ways to pick r elements from the union of these sets is $\binom{m+n}{r}$.

Another way to pick r elements from the union is to pick k elemetns from the first set and then $r - k$ elements from the second set, where k is an integer with $0 \le k \le r$.

This can be done in $\binom{m}{k}\binom{n}{r-k}$ ways, using the product rule. Hence, the total number of ways to pick r elements from the union also equals

$$\binom{m+n}{r} = \sum_{k=0}^{r}\binom{m}{r-k}\binom{n}{k}$$

This proves vander Monde's identity.

Corollary. *If n is a non negative integer, then*

$$\binom{2n}{n} = \sum_{k=0}^{n}\binom{n}{k}^2$$

Proof. We use vander Monde's identity with $m = r = n$ to obtain

$$\binom{2n}{n} = \sum_{k=0}^{n}\binom{n}{n-k}\binom{n}{k} = \sum_{k=0}^{n}\binom{n}{k}^2$$

The last equality was obtained using identity $\binom{n}{k} = \binom{n}{n-k}$.

Remark. This identity was discovered by mathematician Alexander-Theophile Vander Monde in the eighteenth century.

Theorem 1.19. *Let n and r be non negative integer with $r \le n$. Then*

$$\binom{n+1}{r+1} = \sum_{j=r}^{n}\binom{j}{r}.$$

Proof. We use a combinatorial proof.

The left-hand side, $\binom{n+1}{r+1}$, counts the bit strings of length $n + 1$ containing $r + 1$ ones.

We show that the right-hand side counts the same objects by considering the cases corresponding to the possible locations of the final 1 in a string with $r + 1$ ones.

This final one must occur at position $r + 1, r + 2, ...$ or $n + 1$.

Furthermore, if the last one is the kth bit there must be r ones among the first $k - 1$ positions.

Consequently, there are $\binom{k-1}{r}$ such bit strings.

Summing over k with $r + 1 \le k \le n + 1$, we find that there are $\sum_{k=r+1}^{n+1} \binom{k-1}{r} = \sum_{j=r}^{n} \binom{j}{r}$ bit strings of length n containing exactly $r + 1$ ones.

Note that the last step follows from the change of variables $j = k - 1$.

Since the left-hand side and right-hand side count the same objects, they are equal.

This completes the proof.

Problem 1.148. *What is the coefficient of x^2y^4 in $(x + y)^6$?*

Solution. Here $(x + y)^6 = (x + y) . (x + y) ...... (x + y)$ [6 factors]. We have to choose an x from two factors and ay from the remaining four factors.

The x's can be chosen in C(6, 2) = 15 ways and each choice for the x's leaves just one choice for the y's.

Hence the coefficient of x^2y is 15.

Problem 1.149. *What is the coefficient of $x^3y^2z^2$ in $(x + y + z)^9$?*

Solution. This is the same as how many ways one can choose x from three brackets, ay from two brackets and az from two brackets in the expansion $(x + y + z)(x + y + z) ...... (x + y + z)$ (9 factors)

This can be done in $\binom{9}{3\,2\,2} = \dfrac{9!}{3!2!2!} = 15120.$

Problem 1.150. *Find the number of arrangement of the letters in the word ACCOUNTANT.*

Solution. Total number of letters in the word ACCOUNTANT is 10. Out of which A occurs twice, C occurs twice, N occurs twice, T occurs twice and the rest are all different.

Since some of the letters are repeated we need to apply multimonial theorem.

Hence the number of arrangements is

$$\frac{10!}{2!2!2!2!} = 226800.$$

Problem 1.151. *Using the Binomial theorem, expand $(3x^2 + 2y)^5$ and simplify.*

Solution. $(3x^2 + 2y)^5 = (3x^2)^5 + \binom{5}{1}(3x^2)^4 (2y) + \binom{5}{2} (3x^2)^3 (2y)^2 + \binom{5}{3} (3x^2)^2 (2y)^3$

$$+ \binom{5}{4} (3x^2)(2y)^4 + (2y)^5.$$

$$= 243x^{10} + 5(81x^8)(2y) + 10(27x^6)(4y^2) + 10(9x^4)(8y^3) + 5(3x^2)(16y^4) + 32y^5.$$
$$= 243x^{10} + 810x^8y + 1080x^6y^2 + 720x^4y^3 + 240x^2y^4 + 32y^5.$$

Problem 1.152. *Find the coefficient of x^{16} in the expansion of* $\left(2x^2 - \frac{x}{2}\right)^{12}$.

Solution. The general term in the expansion of this expression by the Binomial Theorem is

$$\binom{12}{k}(2x^2)^{12-k}\left(-\frac{x}{2}\right)^k = \binom{12}{k}2^{12-k}\left(-\frac{1}{2}\right)^k x^{24-k}.$$

We want $24 - k = 16 \Rightarrow k = 8$.

The coefficient is $\binom{12}{8}2^4\left(-\frac{1}{2}\right)^8 = \frac{1}{16}\binom{12}{8} = \frac{495}{16}$.

Problem 1.153. *Using the Binomial Theorem, expand* $\left(x-\frac{4}{x}\right)^6$ *and simplify.*

Solution. $$\left(x-\frac{4}{x}\right)^6 = x^6 + \binom{6}{1}x^5\left(-\frac{4}{x}\right) + \binom{6}{2}x^4\left(-\frac{4}{x}\right)^2 + \binom{6}{3}x^3\left(-\frac{4}{x}\right)^3$$

$$+\binom{6}{4}x^2\left(-\frac{4}{x}\right)^4 + \binom{6}{5}x\left(-\frac{4}{x}\right)^5 + \left(-\frac{4}{x}\right)^6$$

$$= x^6 + 6x^5\left(-\frac{4}{x}\right) + 15x^4\left(\frac{16}{x^2}\right) + 20x^3\left(-\frac{64}{x^3}\right) + 15x^2\left(\frac{256}{x^4}\right) + 6x\left(-\frac{1024}{x^5}\right) + \frac{4096}{x^6}$$

$$= x^6 - 24x^4 + 240x^2 - 1280 + \frac{3840}{x^2} - \frac{6144}{x^4} + \frac{4096}{x^6}.$$

Problem 1.154. *Prove that* $\binom{n}{0}+\binom{n}{1}+\binom{n}{2}+\ldots\ldots+\binom{n}{n} = 2^n$ *for all natural numbers n.*

Solution. Consider $(x+y)^n = \sum_{k=0}^{n}\binom{n}{k}x^{n-k}y^k$.

Setting $x = y = 1$, we obtain

$$(1+1)^n = \sum_{k=0}^{n}\binom{n}{k}1^{n-k}1^k.$$

that is, $$2^n = \sum_{k=0}^{n}\binom{n}{k},$$

$$= \binom{n}{0}+\binom{n}{1}+\binom{n}{2}+\ldots\ldots+\binom{n}{n} \text{ as desired.}$$

Problem 1.155. *What is the coefficient of $x^{12}y^{13}$ in the expansion of $(2x - 3y)^{25}$?*

Solution. First, note that this expression equals $(2x + (-3y))^{25}$.

By the Binomial theorem, we have

$$(2x + (-3y))^{25} = \sum_{j=0}^{25} \binom{25}{j} (2x)^{25-j} (-3y)^j$$

Consequently, the coefficient of $x^{12}y^{13}$ in the expansion is obtained when $j = 13$, namely

$$\binom{25}{13} 2^{12} (-3)^{13} = -\frac{25!}{13!12!} 2^{12} 3^{13}.$$

Problem 1.156. *What is the coefficient of $x^{12}y^{13}$ in the expansion of $(x + y)^{25}$?*

Solution. From the Binomial theorem, it follows that this coefficient is

$$\binom{25}{13} = \frac{25!}{13!12!} = 5,200,300.$$

Problem 1.157. *What is the expansion of $(x + y)^4$?*

Solution. From the Binomial theorem it follows that

$$(x + y)^4 = \sum_{j=0}^{4} \binom{4}{j} x^{4-j} y^j$$

$$= \binom{4}{0} x^4 + \binom{4}{1} x^3y + \binom{4}{2} x^2y^2 + \binom{4}{3} xy^3 + \binom{4}{4} y^4$$

$$= x^4 + 4x^3y + 6x^2y^2 + 4xy^3 + y^4.$$

1.5.3 The multinational Theorem 1.20

Let n be a positive integer. Then for all $x_1, x_2 \ldots\ldots x_t$ we have $(x_1 + x_2 + \ldots\ldots + x_t)^n = \Sigma P(n\,;\,q_1, \ldots\ldots q_t)\, x_1^{q_1}\ x_2^{q_2} \ldots\ldots x_t^{q_t}$. Where the summation extends over all sets of non negative integers $q_1, q_2, \ldots\ldots q_t$, where $q_1 + q_2 + \ldots\ldots + q_t = n$. There are $C(n + t - 1, n)$ terms in the expansion of $(x_1 + x_2 + \ldots\ldots + x_t)^n$.

Proof. The coefficient of $x_1^{q_1}\ x_2^{q_2} \ldots\ldots x_t^{q_t}$ is the number of ways of arranging the n letters $\{q_1 \cdot x_1, q_2 \cdot x_2, \ldots\ldots q_t \cdot x_t\}$

Therefore, it is $P(n\,;\,q_1, q_2, \ldots\ldots, q_t)$

The number of terms is determind as follows, each term of the form $x_1^{q_1}\ x_2^{q_2} \ldots\ldots x_t^{q_t}$ is a selection of n objects with repetitions from t distinct types. Hence there are $C(n + t - 1, n)$ ways to do this.

Corollary. For any positive integer t, we have $t^n = \Sigma P(n\,;\,q_1, q_2 \ldots\ldots q_t)$ where the summation extends over all sets of non negative integers $q_1, q_2, \ldots\ldots, q_t$ where $q_1 + q_2 + \ldots\ldots + q_t = n$.

Example

(1) In $(x_1+x_2+x_3+x_4+x_5)^{10}$ the coefficient of $x_1^2x_3\,x_4^3\,x_5^4$ is $P(10\,;2,0,1,3,4)=\dfrac{10!}{2!0!1!3!4!}$ = 12,600.

There are C(10 + 5 – 1, 10) = C(14, 10) = 1,001 terms in the expansion $(x_1+x_2+x_3+x_4+x_5)^{10}$.

(2) In $(2x-3y+5z)^8$, we let $x_1=2x$, $x_2=-3y$, $x_3=5z$ and then the coefficient of $x_1^3x_2^3x_3^2$ is

$$P(8\,;3,3,2)=560.$$

Thus, the coefficient of $x^3y^3z^2$ is

$$2^3(-3)^3(5)^2\,P(8\,;3,3,2)=(2)^3(-3)^3(5^2)(560)$$

(3)
$$\begin{aligned}(x-2y+x)^3 &= P(3\,;3,0,0)x^3+P(3\,;0,3,0)(-2)^3y^3\\ &+P(3\,;0,0,3)\,z^3+P(3\,;2,0,1)\,x^2z\\ &+P(3\,;2,1,0)\,x^2(-2y)+P(3\,;1,2,0)\,x(-2y)^2\\ &+P(3\,;1,0,2)\,xz^2+P(3\,;0,2,1)(-2y)^2z\\ &+P(3\,;0,1,2)(-2y)\,z^2+P(3\,;1,1,1)\,x(-2y)\,z\\ &=x^3-8y^3+z^3+3x^2z-6x^2y+12xy^2+3xz^2+12y^2z-6yz^2-12xyz.\end{aligned}$$

Problem 1.158. *Find the number of 3-part unordered partitions of a set S with n distinct elements.*

Solution. We know that the number of 3-part ordered partitions of S is 3^n.

Let $P_n(t)$ denote the number of t-part unordered partitions of a set with n elements. We are asked to find $P_n(3)$ in this problem.

One 3-part unordered partition of S is {S, ϕ, ϕ} of which there are 3 orderings : (S, ϕ, ϕ), (ϕ, S, ϕ), and (ϕ, ϕ, S).

From each of the other $P_n(3)-1$ unordered partitions there are 3 ! orderings.

Thus, $3!\,(P_n(3)-1)+3=3^n$ and $P_n(3)=(3^{n-1}+1)/2$.

1.6. RAMSEY NUMBER

Let p and q be 2 positive integers. A positive integer r is said to have the {P, q) – **Ramsey property** if in any group of r people either there is a subgroup of p people known to one another or there is a subgroup of q people not known to one another.

(by Ramsey's theorem all sufficiently large integers r have the (p, q) – Ramsey Property).

The smallest r with the (p, r) – **Ramsey Property** is called the **Ramsey number, R(*p*, *q*).**

$R(p, q) = R(q, p)$ and $R(p, 1) = 1$, $R(p, 2) = p$

Let k_i (i = 1, 2, ..., t) and m be positive integers, with each $k_i \geq m$ and $t \geq 2$.

Let < C_1, C_2, ... C_t > be an ordered partition of the class C of all m-element subsets of an n-element set X. (there are thus C(n, m) elements in C). Then the positive integer n has the generalized (k_1, k_2, ..., k_t ; m) – Ramsey Property if, for some value of i in the range 1 to t, X possesses a k_i – element subset B such that all m-element subsets of B belong to C_i. The smallest such n is the **generalized Ramsey number,** R(k_1, k_2, ..., k_t ; m).

Problem 1.159. *Show that if m and n are integers both greater than 2, then*

$$R(m, n) \leq R(m - 1, n) + R(m, n - 1).$$

Solution. Let $P \equiv R(m-1, n)$, $q \equiv R(m, n-1)$, and $r \equiv p + q$.

Consider a group $\{1, 2, ..., r\}$ of r people.

Let L be the set of people known to person 1 and M be the set of people not known to person 1.

The 2 sets together have $r - 1$ people, so either L has at least P people or M has at least q people.

(*a*) If L has $P = R(m-1, n)$ people, then, by definition, it contains a subset of $m - 1$ people known to one another or it contains a subset of n people unknown to one another.

In the former case the $m - 1$ people and person 1 constitute m people known to one another.

Thus, in this case, a group of $R(m-1, n) + R(m, n-1)$ people necessarily includes m mutual acquaintances or n mutual strangers.

i.e., $R(m, n) \leq R(m-1, n) + R(m, n-1)$.

(*b*) By the usual symmetry argument the same conclusion follows when M contains q people.

Problem 1.160. *If R(m – 1, n) and R(m, n – 1) are both even the greater than 2, prove that*

$$R(m, n) \leq R(m-1, n) + R(m, n-1) - 1.$$

Solution. Let $P \equiv R(m-1, n)$, $q \equiv R(m, n-1)$, and $r = p + q$.

It suffices to establish that in any group $X = \{1, 2, ..., r - 1\}$ of $r - 1$ people there is either a subgroup of m people who know one another or a subgroup of n people who do not know one another.

Let d_i be the number of people known to person i, for $i = 1, 2, ... r - 1$.

Since knowing is mutual, $d_1 + d_2 + ... + d_{r-1}$ is necessarily even. But $r - 1$ is odd, so d_i is even for atleast 1 i, which we may take to be $i = 1$.

Let L be the set of people known to person 1 and let M be the set of people not known to person 1.

Since there are an even number of people in L, there must be an ever number of people in M as well.

Now either L has atleat $p - 1$ people or M has atleast q people. But $p - 1$ is odd.

So either L has at least p people or M has at least q people.

(*a*) Suppose L has at least P people. Because $P = R(m-1, n)$, L must contain either $m - 1$ people known to one another or n people not known to one another (in which case the theorem holds.)

In the former case these $m - 1$ people and person 1 will constitute m people known to one another (and the theorem holds.)

(*b*) The case of q or more people in M is handled by symmetry.

Problem 1.161. *Show that if m and n are integers greater than 1, then*

$$R = (m, n) \leq C(m + n - 2, m - 1) \quad ...(1)$$

Solution. When $m = 2$ or $n = 2$, (1) holds with equality.

The proof is by induction on $k = m + n$.

As we have just seen, the result is true when $k = 4$.

Assume the result true for $k - 1$, then

$R(m-1, n) \le C(m+n-3, m-2)$ and

$R(m, n-1) \le C(m+n-3, m-1)$.

Now pascal's identity gives $C(m+n-3, m-2) + C(m+n-3, m-1) = X(m+n-2, m-1)$

So that $R(m-1, n) + R(m, n-1) \le C(m+n-2, m-1)$

But, $R(m, n) \le R(m-1, n) + R(m, n-1)$.

Problem 1.162. *Show that*

(i) $R(4, 4) = 18$

(ii) $R(4, 3) = 9$

(iii) $R(5, 3) = 14$

(iv) $R(3, 3) = 6$.

Solution. *(i)* $R(4, 4) \le R(3, 4) + R(4, 3) = 9 + 9 = 18$

To show that $R(4, 4) > 17$

Consider an arrangement of 17 people about a round table such that each person knows exactly 6 people, the first, second, and fourth persons on one's right and first, and the fourth persons on one's left.

It can be veririfed that in this arrangement these is no subgroup of 4 mutual acquaintances or of 4 mutual strangers.

(ii) $R(4, 3) \le R(3, 3) + R(4, 2) - 1 = 9$

To prove that $R(4, 3) = R(3, 4) > 8$.

We exhibit a group of 8 people which has no subgroup of 3 people known to one another and no subgroup of 4 people not known to one another.

Here is a scenario, 8 people sit about a round table. Each person knows exactly 3 people, the 2 people sitting on either side of him and the person sitting farthest from him.

(iii) $R(5, 3) \le R(4, 3) + R(5, 2) = 9 + 5 = 14$

To see that $R(5, 3) = R(3, 5) > 13$.

Consider a group of 13 people sitting at a round table such that each person knows only the fifth person on his right and the fifth person on his left.

In such a situation there is no subgroup of 3 mutual acquaintances and no subgroup of 5 mutual strangers.

(iv) $R(3, 3) \le 6$.

To show that $R(3, 3) > 5$.

It is enough to consider a seating arrangement of 5 people about a round table in which each person knows only the 2 people on either side.

In such a situation there is no set of 3 mutual acquaintances and no set of 3 people known to one another.

Problem 1.163. *If n points are located in general position in the plane, and if every quadrilateral formed from these n points is convex, then the n points are the vertices of a convex n-gon.*

Solution. Suppose the n points do not form a convex n-gon. Consider the smallest convex polygon that contains the n points. At least one of the n points (say, the point P) is in the interior of this polygon.

Let Q be one of the vertices of the polygon.

Divide the polygon into triangles by drawing line segments joining Q to every vertex of the polygon.

The point P then will be in the interior of one of these triangles, which contradicts the convexity hypothesis.

Problem 1.164. *Show that in any group of 10 people there is always (a) a subgroup of 3 mutual strangers or a subgroup of 4 mutual acquaintances and (b) a subgroup of 3 mutual acquaintances or a subgroup of 4 mutual strangers.*

Solution. (*a*) Let A be 1 of the 10 poeple, the remaining 9 people can be assigned to 2 rooms, those who are known to A are in room Y and those who are not known to A are in room Z.

Either room Y has at least 6 people or room Z has at least 4 people.

(*i*) Suppose room Y has atleast 6 people, then, there is either a subgroup of 3 mutual acquaintances or a subgroup of 3 mutual strangers in this room.

In the former case, A and these 3 people constitute 4 mutual acquaintances.

(*ii*) Suppose room Z has atleast 4 people.

Either these 4 people know one another or at least 2 of them, B and C, do not know each other.

In the former case we have a sub group of 4 mutual acquaintances.

In the later case A, B and C constitute 3 mutual strangers.

(*b*) In the previous scenario, let people who are strangers become acquaintances, and let people who are acquaintances pretend they are strangers. The situation is symmetric.

Problem 1.165. *Let A be any $n \times n$ matrix. Matrix P is an $m \times m$ Principal submatrix of A if P is obtained from A by removing any n-m rows and the same n-m columns. Show that for every positive integer m, there exists a positive integer n such that every $n \times n$ binary matrix A has an $m \times m$ Principal submatrix P in one of the following four categories :*

(i) P is diagonal (ii) Ever non diagonal entry of P is 1.

(iii) P is lower triangular and every element in the lower triangle is 1

(iv) P is upper triangular and every element in the upper triangle is 1.

Solution. Let n be any positive integer greater than R(m, m, m, m ; 2) and let A = $[a_{ij}]$ be any $n \times n$ binary matrix, the rows of which constitute the set X = $\{r_1, r_2, \ldots\ldots, r_n\}$.

The class C of all 2-element subsets of X is partitioned into 4 classes, as follows :

$$C_1 = \{\{r_i, r_j\} : a_{ji} = 0, a_{ij} = 0\}$$

$$C_2 = \{\{r_i, r_j\} : a_{ji} = 1, a_{ij} = 1\}$$

$$C_3 = \{\{r_i, r_j\} : a_{ji} = 0, a_{ij} = 1\}$$

$$C_4 = \{\{r_i, r_j\} : a_{ji} = 1, a_{ij} = 0\}.$$

Since $n \geq$ R(m, m, m, m ; 2), there exists a subset X′ of X with m elements (rows) such that all 2-elements subsets of X′ are contained in one of these 4 classes.

This implies the existence of an $m \times m$ principal submatrix in one of the categories (*i*) through (*iv*).

Problem 1.166. *An arithmetic progression of length n is a sequence of the form* $< a, a + d, a + 2d, \ldots\ldots, a + (n - 1)d >$

Show that in any partition of $X = \{1, 2, \ldots\ldots, 9\}$ *into 2 subsets, at least 1 of sets contains an arithmetic progression of length 3.*

Solution. Suppose that the theoerm is false.

Let X be partitioned into P and Q, and let 5 be an element of P.

Obviously both 1 and 9 [$d = 4$] cannot be in P, so that there are 3 cases to consider.

Case 1. 1 is in P and 9 is in Q.

Since 1 and 5 are in P, 3 is in Q. Since 3 and 9 are in Q, 6 is in P. Since 5 and 6 are in P, 4 is in Q.

Since 3 and 4 are in Q, 2 is in P. Since 5 and 6 are in P, 7 is in Q. Since 7 and 9 are in Q, 8 is in P.

But then P contains the arithmetic Progression 2, 5, 8, *a* contradiction.

Case 2. 9 is in P and 1 is in Q. Set X is invariant when each element is replaced by its tens-complement.

Under this transformation the persent case becomes case 1, which has already been disposed of.

Case 3. 1 and 9 are in Q. The number 7 is either in P or in Q suppose it is in P. Since 5 and 7 are in P, both 3 and 6 are in Q. That means Q has the arithmetic progression 3, 5, 9.

On the otherhand, if 7 is in Q, then 8 is in P.

Since 1 and 7 are in Q, 4 is in P. Since 4 and 5 are in P, 3 is in Q. Since 1 and 3 are in Q, 2 is in P.

Then P has the arithmetic progression 2, 5, 8.

Problem 1.167. *Show that in any group of 20 people there will always be either a subgroup of 4 mutual acquaintances or a subgroup of 4 mutual strangers.*

Solution. Suppose A is one of these 20 people.

People known to A are in room Y and people not known to A are in room Z.

Either room Y has atleast 10 people or room Z has at laest 10 people.

(*i*) If Y has at least 10 people, then there is either a subgroup of 3 mutual acquaintances or a subgroup of 4 mutual strangers, as asserted, in this room.

In the former case A and these mutual acquaintances will form a subgroup of 4 mutual acquaintances.

(*ii*) Inter change "acqaintances" and "strangers" in (*i*).

1.7 THE CATALAN NUMBERS

A point in the cartesian plane whose coordinates are integers is called a **lattice point.** Consider a path from the origin to the lattice point A(m, n), where m and n are non negative, that

(*i*) starts from the origin

(*ii*) is always parallel to the x-axis or the y-axis

(*iii*) makes turns only at a lattice point, either along the positive x-axis or along the positive y-axis,

(*iv*) terminates at A.

A typical path is a sequence of $m + n$ unit steps, m of them horizontal and n of them vertical. The number of paths is $C(m + n, m) = C(m + n, n)$, the number of ways of reserving positions in the sequence for one or the other kind of step.

A path from P_0 to P_m in the cartesian plane is a sequence $<P_0, P_1, ..., P_m>$ of lattice points, $P_i(x_i, y_i)$, such that for each $i = 0, 1, ..., m - 1, x_{i+1} = x_i + 1, y_{i+1} = y_i$ or $x_{i+1} = x_i, y_{i+1} = y_i + 1$.

This path is **good** if $y_i < x_i$ $(i = 0, 1, ... m)$, otherwise it is **bad.**

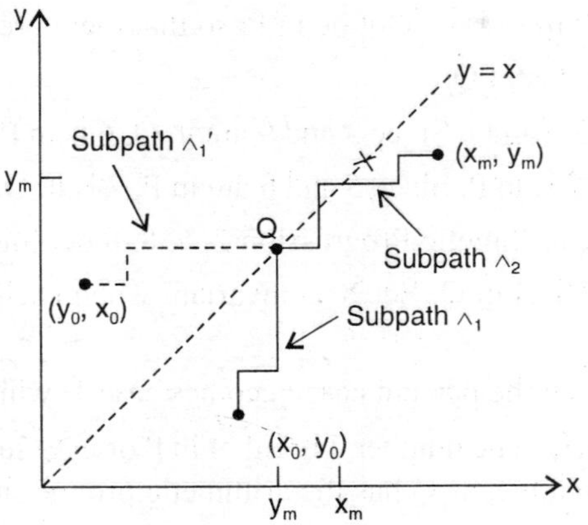

Fig. 1.6.

Set $m = x_m - x_0$ and $n = y_m - y_0$ to obtain the required number as $C(x_m - x_0 + y_m - y_0, x_m - x_0)$.

A **good path** is one that lies entirely below the 45° line (see Fig. 1.6). Thus the conditions $y_0 < x_0$ and $y_m < x_n$ are necessary for a good path, to which may be adjoined $x_0 \le x_m$ and $y_0 \le y_m$ (the x and y coordinates can never decrease along the path). Under these 4 conditions all paths will be good, unless it is possible for a path to intersect the 45° line at some ordinate less than or equal to y_m, *i.e.*, unless $x_0 \le y_m$. The desired criterion is $y_0 < x_0 < y_m < x_m$.

Fig. 1.6 shows a **bad path** from (x_0, y_0) to (x_m, y_m), it first intersects the line $y = x$ in the lattice point Q. If subpath $\wedge_1$, from (x_0, y_0) to Q, is reflected in the 45° line, $\wedge_1' + \wedge_2$ is a path from (y_0, x_0) to (x_m, y_m).

Any path from (y_0, x_0) to (x_m, y_m) defines by partial reflection a **bad path** from (x_0, y_0) to (x_m, y_m). There are $C(x_m - y_0 + y_m - x_0, x_m - y_0)$ **bad paths,** $C(x_m - x_0 + y_m - y_0, x_m - x_0) - C(x_m - x_0 + y_m - y_0, x_m - y_0)$ **good paths** from (x_0, y_0) to (x_m, y_m).

The nth **catalan number**, C_n, is defined as the number of good paths from $(1, 0)$ to $(n, n - 1)$, then

$$C_n = \frac{1}{n} C(2n - 2, n - 1).$$

[We have $C_n = C(2n - 2, n - 1) - C(2n - 2, n)$

$$= C(2n - 2, n - 1)\left[1 - \frac{n-1}{n}\right]$$

$$= \frac{1}{n} C(2n - 2, n - 1)]$$

Problem 1.168. *Find the number of sequences of the form* $<u_1 u_2 \ldots\ldots u_{2n}>$ *such that*

(*i*) u_i *is either* -1 *or* $+1$, *for every i,*

(*ii*) $u_1 + u_2 + \ldots\ldots + u_k \geq 0$, *for* $1 \leq k \leq 2n-1$, *and*

(*iii*) $u_1 + u_2 + \ldots\ldots + u_{2n} = 0$.

Solution. Consider a path from (0, 0) to (n, n) as traced by a particle which makes unit steps in the x and y directions.

Let the particles location after i steps be (x_i, y_i) and define $u_i \equiv (x_i - x_{i-1}) - (y_i - y_{i-1})$.

Then, if the particle never rise above the line $y = x$, the integers u_i ($i = 1, 2, \ldots\ldots, 2n$) statisfy (*i*), (*ii*), (*iii*) above.

Conversely, every sequence $<u_i>$ that obeys (*i*), (*ii*) and (*iii*) defines a path from (0, 0) to (n, n) that never rises above $y = x$.

Hence, the number of such sequences is C_{n+1}.

Problem 1.169. (The Ballot Problem)

Suppose P and Q are 2 candidates for a public office who secured p votes and q votes, respectively. If $p > q$, *find the probability that P stayed ahead of Q throughout the counting of votes.*

Solution. In the cartesian plane let x and y, respectively, denote the votes accumulated by P and Q at any stage.

Every path from (0, 0) to (p, q) repersents a possible history of the voting, and conversely.

Thus, the number of ways the voting could have gone is $C(p + q, p)$, out of which P leads continually in $C(p + q - 1, p - 1) - C(p + q - 1, p)$.

This is the number of good paths from (1, 0) to (p, q).

The desired probability is therefore

$$\frac{C(p+q-1, p-1) - C(p+q-1, p)}{C(p+q, p)} = \frac{p-q}{p+q}.$$

Problem 1.170. *Find the number of paths from (0, 0) to (n, n) such that*

(*a*) *either* $x > y$ *at all interior lattice points or* $y > x$ *at all interior lattice points ; and*

(*b*) $y \leq x$ *at every lattice point on the path, and*

(*c*) *the path never crosses the line* $y = x$.

Solution. (*a*) The number of paths of this type will be twice the number of good paths from (1, 0) to $(n, n-1)$, or $2C_n$.

(*b*) Let A be the point (n, n).

Suppose the origin O(0, 0) is transferred to O′(– 1, 0). The new coordinates are O′(0, 0), O(1, 0), and A$(n + 1, n)$. The number of good paths from O to A, namely, C_{n+1} is equal to the number of paths from O to A in which $y \leq x$ at every lattice point.

(*c*) By reflectional symmetry, the required number is twice the number found in (*b*), or $2C_{n+1}$.

Problem 1.171. *Let X = {1, 2, ..., n}. A function f from X to X is monotonic increasing i,' f(i) ≤ f(i) whenever i < j. Find the number of monotonic increasing functions f from X to X such that f(i) ≤ i for every i in X.*

Solution. Any function of this type will have a graph, $y = f(x)$, consisting of lattice points that can be embedded in a unique way in a path from (0, 0) to (n, n) that does not rise above the line $y = x$.

Conversely, any path from (0, 0) to (n, n) that dose not rise above $y = x$ defines a function of this type.

Thus, the answer is C_{n+1}.

Problem 1.172. A diagonal triangulation of a convex polygon is a division of the polygon into triangles by diagonals which do not intersect except at vertices of the polygon. Show that the number of triangles and the number of diagonals in any diagonal triangulation of a convex polygon with n vertices are $n - 2$ and $n - 3$, respectively.

Solution. Suppose that the numbers of triangles in 2 different triangulations are p and q, then $p\pi = q\pi$ = sum of the interior angles of the polygon. When $p = q$, with the n vertices of the polygon marked 1, 2, ..., n, the diagonals 13, 14, ..., $\overline{1n-1}$ yield a particular triangulation.

This and therefore every-triangulation invovles $n - 2$ triangles. In counting the edges of the $n - 2$ triangles in any triangulation, each diagonal involved is counted twice, since it is an edge of exactly 2 triangles.

Each side of the polygon is counted once, since it is an edge of exactly 1 triangle.

Therefore, with $x \equiv$ number of diagonals, we have $3(n - 2) = 2x + n$, or $x = n - 3$.

Problem 1.173. *Find the number of binary vectors in which the number of 1s is m and the number of 0 s in n, such that every component after the first is preceded by moer 0 s than 1s.*

Solution. Obviously m cannot exceed n. Also, the first 2 components are both 0.

We distinguish between two cases, $m < n$ and $m = n$. where F corresponds to 0 and T corresponds to 1. Of the two 0s at the beginning, we do not take the first into consideration.

We then have a feasible vector (with $m + n - 1$ components) in which the number of 1 s is m.

There are $\dfrac{n-m}{n}$C$(m + n - 1, m)$ such vectors.

If $m = n$, let *a 0* represent a unit step in the x direction and *a* 1 represent a unti in the y direction. Then there are just as many binary vectors of the given type as there are paths from (0, 0) to (n, n) with the property that $x > y$ at each interior lattice point.

The number of these paths is C_n.

Problem 1.174. *Prove combinatorially that the catalan numbers satisfy the non linear recursion relation*

Solution. There are C_{n+1} paths from (0, 0) to (n, n) with $y \leq x$ at each lattice point. (See Fig. 1.7 below)

Let X_i ($i = 1, 2, ..., n$) denote the subset of these paths that have (i, i) as their second lattice point on the line $y = x$.

There are C_i ways of going from (0, 0) to (i, i) (with the origin of coordinates translated to (i, i)), there are C_{n-i+1} ways of going from (i, i) to (n, n).

Hence, by the product rule, X_i contains $C_i\, C_{n-i+1}$ paths.

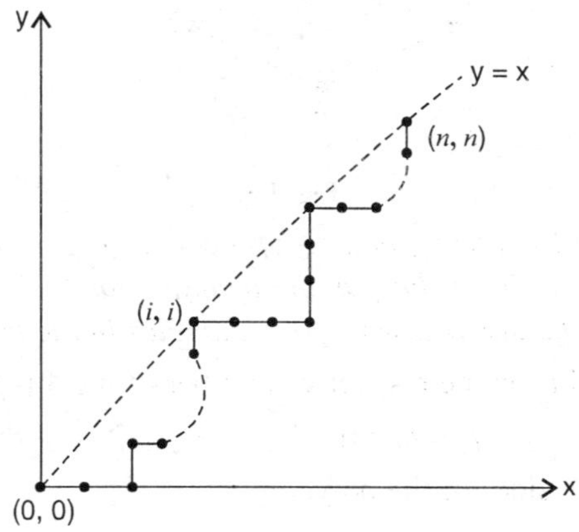

Fig. 1.7.

Problem 1.175. *Find the number of triangulations of a convex polygon with $n \geq 4$ vertices such that every triangle is type 1 or type 2.*

Solution. The triangulation must involve $n - 3$ diagonals, and there must be precisely two type 2 triangles.

Number the vertices consecutively in such fashion that diagonal $d_1 = 2n$ cuts off one of these type 2 triangles. (See Fig. below)

Now our triangulation must induce a triangulation of the convex $(n - 1) - gon$ 23 ...... $n2$ (interior shaded in Fig.) that also has the property $f_0 = 0$.

This implies that side d_1 must be covered by a type 2 triangle (else the n-gon would have three type 2 triangles).

Hence there are 2 choices for d_2 : $d_2 = 3n$ and $d_2 = \overline{2n-1}$. Repeating the argument for the residual $(n - 2)$-gon etc. We obtain (1)(2)(2) ...... (2) $= 2^{n-4}$ triangulations of the n-gon, in each of which the chosen d_{n-3} cuts off the second type 2 triangle.

Because vertex 1 can be any vertex of the n-gon, we would seem to have altogether $n2^{n-4}$ triangulations. But in this grand total each triangulation is counted twice-once according to its beginning type 2 triangle and once according to its ending type 2 triangle our answer is therefore $n2^{n-5}$.

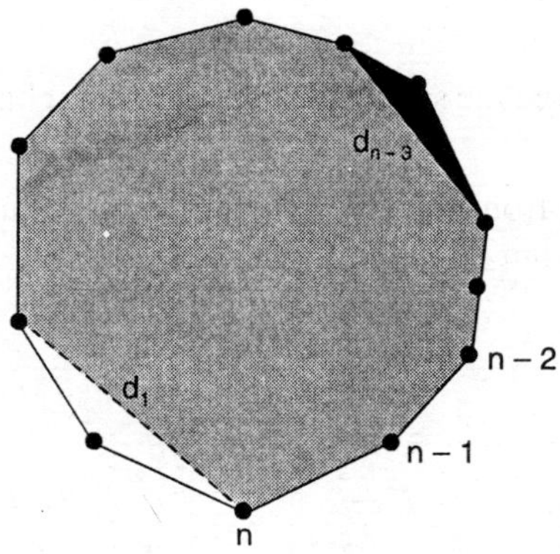

Fig. 1.8.

Problem 1.176. *Consider a triangulation of a convex n-gon ($n \geq y$). Let us call a constituent triangle type 0, type 1, or type 2, according as the triangle has 0, 1 or 2 sides in common with the polygon. Prove that there must be at least two type 2 triangles in the triangulation.*

Solution. Let f_0, f_1 and f_2 be the respective number of the 3 types. We have

$$f_0 + f_1 + f_2 = n - 2 \quad ...(1)$$

and, by a double counting of the sides of the polygon.

$$f_1 + 2f_2 = n \quad ...(2)$$

Subtract (1) and (2),

$$f_2 - f_0 = 2$$

or

$$f_2 = 2 + f_0 \geq 2.$$

Problem 1.177. *Find the number of ways of arranging 2n distinct real numbers as two n-vectors, $u = [u_1\ u_2 \ldots\ldots u_n]$ and $v = [v_1\ v_2 \ldots\ldots v_n]$, such that*

(i) in each vector the components are in strict decreasing order and

(ii) $u_i > v_i$ for all i.

Solution. Imagine that the $2n$ numbers correspond to $2n$ customers (of distinct heights) waiting in front of a box office to buy a \$ 5 ticket, such that there are n customers who have exactly one \$5 bill and n who have exactly one \$ 10 bill.

Let each u_i correspond to a person with a \$ 5 bill (an F) and each v_i to a T.

Now assume that in the waiting line the $2n$ customers stand in order of decreasing heights (first in line is tallest). Then we have a vector of F's and T's with $2n$ components out of which the number of F's is n.

In fact, consider the T customer who corresponds to v_i in the vector v.

The number of T's ahead of this customer in the line is $i - 1$, where as, by condition (*ii*) the number of F's ahead of him is atleast i. So this vector is indeed feasible.

Conversely, every feasible vector W can be decomposed into 2 vector, u and v as above : the vector v is the subline of \$ 5 customers and v is the subline of \$ 10 customers, each arranged in order of decreasing heights.

In the feasible vector W each T is preceded by more F's than T's, hence u and v must obey condition (*ii*).

This one-to-one correspondence shows that the required number of ways is C_{n+1}.

1.8 STIRLING NUMBERS AND BELL NUMBERS

Define the **falling factorial** polynomials by $[x]_0 \equiv 1$ and $[x]_n \equiv x(x-1)(x-2) \ldots\ldots (x-n+1)$, $n = 1, 2, 3, \ldots\ldots$

The coefficient of x^r in $[x]_n$ is known as the **stirling number** of the **first kind**, $S(n, r)$,

$$\text{Thus, } [x]_n \equiv \sum_{r=0}^{\infty} S(n, r)\, x^r \text{ where } S(n, r) = 0 \text{ for } r > n.$$

The recurrence relation

$$S(n+1, r) = S(n, r-1) - nS(n, r).$$

The absolute value of $S(n, r)$ is called the **signless stirling number** of the first kind, we denote it as $S'(n, r)$.

$$S(n, r) = (-1)^{n-r}\, S'(n, r).$$

Define the **rising factorial** polynomials by $[x]^0 \equiv 1$ and

$$[x]^n = x(x+1)(x+2) \ldots\ldots (x+n-1),\ n = 1, 2, 3, \ldots\ldots$$

$$\text{that is, } [x]^n = \sum_{r=0}^{\infty} S(n, r)\, x^r \text{ where } S'(n, r) = 0 \text{ for } r > n.$$

The number ways of partitioning a set of n elements into m (non empty) subsets is denoted by $S(n, m)$ and is known as the **stirling number** of the **second kind.**

By definition, $S(0, 0) = 1$, also, $S(n, m) = 0$ if $m > n$.

The number of surjections (onto mappings) from a set of n elements to a set of m elements is $m!\, S(n, m)$. The number of partitions of a set with n elements is the **Bell number, B_n** ; by fiat $B_0 = 1$.

$$\text{i.e., } B_n = \sum_{m=1}^{n} S(n, m) \text{ and } B_n = \sum_{k=0}^{n-1} C(n-1, k)\, B_k.$$

Problem 1.178. *Find the number of ways of putting n distinct objects into m distinct boxes, if the (left-to-right) order of objects within a box is significant and if empty boxes are permitted. (Note that if m > n, at least m – n boxes must be empty).*

Solution. Indicate the desired number by $f(n, m)$.

Suppose that a distribution of $n-1$ of the objects-there are $f(n-1, m)$ such distributions-brings i_1 objects into box 1, i_2 objects into box 2, ..., i_m objects into box m.

Here, $i_k \geq 0$, $k = 1, 2, \ldots\ldots, m$, and $i_1 + i_2 + \ldots\ldots + i_m = n-1$.

Then the nth object can go into box k in $i_k + 1$ ways (leftmost, second from left, ..., $(i_k + 1)$st from left), for a total of $(i_1 + 1) + (i_2 + 1) + \ldots\ldots (i_m + 1) = n - 1 + m$ arrangements.

Since this number is independent of the particular distribution of the $n - 1$ objects, we have the relation

$$f(n, m) = (n - 1 + m) f(n - 1, m) \text{ from which } f(n, m) = (m + n - 1)(m + n - 2) \text{ ...... } m = [m]^n.$$

Problem 1.179. *If m and n are positive integers prove that the equation*

$x_1 + x_2 + \text{......} + x_m = n$ ...(1), *has exactly $[m]^n/n\,!$ solutions in non negative integers x_k.*

(The result also holds for n = 0).

Solution. This is a matter of putting n identical objects (1 s) into m distinct boxes (an x_1 box, an x_2 box, ..., x_m box) empty boxes being allowed.

If we temporarily make the 1 s distinct by labelling then $1_1, 1_2, \text{......}, 1_n$ then $[m]^n$ arrangements.

However, arrangements that differ only with regard to the labels carried by the 1 s given the same solution to (1).

Thus, the answer is $[m]^n/n\,!$, as stated.

Problem 1.180. *Find the number of ways of selecting r distinct integers out of the first n positive integers such that the selection does not contain 2 consecutive integers.*

Solution. Arrange the first n positive integers in a row, in increasing order starting from 1.

If a number is chosen, put the symbol Y under that number, otherweise, put the symbol N under the number.

Let x_1 be the number of N's preceding the first Y, x_2 the number of N's between the first and second Y's ......, x_r the number of N's between the $(r - 1)^{st}$ and r^{th} Y's, and x_{r+1} the number of N's following the rth Y.

Then there is an one-to-one correspondence between acceptable selections and integral solutions of

$$x_1 + x_2 + \text{......} + x_{r+1} = n - r \text{ with}$$

$$x_1 \geq 0, x_2 \geq 1, \text{......}, x_r \geq 1, x_{r+1} \geq 0$$

Therefore, the desired number as $C(n - r + 1, r)$.

Problem 1.181. *Establish the following analogue to the binomial theorem*

$$[x + y]^n = \sum_{r=0}^{n} C(n, r)\, [x]^{n-r}\, [y]^r.$$

Solution. Perform an induction on n.

The formula is true for $n = 1$, because

$$[x + y]^1 = x + y = C(1, 0)\, [x]^1\, [y]^0 + C(1, 1)\, [x]^0\, [y]^1$$

Assume the formula true for $n = m - 1$. Then

$$[x + y]^m = (x + y + m - 1)\, [x + y]^{m-1}$$

$$= (x + y + m - 1) \sum_{r=0}^{m-1} C(m - 1, r)\, [x]^{m-1-r}\, [y]^r.$$

$$= \sum_{r=0}^{m} \{(x+m-r-1)+(y+r)\} \, C(m-1, r) \, [x]^{m-1-r} \, [y]^r$$

$$= \sum_{r=0}^{m} C(m-1, r) \, [x]^{m-r} \, [y]^r + \sum_{r=0}^{m} C(m-1, r) \, [x]^{m-1-r} \, [y]^{r+1}$$

$$= \sum_{r=0}^{m} C(m-1, r) \, [x]^{m-r} \, [y]^r + \sum_{r=0}^{m} C(m-1, r-1) \, [x]^{m-r} \, [y]^r$$

$$= \sum_{r=0}^{m} C(m, r) \, [x]^{m-r} \, [y]^r$$

and the formula holds for $n = m$.

Problem 1.182. *Use a combinatorial argument to establish the identity*

$$[x]^n = \sum_{k=1}^{n} u(n, k) \, [x]_k \qquad \text{...(1)}$$

where $u(n, k) = (n\,!/k\,!) \, C(n-1, k-1)$.

Solution. The number of increasing words of length n afforded by an m-letter alphabet is $[m]^n/n\,!$.

Consider the set of all increasing words of length n composed from the k-letter subalphabet $a_{i1} < a_{i2} < ... < a_{ik}$, such that each of the k letters appears atleast once in the word. With m replaced by k and with all $\lambda_i = 1$, the cardinality of this set is $C(n-1, k-1)$.

Now, there are $C(m, k)$ choices for the subalphabet, and k ranges from 1 to m.

Clearly, the corresponding sets constitute a partition of the increasing words of length n.

Therefore, $$\frac{[m]^n}{n!} = \sum_{k=1}^{m} C(m, k) \, C(n-1, k-1)$$

$$= \sum_{k=1}^{n} C(n-1, k-1) \, \frac{[m]_k}{k\,!} \qquad \text{...(2)}$$

Problem 1.183. *Find the number of ways of choosing r positive integers from among the first n positive integers such that no 2 consecutive integers appear in the choice and the choice does not include both 1 and n.*

Solution. **Case 1.** The choice includes 1.

We have $x_1 + x_2 + ...+ x_{r+1} = n - r$ with $x_1 \geq 0, x_2 \geq 1, ..., x_r \geq 1, x_{r+1} \geq 0$.

$x_1 = 0$, (there is a Y under 1) and $x_{r+1} \geq 1$

(there is an N under n).

Thus we count the solutions of

$$x_2 + x_3 + \ldots\ldots + x_{r+1} = n - r \text{ with } x_2 \geq 1, x_3 \geq 1, \ldots\ldots, x_{r+1} \geq 1.$$

and obatin $C(n - r - 1, r - 1)$.

Case 2. The choice does not include 1.

Now $x_1 \geq 1$ (there is an N under 1).

Thus, we count the solutions of

$$x_1 + x_2 + \ldots\ldots + x_{r+1} = n - r \text{ with } x_1 \geq 1, x_2 \geq 1, \ldots\ldots, x_{r+1} \geq 0.$$

and obtain $C(n - r, r)$

The total number of ways is,

$$C(n - r - 1, r - 1) + C(n - r, r) = \left[1 + \frac{n-r}{r}\right] C(n - r - 1, r - 1)$$

$$= \frac{n}{r} C(n - r - 1, r - 1).$$

Problem 1.184. *Show that $S(n + 1, m) = S(n, m - 1) + mS(n, m)$.*

Solution. Let $X \equiv \{x_1, x_2, \ldots\ldots, x_n\}$, $A \equiv \{x_{n+1}\}$, and $X' \equiv X \cup A$.

Then $S(n + 1, m)$ is the number of ways of partitioning X' into m subsets.

There are two means of effecting such a partition : either take a partition of X into $m - 1$ subsets and adjoin the set A or take a partition of X into m subsets and include the element x_{n+1} in any one of these.

There are $S(n, m - 1)$ ways of accomplishing the former objective and $mS(n, m)$ ways of accomplishing the letter. Thus the relation is established.

Problem 1.185. *Show that* $S(n, m) = \sum_k C(n - 1, k)\ S(k, m - 1)$.

Solution. The left-hand side represents the number of partitions of n objects into $m \geq n$ cells.

Let a denote a particular one of the objects, and in each partition distinguished the cell containing a as the a cell.

Partition the partitions (!) according to the composition of the a cell, which must contain r objects ($r = 0, 1, \ldots\ldots, n - m$) besides a.

These r objects may be chosen in $C(n - 1, r)$ ways. For each choice, the other $m - 1$ cells may be filled in $S(n - 1 - r, m - 1)$ ways.

Thus $S(n, m) \sum_r C(n - 1, r)\ S(n - 1 - r, m - 1)$ and the required formula results on changing the summation index from r to $k = n - 1 - r$.

Problem 1.186. *Prove the following relations :*

(a) $C(i + j, i)\ S(n, i + j) = \sum_k C(n, k)\ S(k, i)\ S(n - k, j)$

(b) $C(i + j, i)\ S(n, i + j) = \sum_k C(n, k)\ S(k, i)\ S(n - k, j)$.

Solution. (*a*) By our definition

$$[x+y]_n = \sum_k S(n,k)\ (x+y)^k = \sum_k S(n,k)\ \sum_i C(k,i)\ x^i y^{k-i} \qquad ...(1)$$

But, by an induction, we have

$$[x+y]_n = \sum_k C(n,k)\ [x]_k\ [y]_{n-k} \qquad ...(2)$$

In (1), the coefficient of x^iy^j is $S(n, i+j)\ C(i+j, i)$, in (2), the coefficient of x^iy^j is

$$\sum_k C(n,k)\ S(k,i)\ S(n-k,j).$$

(*b*) We have $x^n = \sum_{k=1}^{n} S(n,k)\,[x]_k$ ...(3)

and formula (2) above,

$$(x+y)^n = \sum_k S(n,k)\,[x+y]_k$$

$$= \sum_k S(n,k)\ \sum_i C(k,i)\ [x]_i\ [y]_{k-j} \qquad ...(4)$$

Next we expand $(x+y)^n$ by the binomial theorem and use (3) again,

$$(x+y)^n = \sum_k C(n,k)\ x^k y^{n-k}$$

$$= \sum_k C(n,k)\ \sum_i S(k,i)\ [x]_i\ \sum_j S(n-k,j)\ [y]_j \qquad ...(5)$$

This time we compare the coefficients of $[x]_i\ [y]_j$ in (4) and in (5) to obtain the results.

Problem 1.187. *A partition of a non empty subset Y of a set X is called a partial partition of X. Show that the number of partial partitions of a set X with n elements is* $B_{n+1} - 1$.

Solution. Let a be an element not in X and let X′ be a set obtained by adjoining a to X.

Excluding the partition of X′ into one part, there are $B_{n+1} - 1$ partitions of X′.

From each of these partitions delete the set which contains a, this accounts for all the partial partitions of X.

Problem 1.188. *Find the number of functions from a set X of n elements to a set Y of m elements such that the ranges of these functions each have exactly r elements.*

Solution. A subset of Y with r elements can be chosen in $C(m, r)$ ways.

Once a set is chosen, there are $r!\ S(n, r)$ surjections from X to that set.

Thus, the total number of functions in this category is

$$C(m,r)\ r!\ S(n,r) = P(m,r)\ S(n,r).$$

Problem 1.189. *Prove that the sequence* $< S(n, 1), S(n, 2), \ldots\ldots, S(n, n) >$ *is unimodal for all* $n > 2$.

Solution. Making an induction, we suppose the result true for every $k \leq n$, *i.e.*, for each $k \leq n$, there exists an integer m such that the sequence < S(k, m)> peaks at $m = m_k$.

Moreover, we include in the induction hypothesis the assertion that m_k is non decreasing in k.

We have $S(n + 1, m) = S(n, m - 1) + mS(n, m)$

and $$S(n + 1, m - 1) = S(n, m - 2) + (m - 1)\,S(n, m - 1)$$

Hence, by subtraction.

$$S(n + 1, m) - S(n + 1, m - 1) = [S(n, m - 1) - S(n, m - 2)] + m[S(n, m) - S(n, m - 1)] + S(n, m - 1)$$

By the induction hypothesis, the right hand side, and therefore the left hand side is positive for all $m \leq m_n$, which implies that

$$S(n + 1, 1) < S(n + 1, 2) < \ldots\ldots < S(n + 1, m_n) \quad \ldots(1)$$

Next, consider the case $m \geq m_n + 2$,

$$S(n + 1, m) - S(n + 1, m - 1) = \sum_{k \leq n} C(n, k)\ [S(k, m - 1) - S(k, m - 2)]\,.$$

By the induction hypothesis, each bracketed term on the right hand side is negative.

Hence, $S(n + 1, m_n + 1) > S(n + 1, m_n + 2) > \ldots\ldots > S(n + 1, n + 1)$ ...(2)

Together, (1) and (2) establish that < S(n + 1, m) > is unimodal, with either

$$m_{n+1} = m_n \text{ or } m_{n+1} = m_n + 1.$$

The induction is complete.

Problem Set 1.1

1. An organizing committee requires one student representative from either the first year or the second year or the third year. If there are 60 first year, 40 second year and 20 third year students, how many different representatives are there ?
2. A student wishes to take a combination of three courses, one from each of three science departments. There are 4 physics, 3 chemistry and 2 Biology courses on offer. How many possible combinations are there ?
3. If E be the event of selecting a prime number less than 10 and F, the event of selecting an even number less than 10. Find the number of ways of happening E or F.
4. A person buying a personal computer system is offered a choice of four models of the basic unit, three model of key board, and two models of printer. How many distinct system can be purchased ?
5. A house has 4 doors and 10 windows. In how many ways can a burglar rob the house by entering through a window and exciting through a door.
6. A team consists of 6 boys and 4 girls and the other has 5 boys and 3 girls, how many single matches can be arranged between the two teams when a boy plays against a boy and a girl plays against a girl ?
7. Prove that, given three natural numbers, we can select two of them with even sum.

8. Assume that there are $n > 1$ people in the room. Prove that atleast two of them have equal number of acquaintances.

9. How many friends must you have to guarantee that at least five of them will have birthdays in the same month ?

10. Prove that for any natural number n there is a number written only by 5's and 0's and divisible by n.

11. (*a*) Roberta needs two courses to complete her university degree. There are three sections of the first course which meet in the morning and four afternoon sections. In how many different ways can she select the two courses she needs ?

(*b*) Suppose Roberta needs just one course to complete her degree. In how many ways can she select this course ?

12. In how many of the three-digit numbers 000-999 are all the digits different ?

13. (*a*) Some license plates in california consists of one of the digits 1-9, followed by three (not necessarily distinct) letters and then three of the digits 0-9 (not necessarily distinct). How many possible license plates can be produced by this method ?

(*b*) Other california license plates consists of one of the digits 1-9 followed by three letters and then three of the digits 0-9 (the same digit or letter can be used moer than one). How many license plates of this type can be made ?

(*c*) What is the maximum number of license plates in california that can be made assuming plates have one of the two types described in (*a*) and (*b*) ?

14. From a gruop of 13 men, 6 women, 2 boys and 4 girls,

(*a*) In how many ways can a man, a woman, a boy, and a girl be selected ?

(*b*) In how many ways can a man or a girl be selected ?

(*c*) In how many ways can one person be selected ?

15. Using only the digits 1, 3, 4 and 7.

(*a*) How many two-digits numbers can be formed ?

(*b*) How many three-digit numbers can be formed ?

(*c*) How many two- or three-digit numbers can be formed ?

16. A company produces combination locks, the combinations consisting of three different numbers in the range 0-59 (inclusive) which must be dialed in order. How many different combinations are possible ?

17. New parents wish to give their baby one, two or three different names. In how many ways can the baby be named if the parents will choose from a book containing 500 names ?

18. How many three-digit numbers contain the digits 2 and 5 but none of the digits 0, 3, 7 ?

19. Your are dealt four cards from a standard deck of 52 playing cards. In how many ways can you get four of a kind (four 2's, or four kings, etc) ?

20. Make a table which shows all the totals which are possible when two dice are rolled and the number of ways in which each total can occur.

21. A coin is tossed four times. Make a list of all the possible outcomes. For example, the sequence HHTH, representing head, head, tail, head, is one possibility. How many possibilities are there altogether ?

22. How many possible license plates can be manufactured if a license plate consists of three letters followed by three digits and the digits must be distinct, the letters can be arbitrary ?

23. Let A = $\{a_1, a_2, \ldots\ldots, a_n\}$ be a set of n elements and B = {0, 1}

(*a*) Show that there are 2^n functions from A to B

(*b*) Show that there are $2^n - 2$ onto functions from A to B.

24. During a local campian, eight Republican and five Democratic candidates are nominated for President of the school board.

(*a*) If the President is to be one of these candidates, how many possibilities are there for the eventual winner ?

(*b*) How many possibilities exist for a pair of candidates (one from each party) to oppose each other for the eventual election ?

(*c*) Which counting principle is used in part (*a*) ? in part (*b*) ?

25. Buick automobiles come in four models, 12 colours, three engine sizes, and two transmission types.

(*a*) How many distinct Buicks can be manufactured ?

(*b*) If one of the available colors is blue, how many different blue Buicks can be manufactured ?

26. A certain "Burger Joint" advertises that a customer can have his or her hamburger with or without any or all of the following : catsup, mustard, mayonnaise, lettuce, tomato, onion, pickle, cheese, or mushrooms. How many different kinds of hamburger orders are possible ?

27. While on a saturday shopping spree Jennifer and Tiffany witnessed two men driving away from the front of a jewellery shop, just before a burglar alarm started to sound. Although everything happened rather quickly, when the two young ladies were questioned they were able to give the police the following information about the license plate (which consisted of two letters followed by four digits) on the gataway car. Tiffany was sure that the second letter on the plate was either on O or a Q and the last digit was either a 3 or on 8. Jennifer told the investigator that the first letter on the plate was either a C or a G and that the first digit was definitely a 7. How many different license plates will the police have to check out ?

28. Patter's pastry parlor offers eight different kinds of pastry and six different kinds of muffins. In addition to bakery items one can purchase small, medium, or large containers of the following beverages ; coffee (black, with cream, with sugar, or with cream and sugar), ten (plain, with cream, with sugar, with cream and sugar, with lemon, or with lemon and sugar), hot cocoa, and orange juice. When carol comes to patter's, in how many ways can she order.

(*a*) One bakery item and one medium-sized beverage for herself ?

(*b*) One backery item and the container of coffee for her self and one muffin and one container of tea for her boss, Mr. Didio ?

(*c*) One piece of pastry and one container of tea for herself, one muffin and a container of orange juice for Mr. Didio and one backery item and one container of coffee for each of her two assistants, Mr. Talbot and Mrs. Gillis ?

Problem Set 1.2

1. How many permutations can be made out of the letters of word 'COMPUTER' ? How many of these (*a*) begin with C ? (*b*) begin with R ? (*c*) begin with C and with R ? (*d*) C and R always together ?
2. Find the number of ways in which 5 ladies and 5 gentlemen may be seated in a row, so that no two ladies are together.
3. How many arrangements can be made of the letters of the word

   (*a*) APPLE (*b*) COMMERCE (*c*) PROGRAMMING
4. Find how many arrangements can be made with the letters of the word 'MATHEMATICS'. In how many of them the vowels occur together ?
5. In how many ways can the letters of the word "ARRANGE" be arranged ? If the two R's do not occur together, then how many arrangements can be made ? If besides the two R's, the two A's also do not occur together, how many permutations can be obtained ?
6. In how many different ways can the letters of the word "SALOON" be arranged if (*a*) the two O's must not come together (*b*) the consonants and vowels must occupy alternate places ?
7. In how many ways can 4 prizes be distributed among 5 girls when (*a*) no girl more than 1 prize

   (*b*) a girl may get any number of prizes.

   (*c*) no girl gets all the prizes.
8. How many different numbers of six digits (without repetitions) can be formed from the digits 3, 1, 7, 0, 9, 5 ?

   (*i*) How many of them will have 0 in the unit place ?

   (*ii*) How many of them are divisible by 5 ?

   (*iii*) How many of them are not divisible by 5 ?
9. Find the number of ways in which n different beads can be arranged to form a necklace ?
10. In how many ways can a garland of 15 different flowers be made ?
11. How many different words, each containing 2 vowels and 3 consonants can be formed with 5 vowels and 17 consonants ?
12. Five balls are to be placed in three boxes. Each can hold all the five balls. In how many different ways can be place the balls so that no box remains empty, if

    (*a*) balls and boxes are all different

    (*b*) balls are identical but boxes are different

    (*c*) balls are different but boxes are identical

    (*d*) balls as well as boxes are identical ?
13. How many cards must you pick up from a standard 52 cards deck to be sure of getting at least 1 red card.
14. Out of 9 girls and 15 boys, how many different committees can be formed each consisting of 6 boys and 4 girls ?
15. In how many ways can a committee of 10 to be selected at least 4 women and 4 men from 9 women and 8 men if

(*a*) Ms. X refuses to work with Mr. Y

(*b*) Ms. X and Mr. Y insist to work together ?

16. How many integers must you pick in order to be sure that atleast two of them have the same remainder when divided by 7 ?

17. Find the number of permutations that can be formed by the letters of the word 'DAUGHTER' (*i*) taking all the letters together without any restriction (*ii*) beginning with D (*iii*) beginning with D and ending with R (*iv*) vowels being always together (*v*) not all vowels together (*vi*) not even two vowels together (*vii*) vowels occupying even places.

18. (*a*) How many arrangements are there of all the letters in SOCIOLOGICAL ?

(*b*) In how many of the arrangements in part (*a*) are A and G adjacent ?

(*c*) In how many of the arragements in part (*a*) are all the vowels adjacent ?

19. Find the value(s) of n in each of the following :

(*a*) $P(n, 2) = 90$, (*b*) $P(n, 3) = 3P(n, 2)$ and (*c*) $2P(n, 2) + 50 = P(2n, 2)$.

20. (*a*) How many distinct paths are there from $(-1, 2, 0)$ to $(1, 3, 7)$ in Euclidean three-space if each move is one of the following types ?

(H) : $(x, y, z) \to (x + 1, y, z)$; (V) : $(x, y, z) \to (x, y + 1, z)$; (A) : $(x, y, z) \to (x, y, z + 1)$.

(*b*) How many such paths are there from $(1, 0, 5)$ to $(8, 1, 7)$?

(*c*) Generalize the results in parts (*a*) and (*b*).

21. (*a*) How many permutations of size 3 can one produce with the letters m, r, a, f and t ?

(*b*) List all the combinations of size 3 that result for the letters m, r, a, f and t.

22. How many arrangements of the letters in MISSISSIPPI have no consecutive S's ?

23. Consider the collection of all strings of length 10 made up from the alphabet 0, 1, 2, and 3. How many of these strings have weight 3 ? How many have weight 4 ? How many have even weight ?

24. Determine the coefficient of x^9y^3 in the expansions of

(*a*) $(x + y)^{12}$ (*b*) $(x + 2y)^{12}$ and (*c*) $(2x - 3y)^{12}$.

25. Determine the sum of all the coefficients in the expansions of

(*a*) $(x + y)^3$ (*b*) $(x + y)^{10}$ (*c*) $(x + y + z)^{10}$ (*d*) $(w + x + y + z)^5$

(*e*) $(2s - 3t + 5u + 6v - 11w + 3x + 2y)^{10}$.

26. Determine the coefficient of

(*a*) xyz^2 in $(x + y + z)^4$ (*b*) xyz^2 in $(w + x + y + z)^4$

(*c*) xyz^2 in $(2x - y - z)^4$ (*d*) xyz^{-2} in $(x - 2y + 3z^{-1})^4$

(*e*) $w^3x^2yz^2$ in $(2w - x + 3y - 2z)^8$.

27. Verify that for each integer $n \geq 1$,

$$\binom{2n}{n} - \binom{2n}{n-1} = \frac{1}{n-1}\binom{2n}{n}.$$

28. In how many ways can one parenthesize the product *abcdef*?

29. Determine the number of integer solutions of $x_1 + x_2 + x_3 + x_4 = 32$, where

(*a*) $x_i \geq 0,\ 1 \leq i \leq 4$ (*b*) $x_i > 0,\ 1 \leq i \leq 4$

(*c*) $x_1, x_2 \geq 5, x_3, x_4 \geq 7$ (*d*) $x_i \geq 8,\ 1 \leq i \leq 4$

(*e*) $x_i \geq -2,\ 1 \leq i \leq 4$ (*f*) $x_1, x_2, x_3 > 0, 0 < x_4 \leq 25$.

30. Determine how many ways 20 coins can be selected from four large containers filled with pennies, nickels, dimes and quarters. (Each container is filled with only one type of coin).

Answers 1.1

1. 120 **2.** 24 **3.** 7 **4.** 24 **5.** 40 **6.** 42

11. (*a*) 12 (*b*) 7 **12.** $10 \times 9 \times 8 = 720$

13. (*a*) $9 \times 26 \times 26 \times 26 \times 10 \times 10 \times 10 = 158{,}184{,}000$.

(*b*) $9 \times 26 \times 10^5 = 23{,}400{,}000$

(*c*) $158{,}184{,}000 + 23{,}400{,}000 = 181{,}584{,}000$

14. (*a*) $13 \times 6 \times 2 \times 4 = 624$ (*b*) $13 + 14 = 17$, (*c*) 25

15. (*a*) $4 \times 4 = 16$, (*b*) $4 \times 4 \times 4 = 64$, (*c*) $16 + 64 = 80$. **16.** $60 \times 60 \times 60 = 216{,}000$

17. $500 + (500)(499) + (500)(499)(498) = 1{,}24501 \times 10^8$. **19.** $52 \times 3 \times 2 \times 1 = 312$.

20.

Total	2	3	4	5	6	7	8	9	10	11	12
No. of ways	1	2	3	4	5	6	5	4	3	2	1

21. HHHH, HHHT, HHTH, HTHH, THHH, HHTT, HTHT, HTTH, THHT, THTH, TTHH, HTTT, THTT, TTHT, TTTH, TTTT. There are 16 possibilities in all.

22. $26 \times 26 \times 26 \times 10 \times 9 \times 8 = 12{,}654{,}720$.

23. (*a*) The easiest way to see that there are 2^n functions from A to B is to note that $\{a_i \mid (a_i, 0) \in f\} \leftrightarrow f$ is a one-to-one correspondence between the (2^n) subsets of A and the set of functions $f : A \to B$.

(*b*) Of the 2^n functions $A \to B$, precisely two are not onto,

$$\{(a_1, 0), (a_2, 0), \ldots\ldots (a_n, 0)\} \text{ and } \{(a_1, 1), (a_2, 1) \ldots\ldots, (a_n - 1)\}$$

Thus, the number of onto functions is $2^n - 2$ as claimed.

24. (*a*) 13 (*b*) 40 (*c*) The rule of sum in part (*a*) ; the rule of product in part (*b*).

25. (*a*) 288 (*b*) 24 **26.** 2^9.

27. $2 \times 2 \times 1 \times 10 \times 10 \times 2 = 800$ different license plates.

28. (*a*) $(14)(12) = 168$ (*b*) $(14)(12)(6)(18) = 18,144$ (*c*) 73, 156, 608.

Answers 1.2

1. 4032 ; 5040 ; 5040 ; 720 ; 10080 **2.** 86400

3. (*a*) 60 (*b*) 5040 (*c*) 9979200 **4.** 4989600 ; 120960 **5.** 1260 ; 900 ; 660

6. 240 ; 36 **7.** 10 ; 625 ; 620 **8.** 600 ; 120 ; 216 ; 384

9. $\frac{1}{2}(n-1)!$ **10.** $\frac{1}{2}(14!)$ **11.** 816000 **12.** 150 ; 6 ; 50 ; 2

13. 27 **14.** 630630 **15.** 5586 ; 10876 **16.** 8

17. 40320 ; 5040 ; 720 ; 4320 ; 3600 ; 14400 ; 2880

18. (*a*) $\frac{12!}{(3!2!2!2!)}$ (*b*) $2\left[\frac{11!}{(3!2!2!2!)}\right]$ (*c*) $\left[\frac{7!}{(2!2!)}\right]\left[\frac{6!}{(3!2!)}\right]$

19. (*a*) $n = 10$; (*b*) $n = 5$; (*c*) $n = 5$

20. (*a*) $\frac{(10!)}{(2!7!)} = 360$ (*b*) 360

(*c*) Let *x, y* and *z* be any real numbers and let *m, n* and *P* be any non negative integers. The number of path from (x, y, z) to $(x + m, y + n, z + p)$, as described in part (*a*), is $\frac{(m+n+P)!}{(m!n!P!)}$.

21. (*a*) P(5, 3) = 60, (*b*) *a, f, m* *a, f, r* *a, f, t* *a, m, r* *a, m, t*
a, r, t *f, m, r* *f, m, t* *f, r, t* *m, r, t*

22. $\binom{8}{4}\left(\frac{7!}{4!2!}\right) = 7350$

23. (*a*) $\binom{10}{3} + \binom{10}{1}\binom{9}{1} + \binom{10}{1} = 220$ (*b*) $\binom{10}{4} + \binom{10}{2} + \binom{10}{1}\binom{9}{2} + \binom{10}{1}\binom{9}{1} = 705$

(*c*) $2^{10}\left(\sum_{i=0}^{5}\binom{10}{2i}\right)$ **24.** (*a*) $\binom{12}{9}$ (*b*) $\binom{12}{9}(2^3)$ (*c*) $\binom{12}{9}(2^9)(-3)^3$

25. (*a*) 2^3 (*b*) 2^{10} (*c*) 3^{10} (*d*) 4^5 (*e*) 4^{10}.

26. (*a*) $\binom{4}{1, 1, 2} = 12$ (*b*) 12 (*c*) $\binom{4}{1, 1, 2}(2)(-1)(-1)^2 = -24$ (*d*) -216

(*e*) $\binom{8}{3, 2, 1, 2}(2^3)(-1)^2(3)(-2)^2 = 161, 280.$

27. $$\binom{2n}{n} - \binom{2n}{n-1} = \frac{(2n)!}{n!n!} - \frac{(2n)!}{(n-1)!(n+1)!}$$

$$= \frac{(2n)!(n+1)}{(n+1)!n!} - \frac{(2n)!n}{n!(n+1)!}$$

$$= \frac{(2n)![(n+1)-n]}{(n+1)!n!} = \frac{1}{(n+1)}\frac{(2n)!}{n!n!} = \left(\frac{1}{n+1}\right)\binom{2n}{n}.$$

28. There are $b_5 = (= 42)$ ways.

29. (*a*) $\binom{35}{32}$ (*b*) $\binom{31}{28}$ (*c*) $\binom{11}{8}$ (*d*) 1 (*e*) $\binom{43}{40}$ (*f*) $\binom{31}{28} - \binom{6}{3}$.

30. $\binom{23}{10}$.

CHAPTER

2

The Principles of Inclusion and Exclusion

INTRODUCTION

Many counting problems can be solved using the principle of inclusion-exclusion. For instance, we can use this principle to find the number of primes less than a positive integer. Many problems can be solved by counting the number of onto functions from one finite set to another. The inclusion-exclusion principle can be used to find the number of such functions.

2.1 THE PRINCIPLE OF INCLUSION-EXCLUSION

Let A and B be sub sets of a finite universal set U. Then

$$|A \cup B| = |A| + |B| - |A \cap B| \qquad ...(1)$$

Proof. If $A = \phi$, then $A \cap B = \phi$, and $|A| = |A \cap B| = 0$

so the result holds because each side of (1) equals $|B|$.

Similarly, the result holds if $B = \phi$, so we assume hence forth that neither A nor B is emtpy.

Suppose $A \cap B = \phi$. Let $A = \{a_1, a_2, ..., a_r\}$ and $B = \{b_1, b_2, ..., b_s\}$.

Then $A \cup B = \{a_1, a_2, ..., a_r, b_1, b_2, ..., b_s\}$

and, since there is no repetition among the elements listed here,

$$|A \cup B| = r + s = |A| + |B| = |A| + |B| - |A \cap B|$$

and the desired formula is true.

On the other hand, if $A \cap B \neq \phi$ and we let

$$A \backslash B = \{a_1, a_2, ..., a_r\}$$

$$B \backslash A = \{b_1, b_2, ..., b_s\}$$

$$A \cap B = \{x_1, x_2, ..., x_t\}$$

then $$A = \{a_1, a_2, ..., a_r, x_1, x_2, ..., x_t\}$$

and $$B = \{b_1, b_2, ..., b_s, x_1, x_2, ..., x_t\},$$

so $$A \cup B = \{a_1, ..., a_r, b_1, ..., b_s, x_1, ..., x_t\}$$

With no repetition among the elements listed in any of these sets.

Thus, $$|A| + |B| - |A \cap B| = (r + t) + (s + t) - t$$
$$= r + s + t$$
$$= |A \cup B|.$$

and the formula holds in this case as well.

Note 1. Suppose, for instance, we have three finite sets A, B and C.

$$\downarrow$$

Then $|A \cup B \cup C| = |A \cup (B \cup C)|$

$$\downarrow$$

$$= |A| + |B \cup C| - |A \cap (B \cup C)|$$

$$\downarrow \qquad \downarrow$$

$$= |A| + |B \cup C| - |(A \cap B) \cup (A \cap C)|$$

$$= |A| + [|B| + |C| - |B \cap C|]$$

$$- [|A \cap B| + |A \cap C| - |(A \cap B) \cap (A \cap C)|]$$

$$= |A| + |B| + |C| - |B \cap C| - |A \cap B| - |A \cap C| + |A \cap B \cap C$$

where, at the spots marked with single arrows, we used the fact that $|X \cup Y| = |X| + |Y| - |X \cap Y|$, and at the last step, that $(A \cap B) \cap (A \cap C) = A \cap B \cap C$.

2. Here is the formula for the cardinality of the union of four finite sets.

$$|A \cup B \cup C \cup D| = |A| + |B| + |C| + |D| - |A \cap B| - |A \cap C| - |A \cap D|$$
$$- |B \cap C| - |B \cap D| - |C \cap D| + |A \cap B \cap C|$$
$$+ |A \cap B \cap D| + |A \cap C \cap D| + |B \cap C \cap D|$$
$$- |A \cap B \cap C \cap D|.$$

Theorem 2.1. Let A and B be subsets of a finite universal set U. Then

(*i*) $|A \cap B| \leq \min \{|A|, |B|\}$, the minimum of $|A|$ and $|B|$

(*ii*) $|A \backslash B| = |A| - |A \cap B| \geq |A| - |B|$

(*iii*) $|A^c| = |U| - |A|$

(*iv*) $|A \oplus B| = |A \cup B| - |A \cap B| = |A| + |B| - 2|A \cap B|$
$= |A \backslash B| + |B \backslash A|$

(*v*) $|A \times B| = |A| \times |B|$.

Problem 2.1. *Of 30 personal computers (PCs) owned by faculty members in a certain university department, 20 run windows, eight have 21 inch monitors, 25 have CD-ROM drives, 20 have atleast two of these features, and six have all three*

(*a*) *How many PCs have at least one of these features ?*

(*b*) *How many have none of these features ?*

(*c*) *How many have exactly one features ?*

Solution. Let W be the set of PCs running under windows, M the set of PCs with 21 inch monitors, and C the set with CD-ROM drives.

We are given that $|W| = 20$, $|M| = 8$, $|C| = 25$,

$|(W \cap M) \cup (W \cap C) \cup (M \cap C)| = 20$, and $|W \cap M \cap C| = 6$.

By the principle of Inclusion-Exclusion,

$$20 = |(W \cap M) \cup (W \cap C) \cup (M \cap C)|$$
$$= |W \cap M| + |W \cap C| + |M \cap C| - |(W \cap M) \cap (W \cap C)|$$
$$- |(W \cap C) \cap (M \cap C)| - |(W \cap C) \cap (M \cap C)|$$
$$+ |(W \cap M) \cap (W \cap C) \cap (M \cap C)|.$$

Since each of the last four terms here is $|W \cap M \cap C|$,

We obtain,

$$20 = |W \cap M| + |W \cap C| + |M \cap C| - 2|W \cap M \cap C|;$$

Therefore, $|W \cap C| + |W \cap M| + |M \cap C| = 20 + 2(6) = 32.$

(*a*) The number of PCs with at least one feature is

$$|W \cup M \cup C| = |W| + |M| + |C| - |W \cap M| - |W \cap C| - |M \cap C| + |W \cap M \cap C|$$
$$= 20 + 8 + 25 - (|W \cap M| + |W \cap C| + |M \cap C|) + 6$$
$$= 59 - 32 = 27.$$

(*b*) If follows that 30 – 27 = 3 PCs have none of the specified features.

(*c*) Since the number of computers with exactly one feature is the number with atleast one feature less the number with at least two, the number with exactly one is 27 – 20 = 7.

Problem 2.2. *Continuing the preceding problem, how many faculty have computers with exactly two of the three features described ?*

Solution. The number of computers with exactly two of the features is the number with atleast two less the number with exactly three ;

that is, 20 – 6 = 14.

Problem 2.3. *Find the number of permutations of the digits 1 thruogh 9 in which (a) none of the blocks 23, 45 and 678 appears, (b) none of the blocks, 34, 45, and 738 appears.*

Solution. Let X be the set of all the permutations, then $n(X) = 9\,!$

(*a*) Let A, B and C be the subsets of permutations which respectively contain 23, 45 and 678 as blocks.

Then $n(A) = 8\,!$

Similarly, $n(B) = 8\,!$, $n(C) = 7\,!$, $n(A \cap B) = 7\,!$

$$n(A \cap C) = n(B \cap C) = 6\,!$$

Thus the answer is

$$9\,! - (8\,! + 8\,! + 7\,!) + (7\,! + 6\,! + 6\,!) = 5\,!$$

(*b*) Let A, B and C be the subsets of X in which 34, 45 and 738 respectively apepar as blocks.

Then $n(A) = n(B) = 8\,!$ and $n(C) = 7\,!$

Observe that $A \cap B$ is the subset of permutations in which 3 4 5 appears as a block

So $n(A \cap B) = 7\,!$. Here $n(A \cap C) = 0 = n(A \cap B \cap C)$ and $n(B \cap C) = 6\,!$

Thus the answer is

$$9\,! - (8\,! + 8\,! + 7\,!) + (7\,! + 0 + 6\,!) - 0.$$

Problem 2.4. *Prove that the stirling number of the second kind may be evaluated from the following inclusion-exclusion formula*

$$m\,!\ S(n, m) = m^n - C(m, 1)(m-1)^n + C(m, 2)(m-2)^n - \ldots\ldots + (-1)^{m-1} C(m, m-1)\, 1^n.$$

Solution. Let M denote the set of all mappings from

$$X = \{x_1, x_2, \ldots\ldots, x_n\} \text{ to } Y = \{y_1, y_2, \ldots\ldots, y_m\}, \text{ and for } i = 1, 2, \ldots\ldots, m$$

Let A_i be the subset of M consisting of those mapping that fail to cover y_i.

We have $n(M) = m^n$ and $S_k = C(m, k)\,(m - k)^n$, $k = 1, 2, \ldots\ldots m - 1$.

There are $C(m, k)$ ways to set apart k elements of Y, and X can be mapped into the remaining elements in $(m - k)^n$ ways.

The number of surjections from X to Y as

$$m^n - C(m, 1)\, 1\, (m - 1)^n + \ldots\ldots + (-1)^{m-1}\, C(m, m - 1)\, 1^n$$

Therefore, the number of surjcetions is $m\,!\, S(n, m)$.

Problem 2.5. *Find the number of solutions in integers of the equation $a + b + c + d = 17$, where $1 \le a \le 3$, $2 \le b \le 4$, $3 \le c \le 5$, $4 \le d \le 6$.*

Solution. Let $a = 1 + \alpha$, $b = 2 + \beta$, $c = 3 + \gamma$, $d = 4 + \delta$.

The transformed system is

$$\alpha + \beta + \gamma + \delta = 7,\ 0 \le \alpha, \beta, \gamma, \delta \le 2.$$

Let X be the set of all solutions in non negative integers of $\alpha + \beta + \gamma + \delta = 7$.

Let A be the subset of X for which $\alpha \ge 3$, B be the subset for which $\beta \ge 3$, C be the subset for which $\gamma \ge 3$, D be the subset for which $\delta \ge 3$.

We have $n(X) = C(10, 3)$, $n(A) = n(B) = n(C) = n(D) = C(7, 3)$

$$n(A \cap B) = n(A \cap C) = \ldots\ldots = n(C \cap D) = C(4, 3)$$

and all higher order intersections are empty.

Thus, $n(X) = 120$, $S_1 = C(4, 1)\, C(7, 3) = 140$

$$S_2 = C(4, 2)\, C(4, 3) = 24 \text{ and}$$

The answer is $120 - 124 + 24 = 4$.

Problem 2.6. *At a university 60% of the teacher play tennis, 50% of them play bridge, 70% jog, 20% play tennis and bridge, 30% play tennis and jog and 40% play bridge and jog. What is the percentage of teachers who jog, play tennis and play bridge ?*

Solution. Let us assume that there are 100 teachers in the university. Let T, B and J be he set of teachers who play tennis, who play bridge and who jog respectively.

Therefore, we have

$$|T| = 60,\ |B| = 50,\ |J| = 70$$

$$|T \cap B| = 20,\ |T \cap J| = 30,\ |B \cap J| = 40$$

$$|T \cup B \cup J| = 100,\ |T \cap B \cap J| = ?$$

From the formula

$$|A \cup B \cup C| = |A| + |B| + |C| - |A \cap B| - |A \cap C| - |B \cap C| + |A \cap B \cap C|$$

We have $\qquad 100 = 60 + 50 + 70 - 20 - 30 - 40 + |T \cap B \cap J|$

or $\qquad |T \cap B \cap J| = 100 - 90 = 10.$

Therefore, there are 10% of the teachers who jog, play tennis and play bridge.

Problem 2.7. *Suppose 18 of the 30 personal computers in problem have pentium III processors, including ten of those running windows, all of those with 21 inch monitors, and 15 of those with CD-ROM drives. Suppose also that every computer has atleast one of the four features now specified. How many have atleast three features ?*

Solution. Let P be the set of PCs with pentium III processors. Then, with W, M and C as before, the question asks for the number.

$$n = |(W \cap M \cap C) \cup (W \cap M \cap P) \cup (W \cap C \cap P) \cup (M \cap C \cap P)|$$
$$= |W \cap M \cap C| + |W \cap M \cap P| + |W \cap C \cap P| + |M \cap C \cap P|$$
$$- 6|W \cap M \cap C \cap P| + 4|W \cap M \cap C \cap P| - |W \cap M \cap C \cap P|$$
$$= |W \cap M \cap C| + |W \cap M \cap P| + |W \cap C \cap P| + |M \cap C \cap P| - 3|W \cap M \cap C \cap P|.$$

We are given that

$$30 = |W \cup M \cup C \cup P|$$
$$= |W| + |M| + |C| + |P| - |W \cap M| - |W \cap C| - |W \cap P|$$
$$- |M \cap C| - |M \cap P| - |C \cap P| + |W \cap M \cap C|$$
$$+ |W \cap M \cap P| + |W \cap C \cap P| + |M \cap C \cap P|$$
$$- |W \cap M \cap C \cap P|$$
$$= |W| + |M| + |C| + |P| - |W \cap M| - |W \cap C| - |W \cap P|$$
$$- |M \cap C| - |M \cap P| - |C \cap P| + n + 2|W \cap M \cap C \cap P|$$

because

$$|W \cap M \cap C| + |W \cap M \cap P| + |W \cap C \cap P| + |M \cap C \cap P| = n + 3|W \cap M \cap C \cap P|.$$

In problem 2.1, we learned that

$|W \cap C| + |W \cap M| + |M \cap C| = 32$ and were told that $|W \cap M \cap C| = 6$.

Here, since $M \subseteq P$,

We have $W \cap M \cap C \cap P = W \cap M \cap C$ and so $|W \cap M \cap C \cap P| = 6$.

Therefore, $30 = 20 + 8 + 25 + 18 - 32 - 10 - 8 - 15 + n + 2(6)$.

So $n = 30 - 71 + 65 - 12 = 12$.

Problem 2.8. *How may integers between 1 and 300 (inclusive) are*

(a) divisible by at least one of 3, 5, 7 ?

(b) divisible by 3 and by 5 but not by 7 ?

(c) divisible by 5 but by neither 3 nor 7 ?

Solution. Let A, B and C be the sets of those integers between 1 and 300 which are divisible by 3, by 5, and by 7, respectively ;

Thus,
$$A = \{n \mid 1 \le n \le 300, 3 \mid n\}$$
$$B = \{n \mid 1 \le n \le 300, 5 \mid n\}$$
$$C = \{n \mid 1 \le n \le 300, 7 \mid n\}.$$

(*a*) To be divisible by either 3 or 5 or 7 is to be in atleast one of the sets A, B or C.

Thus, part (*a*) asks us to find $|A \cup B \cup C|$.

To determine this number, we need $|A|, |B|, |C|, |A \cap B|, |A \cap C|, |B \cap C|$ and $|A \cap B \cap C|$.

The elements of A are 3, 6, 9, 12, ..., 300, so $|A| = 100$.

Note that this numebr is $\left[\dfrac{300}{3}\right]$, where, $[x]$ denotes the floor of the real number x.

In general, for natural numbers a and b, the number of positive integers less than or equal to a and divisible by b is $\left[\frac{a}{b}\right]$. Thus, we also have

$$|\mathrm{B}| = \left[\frac{300}{5}\right] = 60 \text{ and } |\mathrm{C}| = \left[\frac{300}{7}\right] = 42$$

Next we have to find $A \cap B$, the set of integers between 1 and 300 which are divisible by both 3 and 5, since 3 and 5 are relatively prime numbers, any number divisible by each of them must be divisible by their product.

Therefore, $A \cap B$ is just the set of integers between 1 and 300 which are divisible by 15 and, similarly, $A \cap C$, $B \cap C$, and $A \cap B \cap C$ are the sets of integers between 1 and 300 which are divisible by 21, 35 and 105 respectively.

We have $|A \cap B| = \left[\frac{300}{15}\right] = 20$, $|A \cap C| = \left[\frac{300}{21}\right] = 14$,

$$|B \cap C| \left[\frac{300}{35}\right] = 8, \ |A \cap B \cap C| = \left[\frac{300}{105}\right] = 2$$

Hence $|A \cup B \cup C| = 100 + 60 + 42 - 20 - 14 - 8 + 2 = 162$.

(*b*) The numbers divisible by 3 and by 5, but not by 7, are precisely those numbers in $(A \cap B) \backslash C$, a set of cardinality $|A \cap B| - |A \cap B \cap C| \ 20 - 2 = 18$.

(*c*) The numbers divisible by 5 by neither 3 nor 7 are those in $B \backslash (A \cup C)$, a set of cardinality $|B| - |B \cap (A \cup C)|$.

Since $B \cap (A \cup C) = (B \cap A) \cup (B \cap C)$

The principle of Inclusion-Exclusion gives

$$\begin{aligned} |B \cap (A \cup C)| &= |B \cap A| + |B \cap C| - |(B \cap A) \cap (B \cap C)| \\ &= |B \cap A| + |B \cap C| - |B \cap A \cap C| \end{aligned}$$

because $(B \cap A) \cap (B \cap C) = B \cap A \cap C$.

Therefore, $|B \cap (A \cup C)| = 20 + 8 - 2 = 26$ and the number we seek is $|B| - 26 = 60 - 26 = 34$.

Problem 2.9. *How many integers between 1 and 500 are*

(*i*) *divisible by 3 or 5 ?*

(*ii*) *divisible by 3 but not by 5 or 6 ?*

Solution. (*i*) Let A and B be the set of integers between 1 and 500 which are divisible by 3 and 5, respectively.

The question asks for $|A \cup B|$.

This number is $|A| + |B| - |A \cap B| = \left[\frac{500}{3}\right] = 233$.

(*ii*) The question asks for $|A \backslash (B \cup C)| = 66$.

Problem 2.10. *How many primes are less than 200 ? Explain your answer.*

Solution. Let A, B, C, D, E, F be the sets of natural numebrs between 1 and 200 (exclusive) which are not prime and divisible by 2, 3, 5, 7, 11 and 13, respectively.

Let A, B, C, D, E, F be the sets of natural numbers between 1 and 200 (exclusive) which are not prime and divisible by 2, 3, 5, 7, 11 and 13, respectively.

$| A \cup B \cup C \cup D \cup E \cup F | = 152$, so the number of primes less than 200 is $198 - 152 = 46$.

2.2 GENERALIZATION OF THE PRINCIPLE OF INCLUSION-EXCLUSION

If A_i are finite subsets of a universal set U, then

$$| A_1 \cup A_2 \cup \ldots\ldots \cup A_n | = \sum_{i=1}^{n} | A_i | - \sum_{i,\,j} | A_i \cap A_j | + \sum_{i,\,j,\,k} | A_i \cap A_j \cap A_k |$$

$$+ \ldots\ldots + (-1)^{n-1} | A_1 \cap A_2 \ldots\ldots \cap A_n | \qquad \ldots(1)$$

where the second summation is taken over all 2-combinations $\{i, j\}$ of the integers $\{1, 2, \ldots\ldots, n\}$, the third summation is taken over all 3-combinations $\{i, j, k\}$ of $\{1, 2, \ldots\ldots n\}$, and so on.

For $n = 4$ there are $4 + C(4, 2) + C(4, 3) + 1 = 2^4 - 1 = 15$ terms and the theorem states that

$$\begin{aligned} | A_1 \cup A_2 \cup A_3 \cup A_4 | &= | A_1 | + | A_2 | + | A_3 | + | A_4 | - | A_1 \cap A_2 | \\ &- | A_1 \cap A_3 | - | A_1 \cap A_4 | - | A_2 \cap A_3 | - | A_2 \cap A_4 | \\ &- | A_3 \cap A_4 | + | A_1 \cap A_2 \cap A_3 | + | A_1 \cap A_2 \cap A_4 | \\ &+ | A_1 \cap A_3 \cap A_4 | + | A_2 \cap A_3 \cap A_4 | - | A_1 \cap A_2 \cap A_3 \cap A_4 | \end{aligned}$$

In general, there are $\binom{n}{1} + \binom{n}{2} + \binom{n}{3} + \ldots\ldots + \binom{n}{n}$

$= 2^n - 1$ terms on the right-hand side of equation (1)

Proof. The theorem will be proved by induction on the number n of subsets A_i.

The theorem is obviously true for $n = 1$, and we have indicated why the theorem holds for $n = 2$ and $n = 3$.

Assume the theorem holds for any n subsets of U.

Suppose, then, that we have $n + 1$ subsets $A_1, A_2, \ldots\ldots A_n, A_{n+1}$. We show the formula holds with n replaced by $n + 1$.

We will use the result for 2 sets repeatedly in the proof.

Consider $A_1 \cup A_2 \cup \ldots\ldots \cup A_n \cup A_{n+1}$ as the union of the sets $A_1 \cup A_2 \cup \ldots\ldots \cup A_n$ and A_{n+1}. Then

$$\begin{aligned} | A_1 \cup A_2 \cup \ldots\ldots \cup A_n \cup A_{n+1} | &= | (A_1 \cup \ldots\ldots \cup A_n) \cup A_{n+1} | \\ &= | A_1 \cup A_2 \cup \ldots\ldots \cup A_n | + | A_{n+1} | - | (A_1 \cup A_2 \cup \ldots\ldots \cup A_n) \cap A_{n+1} | \qquad \ldots(2) \end{aligned}$$

Use the fact that intersection distributes over unions to get that

$$| (A_1 \cup A_2 \cup \ldots\ldots \cup A_n) \cap A_{n+1} | = | (A_1 \cap A_{n+1}) \cup (A_2 \cap A_{n+1}) \cup \ldots\ldots \cup (A_n \cap A_{n+1}) |$$

Thus (2) becomes

$$|A_1 \cup A_2 \cup ...... \cup A_n| + |A_{n+1}| = |(A_1 \cap A_{n+1}) \cup (A_2 \cap A_{n+1}) \cup ...... \cup (A_n \cap A_{n+1})| \quad ...(3)$$

We can apply the inductive hypothesis to 2 of the 3 sets, in equation (3), namely

$$|A_1 \cup A_2 \cup ...... \cup A_n| = \sum_{i=1}^{n} |A_i| - \sum_{i,\, j \le n} |A_i \cap A_j| + \sum_{i,\, j,\, k \le n} |A_i \cap A_j \cap A_k| + ... + (-1)^{n-1} |A_1 \cap A_2 \cap ...... \cap A_n| \quad ...(4)$$

and $|(A_1 \cap A_{n+1}) \cup (A_2 \cap A_{n+1}) \cup ...... \cup (A_n \cap A_{n+1})|$

$$= \sum_{i=1}^{n} |A_i \cap A_{n+1}| - \sum_{i,\, j} |A_i \cap A_{n+1} \cap (A_j \cap A_{n+1})|$$

$$+ \sum_{i,\, j,\, k \le n} |(A_i \cap A_{n+1}) \cap (A_j \cap A_{n+1}) \cap (A_k \cap A_{n+1})|$$

$$+ ...... + (-1)^{n-1} |(A_1 \cap A_{n+1}) \cap (A_2 \cap A_{n+1}) \cap ...... \cap (A_n \cap A_{n+1})| \quad ...(5)$$

Substituting equations (5) and (4) into (3) and making simplification like

$$(A_i \cap A_{n+1}) \cap (A_j \cap A_{n+1}) \cap (A_k \cap A_{n+1}) = A_i \cap A_j \cap A_k \cap A_{n+1},$$ we have

$$|A_1 \cup A_2 \cup ...... \cup A_n \cup A_{n+1}| = \sum_{i=1}^{n} |A_i| - \sum_{i,\, j \le n} |A_i \cap A_j| + \sum_{i,\, j,\, k \le n} |A_i \cap A_j \cap A_k|$$

$$+ ...... + (-1)^{n-1} |A_1 \cap A_2 \cap ...... \cap A_n| + |A_{n+1}| - \sum_{i=1}^{n} |A_i \cap A_{n+1}|$$

$$+ \sum_{i,\, j \le n}^{n} |A_i \cap A_j \cap A_{n+1}| + ... + (-1)^{n} (A_1 \cap A_2 \cap ...... \cap A_n \cap A_{n+1}| \quad ...(6)$$

Now observe that $\left(\sum_{i,\, j \le n} |A_i \cap A_j| + \sum_{i=1}^{n} |A_i \cap A_{n+1}| \right)$, where the first sum is taken over I, the 2-combinations $\{i, j\}$ of $\{1, 2, ..., n\}$, and the second sum is taken over J, the 2-combinations of the form $\{i, n+1\}$ where $i \in \{1, 2, ..., n\}$, can be simplified to $\Sigma\, A_i \cap A_j$ where this sum is taken over all 2-combinations of $\{1, 2, ..., n, n+1\}$ since $I \cup J$ is the set of all 2-combinations of $\{1, 2, ... n, n+1\}$.

Likewise the two sums

$$\sum_{i,\, j,\, k \le n} |A_i \cap A_j \cap A_k| + \sum_{i,\, j \le n} |A_i \cap A_j \cap A_{n+1}|$$

where the first sum is taken over all 3-combinations of $\{1, 2, \ldots\ldots, n\}$ and the second is taken over all 3-combinations of the form $\{i, j, n + 1\}$ of $\{1, 2, \ldots\ldots, n, n + 1\}$, can be simplified to $\sum_{i, j, k} |A_i \cap A_j \cap A_k|$ where this sum is taken over all 3-combinations of $\{1, 2, \ldots\ldots, n, n + 1\}$.

Other similar simplification can also be made.

Thus, equation (6) becomes

$$|A_1 \cup A_2 \cup \ldots\ldots \cup A_n \cup A_{n+1}| = \sum_{i=1}^{n+1} |A_i| - \sum_{i, j} |A_i \cap A_j| + \sum_{i, j, k} |A_i \cap A_j \cap A_k|$$

$$+ \ldots\ldots + (-1)^n |A_1 \cap A_2 \ldots\ldots \cap A_{n+1}| \quad \ldots(7)$$

where the second sum is taken over all 2-combinations of $\{1, 2, \ldots\ldots, n, n + 1\}$, the third sum is taken over all 3-combinations of $\{1, 2, \ldots\ldots n, n + 1\}$, and so on.

In other words, equation (7) is just equation (1) with n replaced by $n + 1$.

The theorem, then, is proved by mathematical induction.

Corollary. $|\bar{A}_1 \cap \bar{A}_2 \cap \ldots\ldots \cap \bar{A}_n| = |U| - |A_1 \cup A_2 \cup \ldots\ldots \cup A_n|$

$$= |U| - \sum_{i=1}^{n} |A_i| + \sum_{i, j} |A_i \cap A_j|$$

$$- \sum_{i, j, k} |A_i \cap A_j \cap A_k| + \ldots\ldots + (-1)^n |A_1 \cap A_2 \cap \ldots\ldots \cap A_n|.$$

Problem 2.11. *In how many ways can the letters {5.a, 4.b, 3.c} be arranged so that all the letters of the some kind are not in a single block ?*

Solution. Let U be the set of $\dfrac{12!}{[5!4!3!]}$ permutations of these letters.

Let A_1 be the arrangements of the letters where the $5a$'s are in a single block, A_2 the arrangements where the $4b$'s are in a sing block, and A_3 the arrangements where the $3c$'s are in one block. Then,

$$|A_1| = \frac{8!}{4!3!}, |A_2| = \frac{9!}{5!3!}, |A_3| = \frac{10!}{5!4!}$$

$$|A_1 \cap A_2| = \frac{5!}{3!}, |A_1 \cap A_3| = \frac{6!}{4!}$$

$$|A_2 \cap A_3| = \frac{7!}{5!}, |A_1 \cap A_2 \cap A_3| = 3!$$

Thus, $|\bar{A}_1 \cap \bar{A}_2 \cap \bar{A}_3| = \dfrac{12!}{5!4!3!} - \left(\dfrac{8!}{4!3!} + \dfrac{9!}{5!3!} + \dfrac{10!}{5!4!}\right) + \dfrac{5!}{3!} + \dfrac{6!}{4!} + \dfrac{7!}{5!} - 3!$.

Problem 2.12. *Count the number of integral soultions to*

$$x_1 + x_2 + x_3 = 20 \text{ where } 2 \le x_1 \le 5,\ 4 \le x_2 \le 7 \text{ and } -2 \le x_3 \le 9.$$

Solution. Let U be the set of solutions (x_1, x_2, x_3) where $2 \le x_1, 4 \le x_2, -2 \le x_3$.

We know that $| U | = C(20 - 2 - 4 + 2 + 3 - 1, 3 - 1 | = C(18, 2)$

Let $\quad A_1 = \{(x_1, x_2, x_3) \in U \mid x_1 \ge 6\}$

$A_2 = \{(x_1, x_2, x_3) \in U \mid x_2 \ge 8\}$

$A_3 = \{(x_1, x_2, x_3) \in U \mid x_3 \ge 10\}$

We wish to count the number of elements in $\overline{A_1} \cap \overline{A_2} \cap \overline{A_3}$

$$= \overline{(A_1 \cup A_2 \cup A_3)}$$

By the principle of inclusion-exclusion

$$| \overline{(A_1 \cup A_2 \cup A_3)} | = | U | - | A_1 \cup A_2 \cup A_3 |$$

$$= | U | - \{| A_1 | + | A_2 | + | A_3 | - | A_1 \cap A_2 | - | A_1 \cap A_3 | - | A_2 \cap A_3 | + | A_1 \cap A_2 \cap A_3 |\}$$

Now A_1 is the set of solutions of equation (1),

where $6 \le x_1, 4 \le x_2, -2 \le x_3$.

Thus, $| A_1 | = C(20 - 6 - 4 + 2 + 3 - 1, 3 - 1) = C(14, 2)$

Similarly, A_2 is the set of soultions of equation (1)

where $2 \le x_1, 8 \le x_2, -2 \le x_3$.

Therefore, $| A_2 | = C(20 - 2 - 8 + 2 + 3 - 1, 3 - 1) = C(14, 2)$

Likewise, $| A_3 | = C(20 - 2 - 4 - 10 + 3 - 1, 3 - 1) = C(6, 2)$

Now $A_1 \cap A_2$ is the set of solutions where $6 \le x_1, 8 \le x_2, -2 \le x_3$.

Thus $| A_1 \cap A_2 | = C(20 - 6 - 8 + 2 + 3 - 1, 3 - 1) = C(10, 2)$

Like wise, $| A_1 \cap A_3 | = C(20 - 6 - 4 - 10 + 3 - 1, 3 - 1) = C(2, 2)$

and $\quad | A_2 \cap A_3 | = C(20 - 2 - 8 - 10 + 3 - 1, 3 - 1) = C(2, 2)$.

Moreover, $| A_1 \cap A_2 \cap A_3 | = 0$, since 20 does not exceed 6 + 8 + 10.

Therefore, $| \overline{A_1 \cup A_2 \cup A_3} | = C(18, 2) - 2C(14, 2) - C(6, 2) + C(10, 2) + 2C(2, 2)$.

Problem 2.13. *A total of 1232 students have taken a course in a course in Spanish, 879 have taken a course in French, and 114 have taken a coruse in Russian. Further, 103 have taken courses in both Spanish and French, 23 have taken courses in both Spanish and Russian, and 14 have taken courses in both French and Russian. If 2092 students have taken atleast one of Spanish, French, and Russian, how many students have taken a course in all three languages ?*

Solution. Let S be the set of students who have taken a course in Spanish, F the set of students who have taken a course in French, and R the set of students who have taken a course in Russian. Then

$$| S | = 1232, | F | = 879, | R | = 114$$

$$|S \cap F| = 103, |S \cap R| = 23, |F \cap R| = 14$$

and $$|S \cup F \cup R| = 2092.$$

Inserting these quantities into the equation

$$|S \cup F \cup R| = |S| + |F| + |R| - |S \cap F| - |S \cap R| - |F \cap R| + |S \cap F \cap R|$$

gives $2092 = 1232 + 879 + 114 - 103 - 23 - 14 + |S \cap F \cap R|$

Solving for $|S \cap F \cap R|$ shows that $|S \cap F \cap R| = 7$.

Therefore, there are seven students who have taken courses in spanish, French and Russian. (See Fig. 2.1)

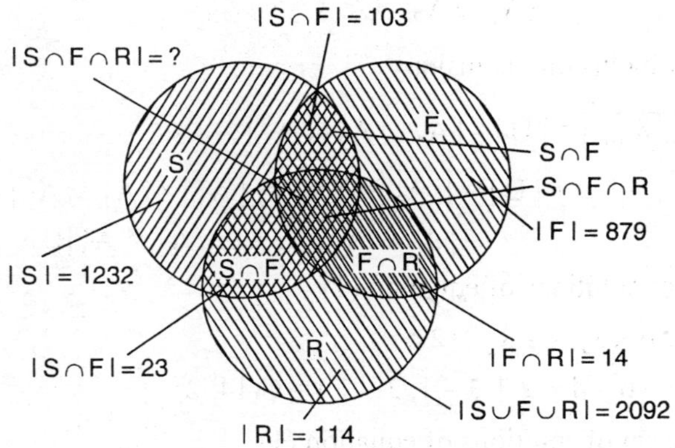

Fig. 2.1. The set of students who have taken courses in Spanish, French, and Russian.

Problem 2.14. *Suppose that there are 1807 freshmen at your school. Of these 453 are taking a course in computer science, 567 are taking a course in mathematics, and 299 are taking courses in both computer science and mathematics. How many are not taking a course either in computer science or in mathematics ?*

Solution. To find the number of freshman who are not taking a course in either mathematics or computer science, subtract the number that are taking a course in either of these subjects from the total number of freshmen.

Let A be the set of all freshmen taking a course in computer science, and let B be the set of all freshmen taking a course in mathematics.

It follows that $|A| = 453$, $|B| = 567$, and $|A \cap B| = 299$.

The number of freshmen taking a course in either computer science or mathematics is

$$|A \cup B| = |A| + |B| - |A \cap B| = 453 + 567 - 299 = 721$$

Consequently, there are $1807 - 721 = 1086$ freshmen who are not taking a course in computer science or mathematics.

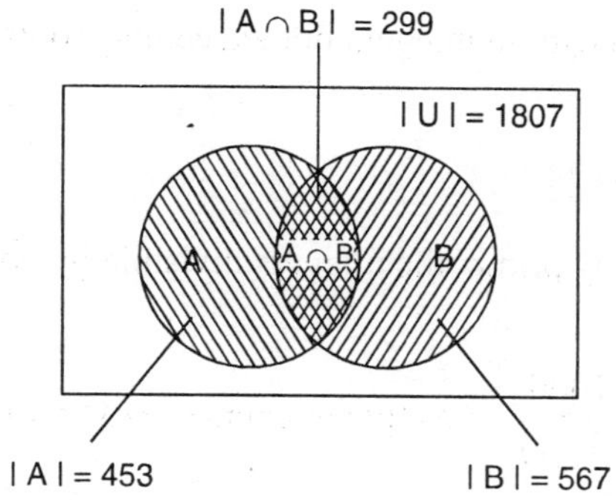

$$| A \cup B | = | A | + | B | - | A \cap B | = 453 + 567 - 299 = 721$$

Fig. 2.2. The set of Freshmen not taking a course in either Computer Science or Mathematics

Problem 2.15. *A discrete mathematics class contains 25 students majoring in computer science, 13 students majoring in Mathematics, and eight joint Mathematics and Computer Science majors. How many students are in this class, if every student is majoring in Mathematics, computer science, or both Mathematics and Computer Science ?*

Solution. Let A be the set of students in the class majoring in computer science and B be the set of students in the class majoring in Mathematics.

Then $A \cap B$ is the set of students in the class who are joint Mathematics and Computer Science majors.

Since every student in the class is majoring in either computer science or mathematics (or both), it follows that the number of students in the class is $| A \cup B |$.

Therefore, $| A \cup B | = | A | + | B | - | A \cap B |$

$$= 25 + 13 - 8 = 30.$$

Therefore, there are 30 students in the class.

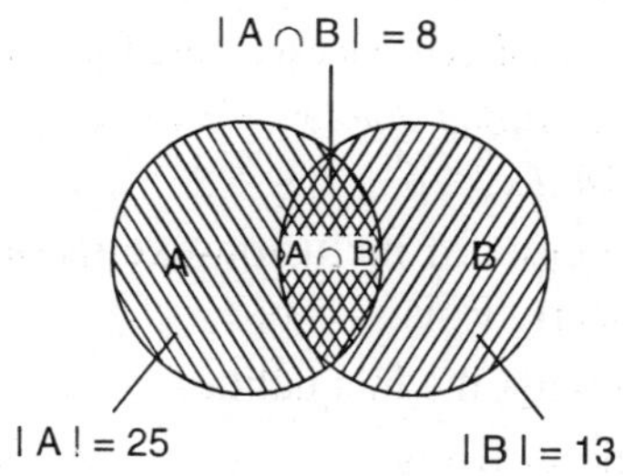

$$| A \cup B | = | A | + | B | - | A \cap B | = 25 + 13 - 8 = 30$$

Fig. 2.3. The set of students in a Discrete Mathematics class.

Problem 2.16. *How many positive integers not exceeding 1000 are divisible by 7 or 11 ?*

Solution. Let A be the set of positive integers not exceeding 1000 that are divisible by 7, and let B be the set of positive integers not exceeding 1000 that are divisible by 11.

Then $A \cup B$ is the set of integers not exceeding 1000 that are divisible by either 7 or 11, and $A \cap B$ is the set of integers not exceeding 1000 that are divisible by both 7 and 11.

We know that among the positive integers not exceeding 1000 there are $\left[\frac{1000}{7}\right]$ integers divisible by 7 and $\left[\frac{1000}{11}\right]$ divisible by 11.

Since 7 and 11 are relatively prime, the integers divisible by both 7 and 11 are those divisible by 7.11.

Consequently, there are $\left[\frac{1000}{(11.7)}\right]$ positive integers not exceeding 1000 that are divisible by both 7 and 11.

It follows that there are

$$|A \cup B| = |A| + |B| - |A \cap B|$$

$$= \left[\frac{1000}{7}\right] + \left[\frac{1000}{11}\right] + \left[\frac{1000}{7.11}\right] = 142 + 90 - 12 = 220$$

Positive integers not exceeding 1000 that are divsible by either 7 or 11.

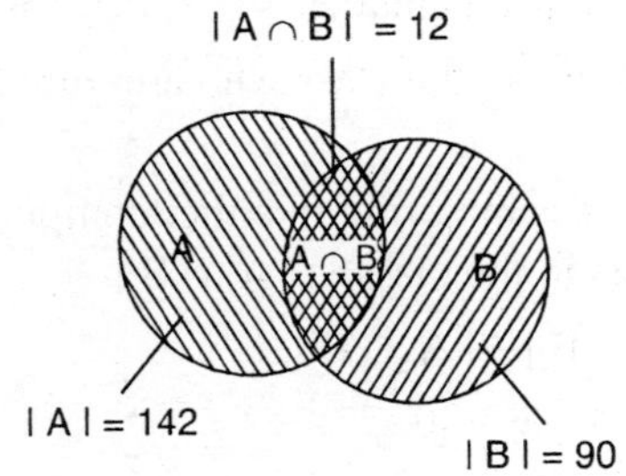

$$|A \cup B| = |A| + |B| - |A \cap B| = 142 + 90 - 12 = 220$$

Fig. 2.4. The set of positive integers not exceeding 1000 divisible by either 7 or 11.

Problem 2.17. *How many solutions does $x_1 + x_2 + x_3 = 11$ have, where x_1, x_2 and x_3 are non negative integers with $x_1 \le 3$, $x_2 \le 4$ and $x_3 \le 6$?*

Solution. To apply the principle of inclusion-exclusion, let a solution have property P_1 is $x_1 > 3$, property P_2 is $x_2 > 4$, and property P_3 is $x_3 > 6$.

The number of solutions satisfying the inequalities

$$x_1 \le 3, x_2 \le 4, \text{ and } x_3 \le 6 \text{ is}$$

$$N(P_1'P_2'P_3') = N - N(P_1) - N(P_2) - N(P_3) + N(P_1P_2) + N(P_1P_3) + N(P_2P_3) - N(P_1P_2P_3)$$

It follows that

$$N = \text{total number of solutions} = C(3 + 11 - 1, 11) = 78$$

$$N(P_1) = (\text{number of solutions with } x_1 \ge 4) = C(3 + 7 - 1, 7) = C(9, 7) = 36$$

$$N(P_2) = (\text{number of solutions with } x_2 \ge 5) = C(3 + 6 - 1, 6) = C(8, 6) = 28$$

$$N(P_3) = (\text{number of solutions with } x_3 \ge 7) = C(3 + 4 - 1, 4) = C(6, 4) = 15$$

$$N(P_1P_2) = (\text{number of solutions with } x_1 \ge 4 \text{ and } x_2 \ge 5)$$

$$= C(3 + 2 - 1, 2) = C(4, 2) = 6.$$

$N(P_1P_3)$ = (number of solutions with $x_1 \geq 4$ and $x_3 \geq 7$)

$= C(3 + 0 - 1, 0) = 1$

$N(P_2P_3)$ = (number of solutions with $x_2 \geq 5$ and $x_3 \geq 7$) = 0

$N(P_1P_2P_3)$ = (number of solutions with $x_1 \geq 4$, $x_2 \geq 5$ and $x \geq 7$) = 0

Inserting these quantities into the formula for $N(P_1'P_2'P_3')$ shows that the number of solutions with $x_1 \leq 3$, $x_2 \leq 4$, and $x_3 \leq 6$ equals

$$N(P_1'P_2'P_3') = 78 - 36 - 28 - 15 + 6 + 1 + 0 - 0 = 6.$$

Problem 2.18. *How many onto functions are there from a set with six elements to a set with three elements ?*

Solution. Suppose that the elements in the codomain are b_1, b_2 and b_3.

Let P_1, P_2 and P_3 be the properties that b_1, b_2 and b_3 are not in the range of the function, respectively.

Note that a function is onto if and only if it has none of the properties P_1, P_2 or P_3.

By the inclusion-exclusion principle it follows that the number of onto functions from a set with six elements to a set with three elements is

$$N(P_1'P_2'P_3') = N - [N(P_1) + N(P_2) + N(P_3)] + [N(P_1P_2) + N(P_1P_3) + N(P_2P_3)] - N(P_1P_2P_3)$$

where N is the total number of functions from a set with six elements to one with three elements.

We will evaluate each of the terms on the right-hand side of this equation, it follows that $N = 3^6$.

Note that $N(P_i)$ is the number of functions that do not have b_i in their range.

Hence, there are two choices for the value of the function at each element of the domain.

Therefore, $N(P_i) = 2^{6.}$

Furthermore, there are C(3, 1) terms of this kind.

Note that $N(P_i\ P_j)$ is the number of functions that do not have b_i and b_j in their range.

Hence, there is only one choice for the value of the function of each element of the domain.

Therefore, $N(P_i\ P_j) = 1^6 = 1$.

Furthermore, there are C(3, 2) terms of this kind.

Also, note that $N(P_1P_2P_3) = 0$, since this term is the number of functions that have none of b_1, b_2 and b_3 in their range.

Clearly, there are no such functions.

Therefore, the number of onto functions from a set wtih six elements to one with three elements is

$$3^6 - C(3, 1)\ 2^6 + C(3, 2)1^6 = 729 - 192 + 3 = 540.$$

Problem 2.19. *How many ways are there to assign five different jobs to four different employees if every employee is assigned at least one job ?*

Solution. Consider the assignment of jobs as a function from the set of five jobs to the set of four employees.

An assignment where every employee gets at least one job is the same as an onto function from the set of jobs to the set of employees.

Hence, it follows that there are

$$4^5 - C(4, 1)3^5 + C(4, 2)2^5 - C(4, 3)1^5 = 1024 - 972 + 192 - 4$$

= 240 ways to assign the jobs so that each employees is assigned atleast one job.

Theorem 2.2. Let m and n be positive integers with $m \geq n$. Then, there are $n^m - C(n, 1)(n-1)^m + C(n, 2)(n-2)^m - \ldots\ldots + (-1)^{n-1} C(n, n-1) \cdot 1^m$ onto functions from a set with m elements to a set with n elements.

Remark. The number of onto functions from a set with m elements to a set with n elements equals $n ! \, S_2 (m, n)$, where $S_2 (m, n)$ is a starting number of the second kind.

Problem 2.20. *In a survey of students at Florida State University the following information was obtained, 260 were taking a statistics course, 208 were taking a mathematics course, 160 were taking a computer programming course, 76 were taking statistics and mathematics, 48 were taking statistics and computer programming, 62 were taking mathematics courses and computer programming, 30 were taking all 3 kinds of courses, and 150 were taking none of the 3 courses.*

Let S = {students taking statistics},
M = {students taking Mathematics}
C = {students taking Computer Programming}

(*a*) *How many students were surveyed ?*

(*b*) *How many students were taking a statistics and a mathematics course but not a computer programming course ?*

(*c*) *How many were taking a statistics and a computer course but not a mathematics course ?*

(*d*) *How many were taking a computer programming and a mathematics course but not a statistics coruse ?*

(*e*) *How many were taking a statistics course but not taking a course in mathematics or in computer programming ?*

(*f*) *How many were taking a mathematics course but not taking a statistics course or a computer programming course ?*

(*g*) *How many were taking a computer programming course but not taking a course in mathematics or in statistics ?*

Solution. We know that $|S| = 260$, $|M| = 208$, $|C| = 160$

$$|S \cap M| = 76, |S \cap C| = 48, |M \cap C| = 62$$

$|S \cap M \cap C| = 30$, and $|\overline{S \cup M \cup C}| = 150$.

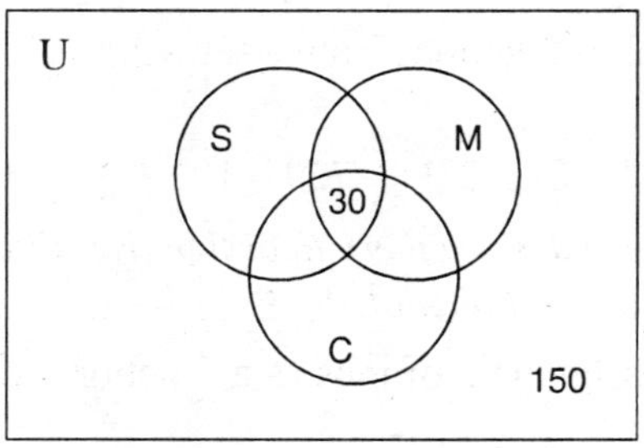

Fig. 2.5.

So we can immediately insert the number of students 2 of the 8 regions of the Venn diagram, the regions $|S \cap M \cap C|$ and $|\overline{S \cup M \cup C}|$.

(*a*) The total number of students surveyed :

$$|U| = |S \cup M \cup C| + |\overline{S \cup M \cup C}|$$

$$= |S| + |M| + |C| - |S \cap M| - |S \cap C| - |M \cap C| + |S \cap M \cap C| + |\overline{S \cup M \cup C}|$$

$$= 260 + 208 + 160 - 76 - 48 - 62 + 30 + 150 = 622.$$

(*b*) We are asked to find :

$$|S \cap M \cap \overline{C}| = |S \cap M - S \cap M \cap C| = |S \cap M| - |S \cap M \cap C|$$

$$= 76 - 30 = 46.$$

(*c*) $|S \cap C \cap \overline{M}| = |S \cap C| - |S \cap C \cap M| = 48 - 30 = 18$

(*d*) $|M \cap C \cap \overline{S}| = |M \cap C| - |M \cap C \cap S| = 62 - 30 = 32.$

Thus, we can insert some more numbers in the appropriate regions of the Venn diagram (See Fig. 2.6).

Only the computation of the number of students taking courses in exactly one of the 3 subjects remains, and that is precisely the content of parts (*e*), (*f*) and (*g*).

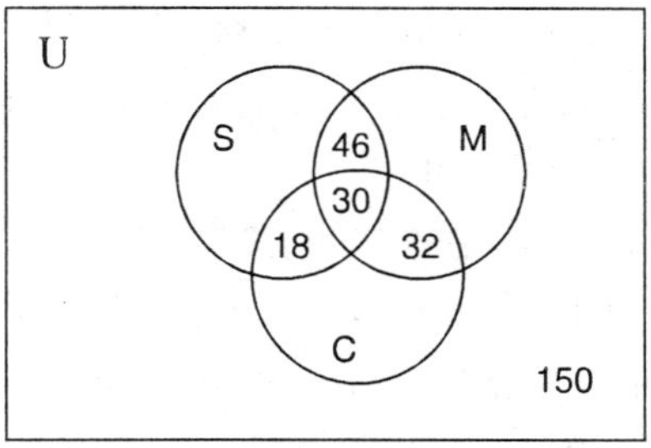

Fig. 2.6.

(*e*) $|S \cap \overline{M} \cap \overline{C}| = |S| - |S \cap M \cap \overline{C}| - |S \cap C \cap \overline{M}| - |S \cap M \cap C|$

$$= 260 - 46 - 18 - 30 = 166$$

(*f*) $|M \cap \overline{S} \cap \overline{C}| = |M| - |M \cap S \cap \overline{C}| - |M \cap C \cap \overline{S}| - |M \cap S \cap C|$

$$= 100$$

(*g*) $|C \cap \overline{S} \cap \overline{M}| = |C| - |C \cap S \cap \overline{M}| - |C \cap \overline{S} \cap M| - |C \cap S \cap M|$

$$= 80.$$

Thus, we can fill in the numbers for all 8 regions in the Venn diagram illustrated in Fig. 2.7.

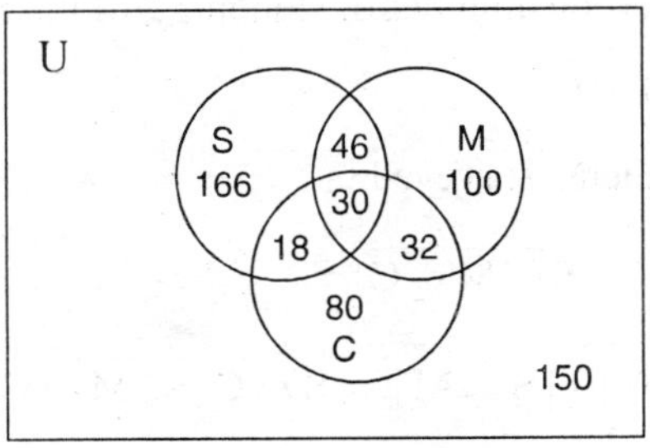

Fig. 2.7.

Problem 2.21. *If there are 200 facutly members that speak French, 50 that speak Russian, 100 that speak Spanish, 20 that speak French and Russian, 60 that speak French and Spanish, 35 that speak Russian and Spanish, while only 10 speak French, Russian, and Spanish, how many speak either French or Russian or Spanish ?*

Solution. Let F be the set of faculty who speak French,

R be the set of faculty who speak Russian,

and S be the set of faculty that speak Spanish.

We know that, $| F | = 200, | R | = 50, | S | = 100$

$$| F \cap R | = 20, | F \cap S | = 60, | R \cap S | = 35 \text{ and } | F \cap R \cap S | = 10$$

Thus, $| F \cup R \cup S | = 200 + 50 + 100 - 20 - 60 - 35 + 10 = 245.$

Problem 2.22. *From a group of 10 professors how many ways can a committee of 5 members be formed so that atleast one of professor A and professor B will be included ?*

Solution. We will give 3 solutions : one using the sum rule, one by counting indirectly, and one using the principle of inclusion-exclusion.

(*i*) The number of committees including both professor A and professor B is C(8, 3) = 56.

The number of committees including professor A but excluding professor B is C(8, 4) = 70.

Since professor A fills one position there are only 4 more position to fill and since Professor B is excluded there are only 8 people from which to choose.

Likewise C(8, 4) is the number of committees including professor B and excluding professor A.

Consequently, by the sum rule the total number of ways of selecting a committee of 5 including Professor A or professor B is C(8, 3) + 2C(8, 4) = 56 + 2.70 = 196.

(*ii*) By counting indirectly we obtain a second solution :

We just need the observation that the total number of committees excluding both professor A and B is C(8, 5) = 56 and the total number of committees is C(10, 5).

Thus we see that C(10, 5) – C(8, 5) = 252 – 56 = 196 is the number of committees including atleast one of the professors.

(*iii*) A third solution uses the principle of inclusion-exclusion.

Among the 252 committees of 5 members, let A_1 and A_2 be the set of committees that include professor A and professor B, respectively.

Since $| A_1 | = C(9, 4) = 126 = | A_2 |$ and $| A_1 \cap A_2 | = C(8, 3) = 56$, it follows that

$$| A_1 \cup A_2 | = 126 + 126 - 56 = 196.$$

Problem 2.23. *Suppose that 200 faculty members can speak French and 50 can speak Russian while only 20 can speak both French and Russian. How many faculty members can speak either French or Russian ?*

Solution. If F is the set of faculty who speak French

and R is the set of faculty who speak Russian,

then, we know that $| F | = 200$, $| R | = 50$ and $| F \cap R | = 20$.

We are asked to compute $| F \cup R |$ which by the formula is $200 + 50 - 20 = 230$.

2.3 THE PIGEON HOLE PRINCIPLE

Suppose that a flock of pigeons flies into a set of pigeon holes to roost. The pigeon hole principle states that if there are more pigeons than pigeon holes, then there must be atleast one pigeon hole with atleast two pigeons in it. Of course, this principle applies to other objects besides pigeons and pigeon-holes.

"If we have six pigeons in five pigeon holes then atleast one pigeonholes would have more than one pigeon".

Or

"If five men (pigeons) are married to four women (pigeon holes) then atleast one women has more than one husband".

Or

"If five men (pigeon holes) are married to six women (pigeons) then atleast one men has more than one wife".

Theorem 2.3. *If n objects (pigeons) are places in m places (pigeon holes) for m < n, then one of the places (pigeon holes) must contains atleast* $\left[\frac{n-1}{m}\right]$ *+ 1* objects (pigeons).

Proof. The proof is very simple. Let us assume that none of the places (pigeon holes) contains more than floor $\left(\frac{(n-1)}{m}\right)$ objects (pigeons). Then there are at the most m^* floor $\left(\frac{(n-1)}{m}\right) \le m^* \frac{(n-1)}{m}$ $= n - 1$ objects. This is contrary to the given fact that there are n objects. This contradiction is because of our assumption that "none of the places contains more than floor $\left(\frac{(n-1)}{m}\right)$ objects, thus our assumption is wrong.

Therefore, one of the pigeon holes must contains atleast

$$\left[\frac{n-1}{m}\right] + 1 \text{ objects.}$$

Theorem 2.4. *If k + 1 or more objects are placed into k boxes, then there is at least one box containing two or more of the objects.*

Proof. Suppose that none of the k boxes contains more than one object.

Then the total number of objects would be atmost k.

This is a contradiction, since there are atleast $k + 1$ objects.

(The pigeon hole principle is also called the Dirichlet drawer principle).

Examples. 1. Among any group of 367 people, there must be at least two with the same brithday, because there are only 366 possible birthdays.

2. In any group of 27 English words, there must be atleast two that begin with the same letter, since there are 26 letters in the English alphabet.

2.4 THE GENERALIZED PIGEON HOLE PRINCIPLE

If N objects are placed into k boxes, then there is atleast one box containing atleast $[N/k]$ objects.

Proof. Suppose that none of the boxes contains more than $[N/k] - 1$ objects.

Then, the total number of objcets is at most

$$k\left(\left[\frac{N}{k}\right]-1\right) < k\left(\left(\frac{N}{k}+1\right)-1\right) = N.$$

where the inequality $[N/k] < (N/k) + 1$ has been used.

This is a contradiction, since there are a total of N objects.

Example. Among 100 people there are atleast $[100/12] = 9$ who were born in the same month.

Theorem 2.5. *Every sequence of $n^2 + 1$ distinct real numbers conatins a subsequence of length $n + 1$ that is either strictly increasing or strictly decreasing.*

Proof. Let $a_1, a_2, \ldots\ldots a_{n^2+1}$ be a sequence of $n^2 + 1$ distinct real numbers.

Associate an ordered pair with each term of the sequence, namely, associate (i_k, d_k) to the term a_k.

where i_k is the length of the longest increasing subsequence starting at a_k and d_k is the length of the longest decreasing subsequence starting at a_k.

Suppose that there are no increasing or decreasing subequences of length $n + 1$.

Then i_k and d_k are both positive integers less than or equal to n, for $k = 1, 2, \ldots\ldots n^2 + 1$.

Hence, by the product rule there are n^2 possible ordered pairs for (i_k, d_k).

By the pigeon hole principle, two of these $n^2 + 1$ ordered pairs are equal.

In other words, there exist terms a_s and a_t, with $s < t$ such that $i_s = i_t$ and $d_s = d_t$.

We will show that this is impossible, because the terms of the sequence are distinct, either $a_s < a_t$ or $a_s > a_t$.

If $a_s < a_t$, then, since $i_s = i_t$, an increasing subsequence of length $i_t + 1$ can be built starting as a_s, by taking a_s followed by an increasing subsequence of length i_t beginning at a_t.

This is a contradiction.

Similarly, if $a_s > a_t$, it can be shown that d_s must be greater than d_t, which is a contradiction.

Example :

The sequence 8, 11, 9, 1, 4, 6, 12, 10, 5, 7 contains ten terms.

Note that $10 = 3^2 + 1$.

There are four increasing subsequences of length four, namely, 1, 4, 6, 12 ; 1, 4, 6, 7 ; 1, 4, 6, 10 ; and 1, 4, 5, 7.

There is also a decreasing subsequence of length four, namely, 11, 9, 6, 5.

Problem 2.24. *How many students must be in a class to guarantee that atleast two students receive the same score on the final exam, if the exam is graded on a scale from 0 to 100 points ?*

Solution. There are 101 possible scores on the final.

The pigeon hole principle shows that among any 102 students there must be atleast 2 students with the same score.

Problem 2.25. *Show that for every integer n there is a multiple of n that has only 0 s and 1 s in its decimal exapnsion.*

Solution. Let n be a positive integer.

Consider the n integers 1, 11, 111, ..., 11, ... 1 (where the last integer in this list is the inetger with $n + 1$ 1s in its decimal expansion).

Note that there are n possible remainders when an integer is divided by n.

Since there are $n + 1$ integers in this list, by the pigeon hole princple there must be two with the same remainder when divided by n.

The difference of these two integers has a decimal expansion consisting entirely of 0s and 1s and is divisible by n.

Problem 2.26. *What is the minimum number of students required in a discrete mathematics class to be sure that atleast six will receive the same grade, if there are five possible grades, A B, C, D and F ?*

Solution. The minimum number of students needed to ensure that atleast six students receive the same grade is the smallest integer N such that [N/5] = 6.

The smallest such integer is N = 5.5 + 1 = 26.

If you have only 25 students, it is possible for there to be five who have received each grade so that no six students have received the same grade.

Thus, 26 is the minimum number of students needed to ensure that atleast six students will receive the same grade.

Problem 2.27. *What is the least number of area codes needed to guarantee that the 25 million phones in a state have distinct ten-digit telephone numbers ?*

(Assume that telephone numbers are of the form NXX-NXX-XXXX, where the first three digits form the area code, N represents a digit from 2 to 9 inclusive, and X represents any digit).

Solution. There are 8 miliion different phone numbers of the form NXX-XXXX.

Hence, by the generalized pigeon hole principle, among 25 million telephones, atleast [25,000,000/ 8,000,000] of them must have identical phone numbers.

Hence, atleast four area codes are required to enure that all ten-digit numbers are different.

Problem 2.28. (*a*) *How many cards must be selected from a standard deck of 52 cards to guarantee that atleast three cards of the same suit are chosen ?*

(*b*) *How many must be selected to guarantee that atleast three hearts are selected ?*

Solution. (*a*) Suppose there are four boxes, and as cards are selceted they are placed in the box reserved for cards of that suit.

Using the generalized pigeon hole principle, we see that if N cards are selected, there is atleast one box containing at least [N/4] cards.

Consequently, we know that atleast three cards of one suit are selected if $[N/4] \geq 3$.

The smallest integer N such that $[N/4] \geq 3$ is $N = 2.4 + 1 = 9$, so nine cards suffice.

Note that if eight cards are selected, it is possible to have two cards of each suit, so that more than eight cards are needed.

Consequently, nine cards must be selected to guarantee that atleast three cards of one suit are chosen.

Note that after 8th card is chosen, there is no way to avoid having a third card of some suit.

(*b*) Since we want to make sure that there are three hearts, not just three cards of one suit.

Note that in the worst case, we can select all the clubs, diamonds, and spades, 39 cards in all, before we select a single heart.

The next three cards will be all hearts, so we may need to select 42 cards to get three hearts.

Problem 2.29. *Suppose that a computer science laboratory has 15 workstations and ten servers. A cable can be used to directly connect a workstation to a server. Only one direct connection to a server can be active at any time. We want to guarantee that at any time any set of ten or fewer workstations can simultaneously access different servers via direct connetions. Although we could do this by connecting every workstation directly to every server (using 150 conncetions), what is the minimum numebr of direct connections needed to achieve this goal ?*

Solution. Suppose that we lebel the workstations $W_1, W_2, ..., W_{15}$ and the servers $S_1, S_2, ...$ S_{10}.

Furthermore, suppose that we connect W_k to S_k for $k = 1, 2, ..., 10$ and each of $W_{11}, W_{12}, W_{13}, W_{14}$ and W_{15} to all ten servers.

We have a total of 60 direct connections.

Clearly any set of ten or fewer workstations can simultaneously access different servers.

We see this by noting that if workstation W_j is included with $1 \leq j \leq 10$, it can access server S_j, and for each workstation W_k with $k \geq 11$ included, there must be a corresponding workstation W_j with $1 \leq j \leq 10$ not included, so that W_k can access server S_j.

(This follows since there are atleast as many available servers S_j as there are workstations W_j with $1 \leq j \leq 10$ not included.

Now suppose there are fewer than 60 direct connections between workstations and servers. Then some server would be connecfed to atmost $[59/10] = 5$ workstations.

(If all servers were connected to atleast six workstations, there would be atleast $6.10 = 60$ direct connections)

This means that the remaining nine servers are not enough to allow the other ten workstations to simultaneously access different servers.

Consequently, at least 60 direct connections are needed. It follows that 60 is the required answer.

Problem 2.30. *During the month with 30 days a baseball team plays atleast one game a day, but no moe than 45 games. Show that there must be a period of some number of consecutive days during which the team must play exactly 14 games.*

Solution. Let a_j be the number of games played on or before the jth day of the month.

Then $a_1, a_2, ... a_{30}$ is an increasing sequence of distinct positive integers, with $1 \le a_j \le 45$.

Moreover, $a_1 + 14, a_2 + 14, ..., a_{30} + 14$ is also an increasing sequence of distinct positive integers, with $15 \le a_j + 14 \le 59$.

Then 60 positive integers $a_1, a_2, ..., a_{30}, a_1 + 14, a_2 + 14, ..., a_{30} + 14$ are all less than or equal to 59.

Hence, by the pigeon hole principle two of these integers are equal.

Since the integers $a_j, j = 1, 2, ..., 30$ are all distinct and the integers $a_j + 14, j = 1, 2, ..., 30$ are all distinct, there must be indices i and j with $a_i = a_j + 14$.

This means that exactly 14 games were played from day $j + 1$ to day i.

Problem 2.31. *Show that among any n + 1 positive integers not exceeding 2n there must be an inetger that divides one of the other integers.*

Solution. Write each of the $n + 1$ integers $a_1, a_2, ..., a_{n+1}$ as a power of 2 times an odd integer.

In other words, Let $a_j = 2^{kj} q_j$ for $j = 1, 2, ..., n + 1$

where k_j is a non negative integer and q_j is odd.

The integers $q_1, q_2, ..., q_{n+1}$ are all odd positive integers less than $2n$.

Since there are only n odd positive integers less than $2n$, it follows from the pigeon hole principle that two of the integers $q_1, q_2, ..., q_{n+1}$ must be equal.

Therefore, there are integers i and j such that $q_i = q_j$.

Let q be the common value of q_i and q_j then $a_i = 2^{ki}q$ and $a_j = 2^{kj} q$.

It follows that if $k_i < k_j$, then a_i divides a_j ; while if $k_i > k_j$, then a_j divides a_i.

Problem 2.32. *Assume that in a group of six people, each pair of individuals consists of two friends or two enemies. Show that there are mutual friends or three mutual enemies in the group.*

Solution. Let A be one of the six people.

Of the five other people in the group, there are either three or more who are friends of A, or three or more who are enemies of A.

This follows from the generalized pigeons hole principle since when five objcets are divided into two sets, one of the sets has atleast [5/2] = 3 elements. In the former cases, suppose that B, C and D are friends of A.

If any two of these three individuals are friends, then these two and A form a group of three mutual friends.

Otherwise, B, C and D form a set of three mutual enemies.

The proof in the later case, when there are three or more enemies of A, proceeds in a similar manner.

Problem 2.33. *Show that if any five numbers from 1 to 8 are chosen, then two fo them will add to 9.*

Solution. Construct four different sets, each containing two nubmers that add upto 9 as follows :

$A_1 = \{1, 8\}, A_2 = \{2, 7\}, A_3 = \{3, 6\}, A_4 = \{4, 5\}$.

Each of the five numbers chosen must belong to one of these sets.

Since there are only four sets, the pigeon-hole principle tells us that two of the chosen numbers belong to the same set. These numbers add up to 9.

Problem 2.34. *Show that if 30 dictionaries in a library contain a total of 61,327 pages, then one of the dictionaries must have atleast 2045 pages.*

Solution. Let the pages be the pigeons and the dictionaries the pigeonholes.

Assign each page to the dictionary in which it appears. Then by the extended pigeonhole principle, one dictionary must contain at least [61,326/30] + 1 or 2045 pages.

Problem. 2.35. *Show that if any 30 people are selected, then one may choose a subset of five so that all five were born on the same day of the week.*

Solution. Assign each person to the day of the week on which she or he was born.

Then 30 pigeons are being assigned to 7 pigeon-holes. By the extended pigeon-hole principle with $n = 30$ and $m = 7$, atleast $[(30 - 1)/7] + 1$ or 5 of the people must have been born on the same day of the week.

Problem 2.36. *Shirts numbered consecutively from 1 to 20 are worn by the 20 members of a bowling league. When any three of these members are chosen to be a team, the league proposes to use the sum of their shirt numbers as a code number for the team. Show that if any eight of the 20 are selected, then from these eight one may form atleast two different teams having the same code number.*

Solution. From the eight selected bowlers, we can form a total of ${}_8C_3$ or 56 different teams.

These will play the role of pigeons. The largest possible team code number is 18 + 19 + 20 or 57, and the smallest possible is 1 + 2 + 3 or 6.

Thus only the 52 code numbers (pigeon-holes) between 6 and 57 inclusive are available for the 56 possible teams. By the pigeon-hole principle, atleast two teams will have the same code number.

The league should use another way to assign team numbers.

Problem 2.37. *Consider the region shown in figure. It is bounded by a regular hexagon whose sides are of length 1 unit. Show that if any seven points are chosen in this region then two of them must be no further apart than 1 unit.*

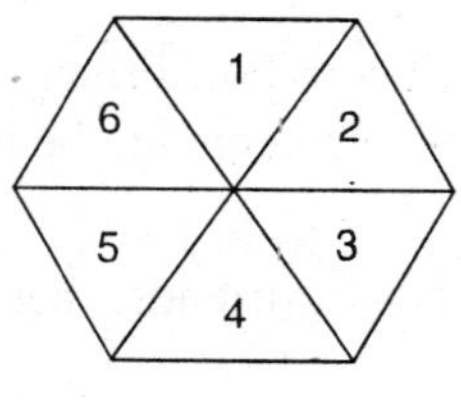

Fig. 2.8.

Solution. Divide the region into six equilateral triangles, as shown in figure.

If seven points are chosen in the region, we can assign each of them to a triangle that contains it.

If the point belongs to several triangles, arbitrarily assign it to one of them.

Then the seven points are assigned to six triangular regions, so by the pigeon-hole principle, atleast two points must belongs to the same region. These two cannot be more than one unit part.

Problem 2.38. *Show that if any 11 numbers are chosen from the set {1, 2, ..., 20}, then one of them will be a multiple of another.*

Solution. The key to solving this problem is to create 10 or fewer pigeon-holes in such a way that each number chosen can be assigned to only one pigeon-hole, and when x and y are assigned to the same pigeon hole we are guaranteed that either $x \mid y$ or $y \mid x$.

Factors are a natural feature to explore. There are eight prime numbers between 1 to 20, but knowing that x and y are multiples of the same prime will not guarantee that either $x \mid y$ or $y \mid x$. We try again.

There are ten odd numbers between 1 to 20.

Every positive integer n can be written as $n = 2^k m$, where m is odd and $k \geq 0$.

This can be seen by simply factoring all powers of 2 (if any) out of n.

In this case let us call m the odd part of n.

If 11 numbers are chosen from the set {1, 2, ..., 20}, then two of them must have the same odd part.

This follows from the pigeon hole principle since there are 11 numbers (pigeons), but only 10 odd numbers between 1 and 20 (pigeon holes) that can be odd parts of these numbers.

Let n_1 and n_2 be two chosen numbers with the same odd part.

We must have $n_1 = 2^{k_1} m$ and $n_2 = 2^{k_2}m$, for some k_1 and k_2. If $k \geq k_2$, then n_1 is a multiple of n_2, otherwise, n_2 is a multiple of n_1.

Problem 2.39. *Prove that in any list of ten natural numebrs $a_1, a_2, \ldots\ldots a_{10}$ there is a string of consecutive items of the list, $a_l, a_{l+1}, a_{l+2}, \ldots\ldots$, whose sum is divisible by 10. (We include the possibility that the "string" consists of just one number).*

Solution. Consider the ten numbers

$$a_1, a_1 + a_2, \ldots\ldots, a_1 + a_2 + \ldots\ldots + a_{10}.$$

If any of these is divisible by 10, we have the desired conclusion, otherwise, each number lies in one of the nine congruence classes $\overline{1}, \overline{2}, \ldots\ldots, \overline{9}$ of integers mod 10.

By the pieon-hole principle, two of them must lie in the same class and hence have a difference divisible by 10.

Again we reach the desired conclusion because, if $s > t$, $(a_1 + a_2 + \ldots\ldots a_s) - (a_1 + a_2 + \ldots\ldots + a_t)$

$$= a_{t+1} + a_{t+2} + \ldots\ldots + a_s.$$

Problem 2.40. *Show that among $n + 1$ arbitrarily chosen integers, there must exist two whose difference is divisible by n.*

Solution. Denote the integers $a_1, a_2, \ldots\ldots, a_{n+1}$.

As we know, the n congruence classes $\overline{0}, \overline{1}, \ldots\ldots, \overline{n-1}$ of integers mod n partition the integers into disjoint sets whose union is Z.

Put each integer a_i into its own congruence class. By the Pigeon-Hole Principle, two integers, a_i and a_j will go to the same class $\overline{k}$.

Since $a_i \in \overline{k}$, $\overline{a_i} = \overline{k}$ and similarly, $\overline{a_j} = \overline{k}$.

Thus, $\overline{a_i} = \overline{a_j}$, so $a_i \equiv a_j \pmod{n}$; that is, $n \backslash (a_i - a_j)$, as desired.

Problem 2.41. *Given five points inside a square whose sides have length 2. Prove that two are within $\sqrt{2}$ of each other.*

Solution. Subdivide the square into four squares with sides of length 1, as shown.

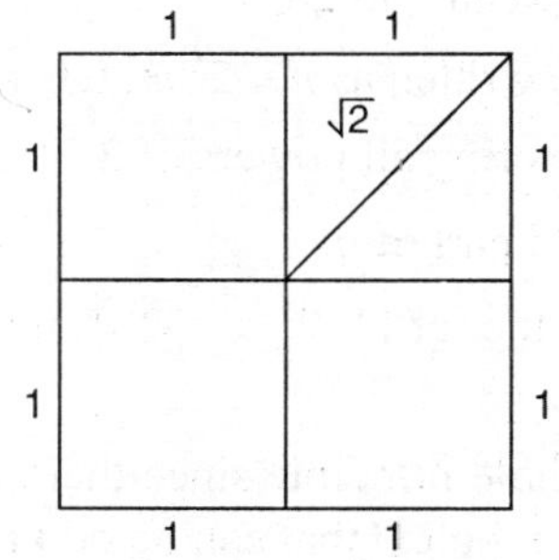

Fig. 2.9.

By the pigeon-hole principle, atleast two of the five chosen points must lie in, or on the boundary of, the same smaller square. But these points are at most $\sqrt{2}$ apart (the length of the diagonal of a smaller square).

Problem 2.42. *Martina has three weeks to prepare for a tennis tournament she decides a play atleast one set every day but not more than 36 sets in all. Show that there is a period of consecutive days during which she will play exactly 21 sets.*

Solution. Suppose Martina plays a_1 sets on day 1, a_2 sets on day 2, and so on, a_{21} sets on the last day of her preparation period.

Consider the 21 natural numbers $a_1, a_1 + a_2, \ldots\ldots, a_1 + a_2 + \ldots\ldots + a_{21}$.

Since each $a_i \geq 1$ and since the sum $a_1 + a_2 + \ldots\ldots + a_{21}$ is the total number of sets Martina will play. We have $1 \leq a_1 < a_1 + a_2 < \ldots\ldots < a_1 + a_2 + \ldots\ldots + a_{21} \leq 36$.

Now the only natural number between 1 and 36 which is divisible by 21 is 21 itself.

Therefore, each of these sums is either 21 or it belongs to one of the 20 non zero congruence classes of integers mod 21.

In the first case, we have $a_1 + a_2 + \ldots\ldots + a_i = 21$ for some i, so Martina would play precisely 21 sets on days through i giving the desired result.

In the second case, by the Pigeon-Hole Principle, two sums lie in the same congruence class and hence have difference divisible by 21.

We obtain $a_{t+1} + a_{t+2} + \ldots\ldots + a_s$ divisible by 21 for some s and t.

Since this is a numebr between 1 and 36, it must equal 21 and we conclude that Martina plays 21 games on days $t + 1$ through s.

Problem 2.43. *In any group of six peopole, atleast three must be mutual friends or at least three must be mutual strangers.*

Solution. Pretend you are one of the six people and put the other $n = 5$ peopole into $m = 2$ "boxes" labeled "my friends" and "strangers to me".

Since $[n/m] = [5/2] = 3$, by the strong form of the pigeon-hole principle, atleast one of these boxes must contain three people.

Suppose three people are your friends. If any two of these three are friends, then together with you, we have a set of three mutual friends.

If no two of these three are friends, then these three people are mutual strangers.

In either case, we have one of the desired conclusions. The remaining possibility, that three of the five people are strangers to you, leads again to the desired conclusion with an argument, which we omit, similar to the one already presented.

Problem 2.44. *Suppose Martina has 11 weeks to prepare for her tournament and that she intends to play atleast one set a day and at most 132 practice sets in all. Draw again the conclusion that during some period of consecutive days, Martina will play precisely 21 sets.*

Solution. The approach used before does not work this time, so we find a slightly different solution.

Let b_i be the number of sets Martina plays on days 1 through i inclusive and consider the 154 numbers

$$b_1, b_2, \ldots\ldots, b_{77}, b_1 + 21, b_2 + 21, \ldots\ldots, b_{77} + 21.$$

The largest number here, $b_{77} + 21$, is at most $132 + 12 = 153$.

By the pigeon-hole principle, we conclude that two are the same. Since $b_i > b_j$ if $i > j$, the only way for two to be equal is for $b_i = b_j + 12$ for some i and j with $i > j$.

Therefore, $b_i - b_j = 21$ and Martina plays 21 sets on days $j + 1$ through i.

Problem 2.45. *If 9 books are to be kept in 4 shelves, there must be atleast one self which contain atleast 3 books.*

Solution. The nine books can be thought of as pigeons and four shelves as the pigeon-holes.

If nine books are to kept in four shelves then average number of books per shelf is 2.25.

Hence it is no longer possible to avoid having at least three books on one shelf.

Problem 2.46. *In a group of 13 children, there must be atleast two children who were born in the same month.*

Solution. The thirteen children can be thought of as the pigeons and 12 months of the year as the pigeon-holes.

Since $13 > 12$, by pigeon-hole principle we can conclude that out of 13 at least 2 were born in the same month.

Problem 2.47. *Find the minimum number of students in a class to be sure that four out of them are born in the same month.*

Solution. Here each month is considered as a pigeon-hole, then $n = 12$ and $k + 1 = 4$ or $k = 3$.

Hence if one takes $k_{n+1} = 37$ students (pigeons), four of them are born in the same month.

Problem 2.48. *Show that in any set of eleven integers, there are two whose difference is divisible by 10.*

Solution. Consider the set A of pigeons is the given set of eleven of integers and the set of pigeon-holes is the set B = {0, 1, ..., 9} of possible right hand digits.

The relevant function $f: A \to B$ takes each integer to its right hand digit and $|A| > |B|$.

Hence, by the pigeon-hole principle, two of inetgers have the same right-hand digit, thus their difference is divisible by 10.

Problem 2.49. *Show that in any room of people who have been doing handshaking there will always be at least two people who have shaken hands the same number of times.*

Solution. Let the pigeon holes be labelled with the different numbers of hands shaken and we put the people (objects) into their correct pigeon-hole.

Suppose there are n people, then, since people only shakehands with each person at most once, the labels on the pigeon holes will go from 0 to $n - 1$.

This is, we have n holes and n people.

It is not possible for the 0th and the $(n - 1)$th holes both to be occupied, because if one person has shaken hands with no body then these cannot be any one person who has shaken hands with every other person.

Thus, we have atmost $n - 1$ holes occupied at any one time.

Hence by the pigeon hole principle at least one of the holes has two occupants, which shows that there are atleast two people who have shaken hands the same number of times.

Problem 2.50. *There are 500 boxer with apples. Each contains no more than x apples. Find the maximum possible value of x, such that one can for sure find 3 boxes containing equal numebr of apples.*

Solution. Let us place these 500 boxes (pigeons) in pigeon-holes marked with number according to the number of apples in a box.

Let there are pigeon-holes marked with 0 for empty boxes, 1 for boxes containing one apple, 2 for boxes containing two apples, and so on up to x.

Thus, we have $x + 1$ pigeon holes, applying the pigeon hole principle we get

$$\left[\frac{500-1}{x+1}\right] + 1 = 3$$

or

$$\left[\frac{500-1}{x+1}\right] = 2$$

$$\Rightarrow \qquad 499 = 2x + 2$$

$$\Rightarrow \qquad x = \left[\frac{499-2}{2}\right] = 248.$$

Therefore, the maximum possible value for x is 248.

Problem 2.51. *15 students wrote a dictation. Tarun made 13 errors. Each of the other students made less than that number of errors. Prove that at least two students made equal number of errors.*

Solution. Since maximum numebr of errors committed by any student is 13, there are students who might have made no errors, some may have made one error and so on.

But, none of the students has made more than 13 errors.

Let these are 14 chairs (pigeon-holes) marked 0, 1, 2, ..., 13. We let the students be seated in the chair having mark equal to the number of errors committed by the student.

After 14 students have been seated, there is one student left who will be sharing chair with any of the 14 students.

This means that there are atleast two students who have made same number of errors.

Here, n = number of pigeons = number of students = 15

m = number of pigeon-holes = 14 (count of number from 0 to 13, possible numebr of errors by different student).

Thus using the pigeon-hole principle we have atleast one pigeon hole containing atleast

$$\left[\frac{n-1}{m}\right]+1=\left[\frac{15-1}{14}\right]+1=1+1=2 \text{ students.}$$

Problem 2.52. *Prove that, given any 12 natural numbers, we can choose two of them such that their difference is divisible by 11.*

Solution. If two numbers a and b have the same remainder upon division by a number C then the difference $a-b$ is divisible by C.

We shall use this property of integers to solve this type of problems. There are 11 possible remainders upon division by 11.

There are 0, 1, 2, ..., 10. Let these be the pigeon holes.

We have 12 pigeons (natural numbers).

If we put these pigeons in the 11 pigeon holes, then we have one pigeon hole containing at least

$$\left[\frac{12-1}{11}\right]+1=2 \text{ integers yielding same remainder when divided by 11.}$$

Using the property of integers, as stated above, we conclude that we can choose two of the 12 natural numbers such that their difference is divisible by 11.

Problem 2.53. *There are pencils in the box, 10 red ones, 8 blue, 8 green 4 yellow. Let us take, with eye closed, some number of pencils from the box. What is the least number of pencils we have to take in order to ensure that we get at least 4 pencils of the same colour ?*

Solution. Let us consider 4 pigeon holes corresponding to four colours. Then, our problem is to find the number of pigeons (pencils) so that when placed in holes, one of the holes must contain at least four pigeons (pencils).

$m = 4$, we have to find n such that

$$\left[\frac{n-1}{4}\right]+1=4.$$

or $$n-1=12 \quad \Rightarrow \quad n=13.$$

Therefore, atleast 13 pencils are to be picked up to ensure that we get atleast 4 pencils of same colour.

Problem 2.54. *There are 33 students in the class, and the sum of their ages is 430 years. Is it true that for some 20 students the sum of their ages is more than 260 years ?*

Solution. Consider the average age of students which is 430/33 = 13.03 years.

Now some students may be of age less than this average, same may of average age and some other may be of age greater than average.

Let there be three pigeon holes L, A and G for less than, average and greater than ages respectively.

If we put the 33 pigeons (students) is these three pigoen-holes then there are three possibilities.

(*i*) all three pigeon-holes get equal number of pigeons.

or (*ii*) the distribution is skewed toward L.

or (*iii*) the distribution is skewed toward G.

In the first case, we can always pick up 20 students (all from G and 9 from A) so that their ages will that more than 260 years.

In the third case too we can pick up 20 students (all from G, remaining from A and, if any more required then from L) so that the ages will total more than 260 years.

In second case, we use the fact that if any number is less than the average then there exists a number which will be greater than the average.

Thus if there are students in L, then these must be some students in G whose age(s) will maintain the average.

By picking up 20 students as in the cases of first and third, theirs total age will exceed 260 years.

Thus, we can always find 20 students such that the sum of their ages is more than 260 years.

Problem 2.55. *There are 50 people in a room. Some of them are acquainted with each other, some not. Prove that there are two persons in the room who have equal nubmers of acquaintances.*

Solution. If there is a person in the room who has no acquaintances at all then each of the other persons in the room may have either 1 or 2, or 3, ... or 48 acquaintances, or do not have acquaintances at all. Therefore we have 49 pigeon holes numbered 0, 1, 2, 3, ..., 48.

If we put 50 people one by one in the pigeon holes according to the number of acquaintances the person has, we have one pigeon hole containing at least.

$$\left[\frac{50-1}{49}\right]+1=1+1=2. \text{ Pigeons.}$$

Thus, there are at laest two persons in the room who have equal number of acquaintances.

Problem 2.56. *51 points where placed in an arbitrary way into the square of side 1. Prove that some 3 of these points can be covered by a circle of radius* $\frac{1}{7}$.

Solution. Let us divide the square into 25 squares of side $\frac{1}{5}$ each.

Then at least one of thee small squares will contain at least $\left[\frac{51-1}{25}\right]+1=3$ points (pigeons).

Now the circle circumvented around the square with three points also contains these three points and has radius $r=\sqrt{\left(\frac{1}{10}\right)^2+\left(\frac{1}{10}\right)^2}=\sqrt{\frac{2}{100}}=\frac{1}{\sqrt{50}}<\frac{1}{\sqrt{49}}=\frac{1}{7}$.

Hence, there are atleast three points inside a circle of radius $\frac{1}{7}$.

Problem 2.57. *Prove that some integral power of 2 has the decimal expansion which starts with the digits 1999, i.e., there exists integer n such that $2^n = 1999$ ......*

Solution. Let there is a number the decimal expansion of which starts with the digits 1999.....

If we prove that there exists a number starting with the digits 1999, ..., such that $\log_2 1999$ ...... is an integer, then the problem is solved.

Problem 2.58. *We are given five arbitrary natural numbers a, b, c, d, e. Prove that either one of them is divisible by 5, or the sum of several numbers in a row is divisible by 5.*

Solution. Let us consider five sums a, $a+b$, $a+b+c$, $a+b+c+d$, $a+b+c+d+e$.

When a number is divided by 5, the possible remainders are 0, 1, 2, 3, 4.

Let the number of pigeon holes is 5.

If we place the five sums one by one is these holes then either each hole will get one pigeon or some of holes will get more than one and some none.

In the first case, the sum corresponding to holes 0 is divisible by 5.

And, in the second case, the difference of the two sums falling int he same-hole is divisible by 5. Also note that the difference of the sums is again one of the natural number or sum of one or more given natural numbers in row.

Thus we have proved that either one of the numbers or the sum of several numbers in a row is divisible by 5.

Problem 2.59. *A chess player wants to prepare for a championship match by playing some practice games in 77 days, she wants to play atleast one game a day but no more than 132 game altogether. Show that there is a period of consecutive days within which she plays exactly 21 games, irrespective of the scheduling of her practice games.*

Solution. Let a_i be the total number of games she plays up through the ith day.

Obviously, the sequence $a_1, a_2, a_3, \ldots\ldots a_{77}$ is a monotonically increasing sequence, where $a_1 \geq 1$ and $a_{77} \leq 132$.

Now let us take another sequence $b_1, b_2, b_3, \ldots\ldots, b_{77}$ such that $b_i = a_i + 21$ and $b_{77} \leq 153$ (notice, here the construction of sequence b).

This sequence b is also monotonically increasing.

We have now 154 numbers (77 from sequence a and 77 from sequence b) and the values of these numbers range from 1 to 153.

Therefore from pigeon-hole principle, we have at least two nubmers having same value.

Since both a and b are monotonically increasing, two numbers of a and two numbers of b cannot be equal.

So, there exists a number in the sequence a, which is equal to a number in the sequence b.

Let a_i is equal to b_j . *i.e.*, $a_i = a_j + 21$.

This proves that there are consecutive days i through j during which she played exactly 21 games.

Problem 2.60. *Prove that there exists a multiple of 1997 whose decimal expansion contains only digits 1 and 0.*

Solution. Let us write 1998 numbers like

$$1, 11, 111, 1111, ..., 111, ... 111.$$

Each of these numbers yields 0 or 1 or 2 or ...... 1996 as remainder when divided by 1997.

Let these 1997 possible remainders ae the pigeon-holes and the above 1998 written numbers are pigeons.

If we place the pigeons one by one in the holes then we have atleast one hole containing at least

$\left[\dfrac{1998-1}{1997}\right] + 1 = 2$ numbers from the 1998 numbers, listed as above.

The difference of these two numbers is a number containing only digits 1 and 0.

2.5 DERANGEMENTS. (Nothing Is In Its Right Place)

The principle of inclusion-exclusion will be used to count the permutations of n objects that leave no objects in their original positions.

A **derangement** is a **permutation** of objects that leaves no object in its original position.

We want to know in how many ways we can arrange the numbers 1, 2, 3, ..., 10 so that 1 is not in first place (its natural position), 2 is not in second place (its natural position), and 10 is not in tenth place (its natural position). These arrangements are called the derangements of 1, 2, ..., 10.

The principle of inclusion-exclusion provide the key to **calculating the number of derangements.** For each $1 \le i \le 10$, an arrangement of 1, 2, 3, ..., 10 is said to satisfy condition C_i of integer i is in the ith place.

The number of derangements of 1, 2, 3, 4 is

$$d_4 = 4\,!\left[1-1+\left(\frac{1}{2}!\right)-\left(\frac{1}{3}!\right)+\left(\frac{1}{4!}\right)\right]$$

$$= 4\,!\left[\left(\frac{1}{2}!\right)-\left(\frac{1}{3}!\right)+\left(\frac{1}{4\,!}\right)\right]$$

$$= (4)(3) - 4 + 1 = 9.$$

These nine derangements are

2143	3142	4123
2341	3412	4312
2413	3421	4321

Among the 24 – 9 = 15 permutations of 1, 2, 3, 4 that are not derangements one finds 1234, 2314, 3241, 1342, 2431, and 2314.

The number of derangements of 1, 2, ..., 10 is

$$d_{10} = N(\bar{C}_1\, \bar{C}_2 \;......\; \bar{C}_{10}) = 10\,! - \binom{10}{1} 9! - \binom{10}{1} 9\,! + \binom{10}{2} 8\,!$$

$$- \binom{10}{3} 7\,! + ... + \binom{10}{10} 0\,!$$

$$= 10\,! \left[1 - \binom{10}{1}\left(\frac{9\,!}{10\,!}\right) + \binom{10}{2}\left(\frac{8\,!}{10\,!}\right) - \binom{10}{3}\left(\frac{7\,!}{10\,!}\right) + ... + \binom{10}{10}\left(\frac{0\,!}{10\,!}\right)\right]$$

$$= 10\,! \left[1 - 1 + \left(\frac{1}{2}!\right) - \left(\frac{1}{3}!\right) + ... + \left(\frac{1}{10\,!}\right)\right]$$

$$= (10\,!)\,(e^{-1}).$$

(In elementary calculus the maclaurin series for the exponential function is given by

$$e^x = 1 + x + \frac{x^2}{2!} + \frac{x^3}{3!} + ... = \sum_{n=0}^{\infty} \frac{x^n}{n\,!}$$

so $$e^{-1} = \sum_{n=0}^{\infty} \frac{(-1)^n}{n!} = 1 - 1 + \frac{1}{2!} - \frac{1}{3\,!} + ...$$

$$= 0.36788$$

for all $k \in Z^+$, if $k \geq 7$, then $\left.\sum_{n=0}^{k} \frac{((-1)^n)}{n!}\right)$.

Theorem 2.6. *The number of derangements of a set with n elements is*

$$D_n = n\,! \left[1 - \frac{1}{1!} + \frac{1}{2!} - \frac{1}{3\,!} + ... + (-1)^n \frac{1}{n\,!}\right].$$

Proof. Let a permutation have property P_i if it fixes element i. The number of derangements is the number of permutations having none of the properties P_i for $i = 1, 2, ..., n$

This means that $D_n = N(P_1'\, P_2' \;......\; P_n')$

Using the principle of inclusion-exclusion, it follows that

$$D_n = N - \sum_{i} N(P_i) + \sum_{i<j} N(P_iP_j) - \sum_{i<j<k} N(P_iP_jP_k) + \ldots\ldots + (-1)^n N(P_1P_2 \ldots\ldots P_n)$$

where N is the number of permutations of n elements.

This equation states that the number of permutations that fix no elements equals the total number of permutations, less the number that fix at least one element, plus the number that fix atleast two elements, less the number that fix atleast three elements, and so on. All the quantities that occur on the right-hand side of this equation will now be found.

First, note that $N = n\,!$, since N is simply the total number of permutations of n elements.

Also, $N(P_i) = (n-1)\,!$. This follows from the product rule, since $N(P_i)$ is the number of permutations that fix element i, so that the ith position of the permutation is determined, but each of the remaining positions can be filled arbitrarily.

Similarly, $N(P_iP_j) = (n-2)\,!$.

Since this is the number of permutations that fix elements i and j, but where the other $n-2$ elements can be arranged arbitrarily.

In general, note $N(P_{i_1} P_{i_2} \ldots\ldots P_{i_m}) = (n-m)\,!$, because this is the number of permutations that fix elements $i_1, i_2, \ldots\ldots i_m$, but where the other $n-m$ elements can be arranged arbitrarily.

Because there are $C(n, m)$ ways to choose m elements from n, it follows that

$$\sum_{i \le j \le k} N(P_i) = C(n, 1)\,(n-1)\,!$$

$$\sum_{1 \le i < j \le n} N(P_iP_j) = C(n, 2)\,(n-2)\,!,$$

and in general,

$$\sum_{1 \le i_1 < i_2 < \ldots\ldots < i_m \le n} N(P_{i_1} P_{i_2} \ldots\ldots P_{i_m}) = C(n, m)\,(n-m)\,!$$

Consequently, inserting these quantities into our formula for D_n gives

$$D_n = n\,! - C(n, 1)\,(n-1)\,! + C(n, 2)\,(n-2)\,! - \ldots\ldots$$

$$\ldots\ldots + (-1)^n\, C(n, n)\,(n-n)\,!$$

$$= n\,! - \frac{n\,!}{1!(n-1)!}(n-1)\,! + \frac{n\,!}{2!(n-2)!}(n-2)\,! - \ldots\ldots + (-1)^n \frac{n\,!}{n\,!0\,!}\,0\,!$$

Simplifying this expression gives

$$D_n = n\,!\left[1 - \frac{1}{1!} + \frac{1}{2!} - \ldots\ldots + (-1)^n \frac{1}{n\,!}\right].$$

Problem 2.61. *Show that $D_n = n\,D_{n-1} + (-1)^n$, for $n \ge 2$ and use this result to evaluate D_n (D_n can be computed without using the inclusion-exclusion principle).*

Solution. The substitution $E_n = D_n - n\, D_{n-1}$ takes the recursion formula of

$$D_n = (n-1)(D_{n-1} + D_{n-2}), \text{ where } n \geq 3 \text{ into } E_n = -E_{n-1}.$$

Since $E_2 = D_2 - 2D_1 = 1$, the latter recursion formula has the solution $E_n = (-1)^n$, for $n \geq 2$.

By interation, $D_n = n((n-1)\, D_{n-2} + (-1)^{n-1}) + (-1)^n$

$$= n(n-1)\, D_{n-2} + n(-1)^{n-1} + (-1)^n$$

$$= n(n-1)(n-2)\, D_{n-3} + n(n-1)(-1)^{n-2} + n(-1)^{n-1} + (-1)^n \;......$$

$$= n(n-1) \;......\; (3) - n(n-1) \;......\; (4) + n(n-1) \;......\; (5) - \;......$$

$$+ (-1)^{n-2}\, n(n-1) + (-1)^{n-1}\, n + (-1)^n$$

$$= n\,! \left[\frac{1}{2!} - \frac{1}{3!} + \;......\; + \frac{(-1)^n}{n!}\right].$$

Problem 2.62. *Show that $D_n = (n-1)(D_{n-1} + D_{n-2})$, where $n \geq 3$.*

Solution. Consider those derangements of X = {1, 2, ... n} in which r occupies the first position. Then either 1 does not occupy the rth position or it does.

There are D_{n-1} derangements of the former type (with respect to X – {r}, position r functions as position 1) and D_{n-2} of the latter type.

The element r can be chosen in $n - 1$ ways.

Problem 2.63. *Out of 30 students in a dormitory, 15 take an art course, 8 take a biology coruse, and 6 take a chemistry course. It is known that 3 students take all 3 courses. Show that 7 or more students take none of the course.*

Solution. Let A be the set of students who take the art course, B be the set for biology, and C be the set for chemistry. Then $S_1 = 15 + 8 + 6 = 29$ and $S_3 = 3$ (given).

Let x be the number who are not in any of these courses.

$$x = 30 - 29 + S_2 - 3 = S_2 - 2$$

Now the intersection of the 3 sets is a subset of the intersection of any 2 of the sets.

Therefore, $S_2 \geq 3 + 3 + 3 = 9$ or $x \geq 9 - 2 = 7$.

Problem 2.64. *Find the number of 4-digit positive integers the sum of the digits of which is 31.*

Solution. We wish to count the solutions in integers of the system

$$a + b + c + d = 31,\; 0 \leq a, b, c, d \leq 9$$

The answer is

$$C(34, 3) - C(4, 1)\, C(24, 3) + C(4, 2)\, C(14, 3) - C(4, 3)\, C(4, 3) + 0.$$

Problem 2.65. *Find the number of injections (one-to-one mappings) from a finite set X of n elements to X such that each has at least 1 fixed point.*

Solution. The total number of injections is $n\,!$

Any derangement of X defines a unique injection which has no fixed point, and vice versa.

Thus, the answer is $n\,! - D_n$.

Problem 2.66. The Hatcheck Problem

A new employee checks the hats of n people at a restaurant, forgetting to put chain check numbers on the hats. When customers return for their hats, the checker gives them back hats chosen at random from the remaining hats. What is the probability that no one receives the correct hat ?

Remark. The answer is the number of ways the hats can be arranged so that there is no hat in its original position divided by n !, the number of permutations of n hats. We will return to this problem after we find the number of permutations of n objects that leave no objects in their original position.

Solution. The probability that no one receives the correct that is $\frac{D_n}{n!}$.

This probability is

$$\frac{D_n}{n!} = 1 - \frac{1}{1!} + \frac{1}{2!} - \ldots\ldots + (-1)^n \frac{1}{n!}.$$

The values of this probability for $2 \le n \le 7$ are displayed in table below.

TABLE : The probability of a derangement

n	2	3	4	5	6	7
$\frac{D_n}{n!}$	0.50000	0.33333	0.37500	0.36667	0.36806	0.36786

Using methods from calculus it can be shown that

$$e^{-1} = 1 - \frac{1}{1!} + \frac{1}{2!} - \ldots\ldots + (-1)^n \frac{1}{n!} + \ldots\ldots \sim 0.368.$$

Since this is an alternating series with terms tending to zero, it follows that as n grows without bound, the probability that no one receives the correct hat.

Converges to $e^{-1} \sim 0.368$.

In fact, this probability can be shown to be within $\frac{1}{(n+1)!}$ of e^{-1}.

Problem 2.67. *Let n books be distributed to n students. Suppose that the books are returned and distributed to the students again later on. In how many ways can the books be distributed so that no student will get the same book twice ?*

Solution. The first time the books are distributed n ! ways. The second time D_n ways. Hence, the total number of ways is given by

$$n!\, D_n = (n!)^2 \left[1 - \frac{1}{1!} + \frac{1}{2!} + \ldots\ldots + (-1)^n \frac{1}{n!}\right].$$

Problem 2.68. *Find the number of derangements of the integers from 1 to 10 inclusive, satisfying the condition that the set of elements in the first 5 places is :*

(*a*) *1, 2, 3, 4, 5, in some order*

(*b*) *6, 7, 8, 9, 10, in some order.*

Solution. (*a*) The integers 1, 2, 3, 4 and 5 can be placed into the first 5 places in D_5 ways, the last 5 integers 6, 7, 8, 9 and 10 can be placed in the last 5 places in D_5 ways. Hence, the answer is $D_5 \cdot D_5 = 1936$.

(*b*) Any arrangement of 6, 7, 8, 9 and 10 in the first 5 places is a derangement so there are 5 ! possibilities the same is true for the integers 1, 2, 3, 4 and 5 in the last 5 places.

Hence there are $(5\,!)^2 = 14{,}400$ such derangements.

Problem 2.69. *There are 6 pairs of children's gloves in a box. Each pair is of a different colour "Suppose the right gloves are distributed at random to 6 children, and then the left gloves also are distributed to them at random. Find the probability that*

(*a*) *no child gets a matching pair,*

(*b*) *everybody gets a matching pair*

(*c*) *exactly 1 child gets a matching pair, and*

(*d*) *atleast 2 children gets matching pairs.*

Solution. The right gloves can be distributed in 6 ! ways, after which the left gloves can be distributed in 6 ! ways.

Thus there are (6 !) (6 !) equiprobable outcomes.

(*a*) For each of the 6 ! distributions of right gloves there are D_6 distributions of left gloves that result in no matching pairs.

The required probability is therefore $(6\,!)\, D_6/(6\,!)^2 = D_6/6\,!$.

(*b*) For each of the 6 ! distributions of right gloves there is 1 distribution of left gloves that yields 6 matching pairs. The required probability is $(6\,!)\,(1)/(6\,!)^2 = 1/6\,!$.

(*c*) For each of the 6 ! distributions of right gloves there are (1) D_5 distributions of left gloves that give Annie-and Annie alone-a matching pair.

Hence the required probability is

$$\frac{6[(6\,!)(1)\, D_5]}{(6\,!)^2} = \frac{D_5}{5\,!}.$$

(*d*) Using the results of (*a*) and (*c*) probability

$$= 1 - \frac{D_6}{6\,!} - \frac{D_5}{5\,!}.$$

Problem 2.70. *Find the number of ways the integers 0, 1, 2, ..., n – 1 can be arranged in a circle so that no arrangement, when read clockwise has a block of 2 consecutive integers or the block [n – 10].*

Solution. Let X be the set of circular permutations of $\{0, 1, 2, \ldots\ldots, n-1\}$,

$$n(X) = (n-1)\,!$$

If we operate modulo n, the n excluded blocks can be notated as $i\ \overline{i+1}$ $(i = 0, 1, 2, \ldots\ldots, n-1)$.

Let A_i be the subset of X in which each circular permutation contains the block $i\ \overline{i+1}$.

To evaluate S_r, for $r < n$, we may suppose the r blocks selected to be disjoint.
Thus leading to $(r + n - 2r - 1)! = (n - r - 1)!$ circular permutations.
i.e., $S_r = C(n, r)(n - r - 1)!$, $r = 1, 2, \ldots\ldots, n - 1$.
Clearly, $S_n = 1$, the answer is

$$\sum_{r=0}^{n-1} (-1)^r\ C(n, r)(n-r-1)! + (-1)^n = n! \sum_{r=0}^{n} \frac{(-1)^r}{r!(n-r)}.$$

Problem 2.71. *Find the number, T_n, of permutations of Z = {1, 2, ……, n} such that no permutation contains a block of 2 consecutive integers.*

Solution. Let X denote the set of all permutations of Z,

$$n(X) = n!$$

Let A_i $(i = 1, 2, \ldots\ldots, n-1)$ represent the subset of X consisting of all permutations in which the block $i\ \overline{i+1}$ occurs.

If one choose r mutually disjoint blocks, these r objets and the remaining $n - 2r$ elements of z together generate $(n - r)!$ permutations.

This same number, $(n - r)!$ is obtained for any selection of r blocks disjoint or not.

Hence, $S_r = C(n - 1, r)(n - r)!$, for $r = 1, 2, \ldots\ldots, n - 1$

and
$$T_n = \sum_{r=0}^{n-1} (-1)^r\ C(n-1, r)\,(n-r)! = \sum_{r=0}^{n-1} (-1)^r \left[\frac{n!}{r!} - \frac{(n-1)!}{(r-1)!}\right]$$

$$= n! \sum_{r=0}^{n-1} \frac{(-1)^r}{r!} + (n-1)! \sum_{s=0}^{n-2} \frac{(-1)^s}{s!}$$

We see that,

$$T_n = [D_n - (-1)^n] + [D_{n-1} - (-1)^{n-1}]$$
$$= D_n + D_{n-1}$$

or by $D_n = (n-1)(D_{n-1} + D_{n-2})$,

$$T_n = \frac{D_{n+1}}{n}.$$

Problem 2.72. *Tabulate D_n and T_n for $n = 1(1)10$.*

Solution. Start with $D_1 = 0$ and $T_1 = 0$

To compute D_n use the relation $D_n = (n-1)\,T_{n-1}$.

To compute T_n use the relation $T_n = D_n + D_{n-1}$ (see table)

n	D_n	T_n
1	0	1
2	1	1
3	2	3
4	9	11
5	44	53
6	265	309
7	1845	2119
8	14833	16687
9	133496	148329
10	1334961	1468457

Problem 2.73. *Each of n woman who attend a banquet checks her coat and hat with the receptionist on arrival. Upon leaving, each woman is given a coat and a hat at random. Find the number of ways these coats and hats may be distributed such that*

(a) nobody gets back either her coat or her hat, and

(b) nobody gets back both her coat and her hat.

Solution. (*a*) The number of derangements for coats is D_n and the numebr of derangements for hats is D_n.

Because the handing out of coats is independent of the handing out of hats, there are $(D_n)^2$ distributions with the specified property.

(*b*) Let A_j ($j = 1, 2, ..., n$) be the subsets of distributions in which woman j gets back both her coat and her hat.

Then, $\quad n(X) = (n\,!)^2,\ S_r = C(n, r)\,[(n-r)\,!]^2$

$r = 1, 2, ..., n$

and the answer is given by

$$n(X) - S_1 + ... + (-1)^n S_n.$$

Problem 2.74. *Use a combinatorial argument to establish the identity*

$$\sum_{r=0}^{n} \frac{D_r}{r\,!\,(n-r)\,!} = 1.$$

Solution. The n ! permutations of an n set may be classified according as they have $n, n-1, n-2, ...\ 0$, fixed points.

Thus, $n\,! = C(n, n)\,D_0 + C(n, n-1)\,D_1 + C(n, n-2)\,D_2 + ... + C(n, 0)\,D_n$

Divide through by n !, to obtain

$$\sum_{r=0}^{n} \frac{D_r}{r\,!\,(n-r)\,!} = 1.$$

2.6 ROOK POLYNOMIALS AND ARRANGEMENTS WITH FORBIDDEN POSITIONS

In the game of chess a **rook** is a piece that can capture another piece or pawn situated in the same row or same column, provided that there is no other piece or pawn between the two. The problem of non taking rooks asks for the number of ways r_k (C), of positioning k identical rooks on a **chessboard** C, with n squares, so that no rook can capture another rook.

The polynomial

$$R(x, C) \equiv 1 + r_1x + r_2x^2 + \ldots\ldots + r_nx^n$$

is called the **rook polynomial of C.**

Given a chessboard C in which a special square has been distinguished. Let D be the board obtained from C by deleting the row and the column of the special square, let E be the board obtained from C by deleting only the special square.

Either there is a rook in the special square or there is no rook in that square. The number of ways of placing k non taking rooks in C, one of which is in the special square, is r_{k-1} (D), the number of ways of placing k non taking rooks so that no rook is in the special square of r_k (E).

$$r_k\,(C) = r_{k-1}\,(D) + r_k\,(E)$$

$$x^k\, r_k\,(C) = x^k\, r_{k-1}\,(D) + x^k\, r_k\,(E)$$

that is $R(x, C) = x\,R(x,D) + R(x, E)$

(expansion formula for the Rook polynomial).

For each element i of $N = \{1, 2, \ldots\ldots, n\}$ let there be specified a (possibly empty) subset, A_i, of N. Required is the number of bijective functions f from N to N such that, for every i, $f(i)$ does not belong to A_i. Because f is a permutation of N, the elements of A_i are called the **forbidden positions for $i \in$ N.**

This problem may be transformed into one of non-taking rooks on a subboard C′ of an $n \times n$ chessboard C, provided that one agrees that 2 rooks on C′ are non taking if and only if they are non taking on C. On this understanding, there are 2 non taking rooks on the subboard of Fig. 2.10.

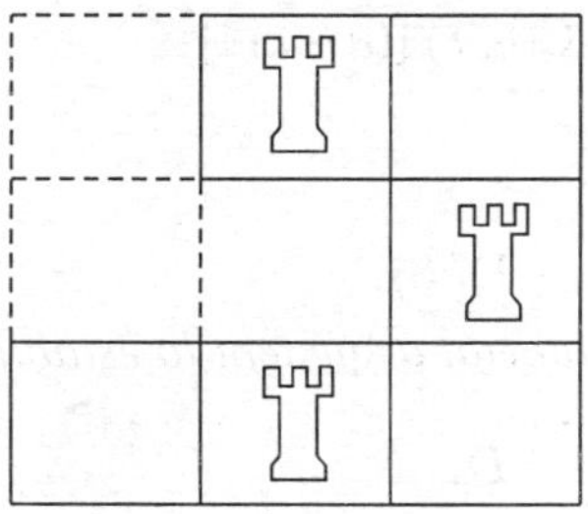

Fig. 2.10.

Let square (i, j) of C be coloured red if $j \in A_i$; otherwise, let it be coloured green. The green squares compose our subboard C′, it is evident that every placement of n nontaking roosk on C′ corresponds to a permutation f of N with the desired property, there exists r_n (C′) such permutations.

A placement of k non taking rooks on C″ corresponds to a family of $(n - k)$! permutations with k elements in forbidden positions. Thus, in the sieve formula,

$$S_k = r_k\,(C'')(n-k)\,!, \qquad k = 0, 1, \ldots\ldots, n$$

and the required number is

$$r_n(C') = \sum_{k=0}^{n} (-1)^k r_k \ (C'')\,(n-k)\,!$$

$$r_n(C'') = \sum_{k=0}^{n} (-1)^k r_k \,(C')\,(n-k)\,!.$$

Problem 2.75. *Find the rook polynomial of the standard 8 × 8 chessboard.*

Solution. One can choose k rows out of the 8 rows in C(8, k) ways.

After that the k rooks can be placed in these k rows so that there is exactly 1 rook in each row and no 2 rooks are in the same column, in P(8, k) ways.

Hence, for $k = 1, 2, \ldots\ldots, 8$ (and also for $k = 0$ and $k > 8$),

$r_k = C(8, k)\, P(8, k)$, giving

$$R(x, C_{8+8}) = \sum_{k \geq 0} (k\,!)\, C(8, k)^2\, x^k \,.$$

Problem 2.76. *Find the rook polynomial of a 2 × 2 board by use of the expansion formula.*

Solution. For any choice of the special square,

$$R(x, D) = 1 + x \text{ and } R(x, E) = 1 + 3x + x^2.$$

Hence,
$$R(x, C) = x(1 + x) + (1 + 3x + x^2)$$
$$= 1 + 4x + 2x^2.$$

Problem 2.77. *Find the number of permutations of a set of n elements such that exactly j elements (j = 0, 1, ..., n) are i_n forbidden positions.*

Solution. We have, the sieve formula

$$S_k = r_k(C'')\,(n-k)\,!,\ k = 0, 1, \ldots\ldots, n$$

$$r_n(C') = \sum_{k=0}^{n} (-1)^k\, r_k\,(C'')\,(n-k)!$$

and
$$e_j = s_j - C(j+1, 1)\, s_{j+1} + C(j+2, 2)\, s_{j+2} - \ldots\ldots + (-1)^{m-j}\, C(m, m-j)\, s_m$$

Replace m by n,

$$e_j = s_j - C(j+1, 1)\, s_{j+1} + C(j+2, 2)\, s_{j+2} - \ldots\ldots + (-1)^{n-j}\, C(n, n-j)\, s_n$$

Here $s_k = r_k\,(n-k)\,!$, r_k being the number of ways of placing k non taking rooks on the forbidden (red) subboard.

Problem 2.78. *Let R(m, n ; x) be the rook polynomial of a m × n chessboard.*

(a) Find the coefficient of x^k in this polynomial.

(b) Establish the recurrence relation

$$R(m, n\,;\, x) = R(m-1, n\,;\, x) + nxR(m-1, n-1\,;\, x)$$
$$= R(m, n-1\,;\, x) + mxR(m-1, n-1\,;\, x).$$

Solution. (*a*) The distinct rows to be occupied by k non taking rooks may be chosen in C(m, k) ways.

The distinct columns may then be chosen in $P(n, k) = k!\, C(n, k)$ ways.

Thus, the required coefficient is

$$r_k(m, n) = k!\, C(m, k)\, C(n, k).$$

(*b*) The number of ways of placing k non taking rooks, with no rook in the first row, is

$$r_k(m-1, n).$$

The number of ways of placing k non taking rooks, with a rook on the jth square of the first row is

$$r_{k-1}(m-1, n-1).$$

Therefore, $r_k(m, n) = r_k(m-1, n) + \sum_{j=1}^{n} r_{k-1}(m-1, n-1)$

$$= r_k(m-1, n) + n r_{k-1}(m-1, n-1)$$

which implies $R(m, n\,; x) = R(m-1, n\,; x) + nxR(m-1, n-1\,; x)$.

Problem 2.79. *Find the number of 6-letter words than can be formed using the letters A, B and C (each twice) in such manner that A does not appear in the first 2 positions, B does not appear in the third position, and C does not appear in the fourth and fifth positions.*

Solution. Temporarily making the letters distinguishable, we have the baord of Fig. 2.11.

	1	2	3	4	5	6
A	X	X				
A′	X	X				
B			X			
B′			X			
C				X	X	
C′	X	X		X	X	

Fig. 2.11.

The rook polynomial of the forbidden subboard is

$$(1 + 4x + 2x^2)^2 (1 + 2x) = 1 + 10x + 36x^2 + 56x^3 + 36x^4 + 8x^5$$

Then, $r_n(C') = \sum_{k=0}^{n} (-1)^k\; r_k(C'')\,(n - K)!$, yields

$$r_6 = (6\,!)\,(1) - (5\,!)\,(10) + (4\,!)\,(36) - (3\,!)\,(56) + (2\,!)\,(36) - (1\,!)\,(8)$$
$$= 112.$$

But A and A′, etc., are really not distinguishable. Thus, the answer is 112/(2 !)(2 !)(2 !) = 14.

2.7 HIT POLYNOMIAL

The generating Polynomial of the numbers e_j,

$$e_j = s_j - C(j+1, 1)\, s_{j+1} + C(j+2, 2)\, s_{j+2} - \ldots\ldots + (-1)^{n-j}\, C(n, n-j)\, s_n.$$

$H(x) = e_0 + e_1x + e_2x^2 + \ldots\ldots + e_n x^n$ is known as the hit polynomial, since an object in a forbidden position constitutes a hit, then

$$H(x) = \sum_{k=0}^{n} S_k\,(x-1)^k = \sum_{k=0}^{n} r_k\,(n-k)!\,(x-1)^k.$$

Problem 2.80. *Let A and B be two disjunct (or disjoint) subboards of a given chessboard C, this means that no row or column of C contains both a square of A and a square of B. Show that the rook polynomial of the subboard A ∪ B is given by*

$$R(x, A \cup B) = R(x, A)\, R(x, B).$$

Solution. If k non taking rooks are placed on A ∪ B, there must exist a non negative integer t such that t (non taking) rooks are on A and $k - t$ rooks are on B.

Since A and B are disjunct, the way the t rooks are placed in independent of way the $k - t$ rooks are placed.

Hence, the number of ways of placing the k rooks is

$$r_k\,(A \cup B) = \sum_{t=0}^{k} r_t\,(A)\, r_{kt}\,(B) \qquad \ldots(1)$$

The left side of (1) is the coefficient of x^k in R(x, A ∪ B), the right side is the coefficient of x^k in R(x, A) R(x, B).

Problem 2.81. *Find the rook polynomial of the board C that is obtained from a 3 × 3 board by deleting the middle square in the first row and the first and the last squares in the third row.*

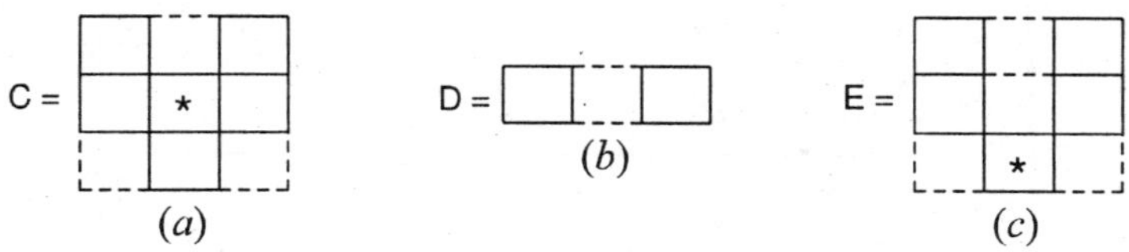

Fig. 2.12.

Solution. Choosing the central square of C as the special square, we have

$$R(x, C) = xR(x, D) + R(x, E) = x(1 + 2x) + R(x, E).$$

Further decomposing E relative to its third-row square,

$$R(x, E) = x(1 + 4x + 2x^2) + (1 + 4x + 2x^2)$$

Thus,
$$R(x, C) = x(1 + 2x) + (x + 1)(1 + 4x + 2x^2)$$
$$= 1 + 6x + 8x^2 + 2x^3.$$

Problem 2.82. *Five workers (P, Q, R, S, T) are to be assigned to 4 jobs (a, b, c, d), with the forbidden jobs in Fig. 2.13. Find the hit polynomial.*

	a	b	c	d	δ
P			X		
Q	X	X			
R				X	
S				X	
T			X		

Fig. 2.13.

Solution. If this 5 × 4 assignment problem is to be viewed as a problem in restricted permutations (bijective mappings), it is necessary to append a fifth, dummy job (δ) which is open to all 5 workers.

Clearly, any hit in the 5 × 5 problem is a hit in the 5 × 4 problem, and conversely.

Transforming the 5 × 5 tableau by a row-c-permutation and Applying

$$R(x, A \cup B) = R(x, A)\, R(x, B),$$

$$R(x) = (1 + 2x)^3 = 1 + 6x + 12x^2 + 8x^3$$

Hence, by $\quad H(x) = e_0 + e_1 x + e_2 x^2 + \ldots\ldots + e_n x^n,$

$$H(x) = 1(5\,!) + 6(4\,!)(x - 1) + 12(3\,!)(x - 1)^2 + 8(2\,!)(x - 1)^3$$

$$= 32 + 48x + 24x^2 + 16x^3.$$

Problem 2.83. *It is evident that the rook polynomial of an arbitrary board is invariant under permutations of rows and/or columns. Use this fact to compute the rook polynomial of the forbidden subboard in Fig. 2.14.*

	1	2	3	4	5	6
1		X		X		
2	X					
3			X			
4	X				X	
5		X		X		
6						X

Fig. 2.14.

Solution. The row-and column-permutation, indicated in Fig. 2.15 transform the forbidden subboard into a union of 4, pairwise distinct boards.

Applying $R(x, A \cup B) = R(x, A)\, R(x, B)$,

$$R(x) = (1 + 4x + 2x^2)(1 + x)(1 + 3x + x^2)(1 + x)$$
$$= 1 + 9x + 30x^2 + 47x^3 + 37x^4 + 14x^5 + 2x^6$$

	2	4	3	1	5	6
1	X	X				
5	X	X				
3			X			
2				X		
4				X	X	
6						X

Fig. 2.15.

Problem Set 2.1

1. In a survey of 800 voters, the following information was found, 300 were college educated, 260 were from high-income families, 325 were registered Democrats, 184 were college educated and from high-income families, 155 were college educated and registered Democrats, 165 were from highincome families and were registered Democrats, 94 were college educated, from high-income families, and were registered Democrats.

   Let E = {Voters who were college educated}

   I = {Voters who were from high-income families}

   D = {Voters who were registered Democrats}

   Draw a Venn diagram and list the number of elements in the 8 different regions of the diagram.

2. A certain computer enter employs 100 computer programmers. Of these 47 cm program in FORTRAN, 35 in pascal and 23 can program in both languages. How many can program in neither of these 2 languages ?

3. An advertising agency has 1,000 clients. Suppose that T is the set of clients that use television advertising, R is the set of clients that use radio advertising and N is the set of clients who use newspaper advertising. Suppose that $|T| = 415$, $|R| = 350$, $|N| = 280$, 100 clients use all 3 types of advertising, 175 use television and radio, 180 use radio and newspapers and $|T \cap N| = 165$.

   (*a*) Find $|T \cap R \cap \overline{N}|$

   (*b*) How many clients use radio and newspaper advertising but not television ?

(*c*) How many use television but do not use newspaper advertising and do not use radio advertising ?

(*d*) Find $|\overline{T} \cap \overline{R} \cap \overline{N}|$.

4. Suppose that, in addition to the information given in Problem 2, there are 20 employees that can program in COBOL, 12 in COBOL and FORTRAN, 11 in pascal and COBOL and 5 in FORTRAN, pascal and COBOL. How many can program in none of these 3 language ?

5. In a group of 15 pizza experts, ten like canadian becon, seven like anchovies and six like both

(*a*) How many people like at least one of these toppings ?

(*b*) How many like canadian bacon but not anchovies ?

(*c*) How many like exactly one of the two toppings ?

(*d*) How many like neither ?

6. In a recent survey of college graduates, it was found that 200 had undergraduate degree in arts, 95 had undergraduate degrees in science, and 120 had graduate degrees. Fifty-five of those with undergraduate arts degrees had also a graduate degree, 40 of those with science degrees had a graduate degree, 25 people had undergraduate degrees in both arts and science, and five people had undergraduate degrees in arts and science and also a graduate degree.

(*a*) How many people had at least on of the types of degrees mentioned ?

(*b*) How many people had an undergraduate degree in science but no the degree ?

7. Of the 2300 delegetes at a political convention 1542 voted in favor of a motion to decrease the deflicit, 569 voted in favor of a motion dealing with environmental issues, and 1197 voted in favor of a motion not to increase taxes. Of those voting in favour of the motion cocerning envirnmental issues, and 327 also voted to decrease the deficit and 92 voted not to increase taxes. Eight hundred and thirty nine people voted to decrease the deficit while also voting against increasing taxes, but of these 839, only 50 voted also in favor of the motion dealing with the environment.

(*a*) How many delegates did not vote in favor of any of the three motions ?

(*b*) How many of these who voted against increasing taxes voted in favor of neither of the other two motion ?

8. Suppose U is a set containing 75 elements and A_1, A_2, A_3, A_4 are subsets of U with the following properties :

each subset contains 28 elements; the intersection of any two of the subsets contains 12 elements ; the intersection of any three of the subsets contains 5 elements ; the intersection of all four subsets contains 1 element.

How many elements belong to none of the four subsets ?

9. How many integers between 1 and 500 are

(*a*) divisible by 3 or 5 ? (*b*) divisible by 3 but not by 5 or 6 ?

10. How many intergers between 1 and 250 are divisible by at least one of the three integers 4, 6 and 15 ?

11. How many integers between 1 and 1000 (inclusive) are not divisible by 2, 3, 5 or 7 ?

12. Let a and b be natural numbers. Show that the number of positive integers less than or equal to a and divisble by b is $\left[\frac{a}{b}\right]$.

13. Let A, B and C be sets. Prove that

$$|A \oplus B \cap C| = |A \cap B| + |B \cap C| - 2|A \cap B \cap C|.$$

14. How many integral solutions are there of $x_1 + x_2 + x_3 + x_4 = 20$ if

$$1 \leq x_1 \leq 6 \,;\, 1 \leq x_2 \leq 7,\, 1 \leq x_3 \leq 8 \text{ and } 1 \leq x_4 \leq 9\ ?$$

15. Find the number of integers between 1 and 1000 inclusive that are divisible by none of 5, 6 and 8. Note that the intersection of the set of integers divisible by 6 with the set of integers divisible by 8 is the set of integers divisible by 24.

16. Show that in any group of eight people, at least two have birthday which fall on the same day of the week in any given year.

17. In a group of 100 people, several will have their birthdays in the same month. At least how many must have birthdays in the same month ? Why ?

18. If 20 processors are interconnected and every processor is connected to at least one other, show that at least two processors are directly connected to the same number of processors.

19. In a 12 day period, a small business mailed 195 bills to customers. Show that during some period of three consecutive days, at least 49 bills were mailed.

20. Of any 26 points within a rectangle measuring 20 cm by 15 cm, show that atleast two are within 5 cm of each other.

Problem Set 2.2

1. Find the number of permutations of the integers 1 to 10 inclusive

(*a*) such that exactly 4 of the integers are in either natural positions (that is, exatly 6 of the inegers are deranged)

(*b*) such that 6 or more of the integers are deranged

(*c*) that do not have 1 in the first place, nor 4 in the fourth place, nor 7 in the seventh place

(*d*) such that no odd integer will be in the natural positon.

(*e*) that so not begin with a 1 and do not end with 10.

2. A simple code is made by permuting the letters of the alphabet with every letter being replaced by a distinct letter. How many different codes can be made is this way ?

3. Eight people enter an elevator at the first floor. The elevator discharges passengers on each successive floor until it empties on the fifth floor. How many different ways can this happen ?

4. In how many way can the letters $\{4.a, 3.b, 2.c\}$ be arranged so that all the letters of the same kind are not in a single block ?

5. A book binder is to bind 10 different books in red, blue and brown cloth. In how many ways can be do this if each colour of cloth is to be used for at least one book ?

6. At a theater 10 men check their hats. In how many ways can their hats be returned so that

(*a*) no man receives his own hat ?

(*b*) at least 1 of the man receives his own hat ?

(*c*) at least 2 of the men receives their own hats ?

7. The squares of a chessboard are painted 8 different colors. The squares of each row are painted all 8 colors and no 2 consecutive squares in one column can be painted the same colour. In how many ways can this be done ?

8. In how many ways can each of 10 people select a left glove and a right glove out of a total of 10 pairs of gloves so that no person selects a matching pair of gloves ?

9. Thirty students take a quiz. Then for the purpose of grading, the teacher asks teh students to exchange papers so that no one is grading his own paper. How many ways can this be done ?

10. How many ways are there to select a 5 card hand from a deck of 52 cards such that the hand contains at least one card in each suit ?

11. How many 13-card bridge hands have at least

(*a*) One card in each suit ?

(*b*) One void suit ?

(*c*) One of each honor card (honor cards are aces, kings, queens and jacks) ?

12. How many ways are there to assign 20 different people to 3 different rooms with at least 1 person in each room ?

13. (*a*) How many integers between 1 and 10^6 inclusive include all of the digits 1, 2, 3 and 4 ?

(*b*) How many of the numbers between 1 and 10^6 inclusive consist of the digits 1, 2, 3, 4 alone ?

14. Find D_6, D_7 and D_8.

15. Eleven books are arranged on a shelf in alphabetical order by author name. In how many can your little sister rearrange these books so that no book is in its original position ?

16. Twenty people check their hats at a theater. In how many ways can their hats be returned so that exactly one person receives his or her own hat ?

17. In how many ways can the integers 1 through 9 be permuted such that no odd integer will be in its natural position ?

18. Without any calculation, prove that

$$n! = D_n + \binom{n}{1} D_{n-1} + \binom{n}{2} D_{n-2} + \ldots + \binom{n}{n-1} D_1 + 1$$

19. Prove that $D_n = nD_{n-1} + (-1)^n$ for all $n \geq 2$.

20. In how many ways can the integers 1 through 9 be permuted such that no even integer is in its natural position ?

21. In how many ways can the integers 1 through 9 be permuted such that exactly four of the nine integers are in their natural positions ?

22. Prove that $D_n \equiv (-1)^n \pmod{n}$.

23. Prove that D_n is even if and only if n is odd.

24. Prove that $D_n = (n-1)(D_{n-1} + D_{n-2})$ for all $n \geq 3$.

25. Let $a, b, c \in Z^+$ with $C = \gcd(a, b)$. Prove that

$$\phi(ab)\,\phi(c) = \phi(a)\,\phi(b)c.$$

26. If $n \in Z^+$, prove that (*a*) $\phi(2n) = 2\phi(n)$ when n is even and (*b*) $\phi(2n) = \phi(n)$ when n is odd.

27. Let D_{18} denote the set of positive divisors of 18. For $d \in D_{18}$ let $S_d = \{n/0. < n \le 18$ and gcd $(n, 18) = d\}$

(*a*) Show that the collection S_d, $d \in D_{18}$, provides a partition of $\{1, 2, 3, 4, ..., 17, 18\}$.

(*b*) Note that $| S_1 | = 6 = \phi(18)$ and $| S_2 | = 6 = \phi(9)$. For each $d \in D_{18}$, express $| S_d |$ in terms of Euler's fit function.

28. (*a*) Find the rook polynomial for the standard 8×8 chessboard.

(*b*) Answer part (*a*) with 8 replaced by n, for $n \in Z^+$.

29. (*a*) Find the rook polynomials for the shaded chessboards in Fig. (2.16).

(*b*) Generalised the chessboard (and rook polynomial) for Fig. (2.16) (*i*).

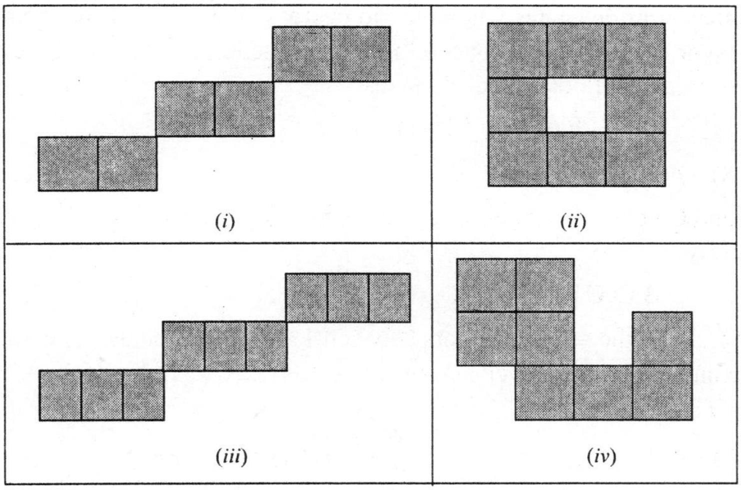

Fig. 2.16.

Answers 2.1

1.

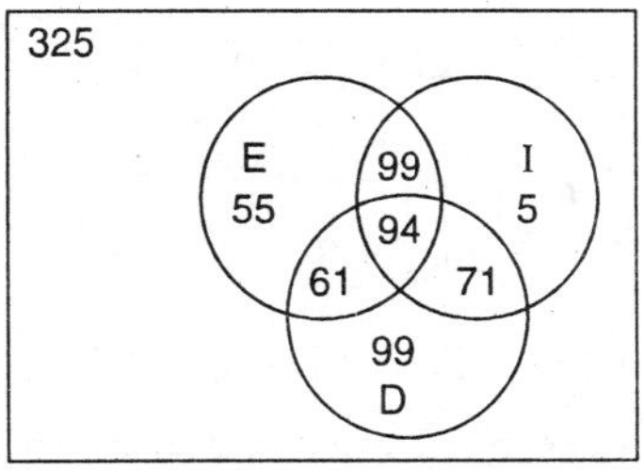

Fig. 2.17.

2. $100 - 47 - 35 + 23 = 41$.

3. (*a*) 75 (*b*) $| R \cap N \cap \overline{T} | = 80$, (*c*) $| T \cap \overline{N} \cap \overline{R} | = 175$

(*d*) $| \overline{T} \cap \overline{R} \cap \overline{N} | = | \overline{T \cup R \cup N} | = 1{,}000 - 625 = 375$

4. $100 - 47 - 35 - 20 + 23 + 12 + 11 - 5 = 39$.

5. Let B be the set of connoisseurs of canadian bacon and A the set of those who like anchovies.

$(a)\ |B \cup A| = |B| + |A| - |B \cap A| = 10 + 7 - 6 = 11$

$(b)\ |B \setminus A| = |B| - |B \cap A| = 4;$

$(c)\ |B \oplus A| = |B \cup A| - |B \cap A| = 5$

$(d)\ |U| - |B \cup A| = 4.$

6. Let A be the set of people with undergraduate degrees in arts S those with undergraduate degrees in science, and G those with graduate degree.

$(a)\ |A \cup S \cup G| = 300, \quad (b)\ \ |S \setminus (A \cup G)| = 35.$

7. Let D be the set of delegates who voted to decrease the deficit, E be the set of delegates who voted in favor of the motion concerning environmental issues, and T be the set of delegates who voted in favor of not increasing taxes.

$(a)\ |(D \cup E \cup T)^c| = 200, \ (b)\ \ |T \setminus (D \cup E)| = 316.$

8. $75 - (4(28) - 6(12) + 4(5) - 1) = 75 - 59 = 16.$

10. Let A, B and C be the sets of integers between 1 and 250 which are divisible by 4, by 6 and by 15 respectevely.

We want $|A \cup B \cup C| = 91$.

11. Let A, B, C, D be the sets of integers between 1 and 1000 (inclusive) which are divisible by 2, by 3, by 5 and by 7, respectively. We want $|A \cup B \cup C \cup D|^c = 1000 - 772 = 228$.

12. Write $a = qb + r$ with $0 \le r < b$. Then $\left\lfloor \frac{a}{b} \right\rfloor = q$. Clearly the positive integers $b, 2b, ..., qb$ are all less than or equal to a and are all divisible by b. On the other hand, if $sb \le a$, then s must belong to the set $\{1, 2, ..., q\}$. Hence, there are exactly q such natural numbers, as required.

13.
$$\begin{aligned} |(A \oplus B \cap C)| &= |((A \setminus B) \cup (B \setminus A)) \cap C| = |((A \setminus B) \cap C) \cup ((B \setminus A) \cap C| \\ &= |(A \cap C) \setminus B| + |(B \cap C) \setminus A| \text{ Since } (A \setminus B) \cap (B \setminus A) = \phi \\ &= |A \cap C| - |A \cap B \cap C| + |B \cap C| - |A \cap B \cap C| \\ &= |A \cap C| + |B \cap C| - 2|A \cap B \cap C|. \end{aligned}$$

14. $\binom{19}{3} - \binom{13}{3} - \binom{12}{3} - \binom{11}{3} - \binom{10}{3} + \binom{6}{3} + \binom{5}{3} + \binom{4}{3} + \binom{4}{3} + \binom{3}{3} = 217$

15. $1000 - (200 + 166 + 125) + (33 + 25 + 41) - 8 = 600.$

16. There are seven days of the week. Hence, we wish to put eight objects (people) into seven boxes (days). By the pigeon-hole principal, same day must have at least two people corresponding to it.

17. $\left\lceil \frac{100}{12} \right\rceil = 9.$

18. Any given processor is connected to atleast one of 19 other processors. There are 20 processors, so the pigeon-hole principal assures us that atleast two are connected to the same number.

19. Let box 1 correspond to days 1, 2 and 3 box 2 to days 4, 5 and 6 box 3 to days 7, 8 and 9 and box 4 to days 10, 11 and 12. Putting each bill into the box corresponding to the day it was mailed, we see that one box must contain at least $\left\lceil \frac{195}{4} \right\rceil = 49$ bills, by the general form of the pigeon hole principal. This gives the desired result.

20. Divide the rectangle into 25 rectangles, each 3×4. By the pigeon hole principal, at least two po .nts are within, or on the boundry of one of those smaller rectangles. This gives the result since the maximum distance between two points of such a rectangle is $\sqrt{3^2 + 4^2} = 5$.

Answers 2.2

1. (*a*) $C(10, 6)\, D_6$

(*b*) $\binom{10}{6} D_6 + \binom{10}{7} D_7 + \binom{10}{8} D_8 + \binom{10}{9} D_9 + \binom{10}{10} D_{10}$

(*c*) $10! - (3)9! + (3)8! - 7!$

(*d*) $10! - \binom{5}{1} 9! + \binom{5}{2} 8! - \binom{5}{3} 7! + \binom{5}{4} 6! - \binom{5}{5} 5!$

(*e*) $10! - (2)9! + 8!$

2. D_{26} **3.** $4^8 - \binom{4}{1} 3^8 + \binom{4}{2} 2^8 - \binom{4}{3} 1^8 = 40,824.$

4. Let A_1 be the arrangements where the 4 *a*'s are one block, A_2 the arrangements with the 3*b*'s in a single block, and A_3 the arrangements with the 2*c*'s in a block.

Then, $|\overline{A} \cap \overline{A_2} \cap \overline{A_3}| = \frac{9!}{4!3!2!} - \left(\frac{6!}{3!2!} + \frac{7!}{4!2!} + \frac{8!}{4!3!} \right) + \frac{4!}{2!} + \frac{5!}{3!} + \frac{6!}{4!} - 3! = 871.$

5. $3^{10} - (3)2^{10} + 3$ ways

6. (*a*) D_{10} (*b*) $10! - D_{10}$ (*c*) $10! - D_{10} - 10D_9$

7. The first row can be painted 8! ways. Each row after the first can be painted D_8 ways. Hence the number of ways is $8!\,(D_8)^7$

8. $(10!)D_{10}$ **9.** D_{10} **10.** $\binom{52}{5} - 4\binom{39}{5} + 6\binom{29}{5} - 4\binom{13}{5}$

11. (*a*) $\binom{52}{13} - \binom{4}{1}\binom{39}{13} + \binom{4}{2}\binom{26}{13} - \binom{4}{3}$ **12.** $3^{20} - 3.2^{20} + 3$

13. (*a*) $10^6 - 4.9^6 + 6.8^6 - 4.7^6 + 6^6$ (*b*) $4 + 4^2 + 4^3 + 4^4 + 4^5 + 4^6 = (4^7 - 4)/3$.

14. $D_6 = 265$, $D_7 = 1854$, $D_8 = 14,833$. **15.** $D_{11} = 11!\left(1 - \frac{1}{1!} + \frac{1}{2!} - \frac{1}{3!} + \ldots - \frac{1}{11!}\right)$

16. $20D_{19}$ **17.** $9! - [5(8!) - 10(7!) + 10(6!) - 5(5!) + 4!] = 205,056$

26. (*a*) If n is even, then by the Fundamental theorem of Arithmetic we may write $n = 2^k m$, where $k \geq 1$ and m is odd. Then $2n = 2^{k+1} m$ and $\phi(2n) = (2^{k+1})\left(1 - \frac{1}{2}\right)\phi(m) = 2^k \phi(m) = 2(2^k)\left(\frac{1}{2}\right)\phi(m)$

$$= 2\left[2^k\left(1 - \frac{1}{2}\right)\phi(m)\right] = 2[\phi(2^k m)] = 2\phi(n).$$

(*b*) When n is odd, we find that $\phi(2n) = (2n)\left(1 - \frac{1}{2}\right)\pi_{p/n}\left(1 - \frac{1}{p}\right)$, where the product is taken over all (odd) primes dividing n. If $n = 1$, then $\pi_{p/n}\left(1 - \frac{1}{p}\right)$ is 1.

But $(2n)\left(1 - \frac{1}{2}\right) \cdot \pi_{p/n}\left(1 - \frac{1}{p}\right) = n\,\pi_{p/n}\left(1 - \frac{1}{p}\right) = \phi(n)$.

27. (*a*) $S_1 = \{1, 5, 7, 11, 13, 17\}$, $S_2 = \{2, 4, 8, 10, 14, 16\}$

$S_3 = \{3, 15\}$, $S_6 = \{6, 12\}$, $S_9 = \{9\}$, $S_{18} = \{18\}$

(*b*) $|S_1| = 6\ \phi(18)$ $|S_3| = 2\ \phi(6)$ $|S_9| = 1 = \phi(2)$

$|S_2| = 6 = \phi(9)$ $|S_6| = 2 = \phi(3)$ $|S_{18}| = 1 = \phi(1)$

28. (*a*) $\binom{8}{0} + \binom{8}{1}8x + \binom{8}{2}(8.7)x^2 + \binom{8}{3}(8.7.6)x^3 + \binom{8}{4}(8.7.6.5)x^4 + \ldots + \binom{8}{8}(8!)x^8 = \sum_{i=0}^{8}\binom{8}{i} P(8, i)x^i$ (*b*) $\sum_{i=0}^{n}\binom{n}{i} P(n, i)x^i$.

29. (*a*) (*i*) $(1 + 2x)^3$ (*ii*) $1 + 8x + 14x^2 + 4x^3$ (*ii*) $1 + 9x + 25x^2 + 21x^3$ (*iv*) $1 + 8x + 16x^2 + 7x^3$

(*b*) If the board C consists of n steps, and each step has k blocks, then $r(C, x) = (1 + kx)^n$.

CHAPTER

3

Generating Functions

INTRODUCTION

An important idea in Mathematics is to establish connection between two fields in order to apply knowledge in one field to the other field, or at least take a problem of one field and transform it to a problem in the other field. This idea motivatives the idea of a generating function, which establishes a connection between functions of a real variable and sequences of numbers. De Moivre introduced the notion of generating function in 1730 AD.

3.1. GENERATING FUNCTION : DEFINITION

A generating function can be defined as "Given a numeric sequence $a = a_0, a_1, a_2, ..., a_n ......$ the series.

$$f(x) = a_0 + a_1x + a_2x^2 + ... + a_nx^n + ...$$

is called the **generating function** of the sequence".

Suppose $h = h_0, h_1, h_2, ..., h_n, ...$ is a sequence of numbers. We write it as an infinite sequence, but we mean to include finite sequences also as well.

If $g(x) = h_0 + h_1x + h_2x^2 + ... + h_nx^n + ...$

$$= \sum_{n=0}^{\infty} h_n x^n$$

$h_0, h_1, h_2, ..., h_n$ is finite sequnce, we make it of finite length, by setting $h_r = 0$ for $r > n$.

Thus, the function generates the sequence as its sequence of coefficients. If the sequence is finite then there is an m for which $h_r = 0$ for $r > m$.

In this case $g(x)$ is an ordinary polynomial in x of degree m. The function $g(x) = (1 + x)^m$ generates the binomial coefficients $h_r = C(m, r)$.

Therefore a generating function in which coefficients of x^n are sequence terms of the sequence h, is **called the Binomial generating function** of the sequence h.

The binomial coefficient $C(m, r)$ gives us the total number of combinations of r selections from m objects.

We know that $P(m, r) = r ! * C(m, r)$. This means if $C(m, r)$ is coefficient of x^r then $P(m, r)$ is coefficient of $\frac{x^r}{r!}$.

Let $h(x)$ is a generating function for a sequence a given in series form as below

$$h(x) = a_0 + a_1 x + a_2 \frac{x^2}{2!} + a_3 \frac{x^3}{3!} + \ldots\ldots$$

$$= \sum_{n=0}^{\infty} a_n \frac{x^n}{n!}, \text{ this series is exponential.}$$

The generating function $h(x)$ defined as above for a sequence $a = a_0, a_1, a_2, \ldots\ldots$ is called **Exponential generating function.**

3.2. INTRODUCTORY EXAMPLES

Problem 3.1. *Find a generating function for a_r = the number of non negative integral solutions of $e_1 + e_2 + e_3 + e_4 + e_5 = r$.*
where $0 \le e_1 \le 3$, $0 \le e_2 \le 3$, $2 \le e_3 \le 6$, $2 \le e_4 \le 6$, e_5 is odd, and $1 \le e_5 \le 9$.

Solution. Let $A_1(x) = A_2(x) = 1 + x + x^2 + x^3$

$$A_3(x) = A_4(x) = x^2 + x^3 + x^4 + x^5 + x^6$$

and $$A_5(x) = x + x^3 + x^5 + x^7 + x^9.$$

Thus, the generating function we want is

$$A_1(x)\, A_2(x)\, A_3(x)\, A_4(x)\, A_5(x)$$
$$= (1 + x + x^2 + x^3)^2\, (x^2 + x^3 + x^4 + x^5 + x^6)^2\, (x + x^3 + x^5 + x^7 + x^9).$$

Problem 3.2. *Find a generating function for a_r = the number of non negative integral solutions to $e_1 + e_2 + \ldots\ldots + e_n = r$, where $0 \le e_i \le 1$.*

Solution. Let $A_i(x) = 1 + x$ for each $i = 1, 2, \ldots\ldots, n$.

Thus, the generating function we want is

$$A_1(x)\, A_2(x) \ldots\ldots A_n(x) = (1 + x)^n.$$

The binomial theorem gives all the coefficients and thus we know the number of solutions to be above equation is C(n, r).

Problem 3.3. *Find a generating function for a_r = the number of non negative integral solutions to $e_1 + e_2 + \ldots\ldots + e_n = r$, where $0 \le e_i$ for each i.*

Solution. Here since there is no upper bound constraint on the e_i's.

Let $A_1(x)\, A_2(x) \ldots\ldots A_n(x) = (1 + x + x^2 + \ldots\ldots + x^k \ldots\ldots)^n$. We know that $\sum_{r=0}^{\infty} C(n-1+r, r)x^r$

must be another expressions for this same generating function

that is, $$\left(\sum_{k=0}^{\infty} x^k\right)^n = \sum_{r=0}^{\infty} C(n-1+r, r)x^r.$$

In particular, $\left(\sum_{k=0}^{\infty} x^k\right)^2 = \sum_{r=0}^{\infty} (r+1)x^r$ and $\left(\sum_{k=0}^{\infty} x^k\right)^3 = \sum_{r=0}^{\infty} \frac{(r+2)(r+1)}{2} x^2$

Since for $n = 2$, $C(n-1+r, r) = r + 1$ and

for $n = 3$, $C(n-1+r, r) = (r+2)(r+1)/2$.

Problem 3.4. *Find a generating function for a_r = the number of ways of distributing r similar balls into n numbered boxes where each box is non empty.*

Solution. First we model this problem as an integral solution of an equation problem, namely, we are to count the number of integral solutions to $e_1 + e_2 + ... + e_n = r$, where each $e_i \geq 1$.

Then, in turn, we build the generating function

$$(x + x^2 + ...)^n = \left(\sum_{r=1}^{\infty} x^r\right)^n, \text{ must equal } \sum_{r=n}^{\infty} C(r-1, n-1)x^r.$$

Problem 3.5. *Find a generating function for a_r = the number of ways the sum r can be obtained when :*

(a) 2 distinguishable dice are tossed.

(b) 2 distinguishable dice are tossed and the first shows an even number and the second shows an odd number.

(c) 10 distinguishable dice are tossed and 6 specified dice show an even number and the remaining 4 show an odd number.

Solution. In (*a*) We are to count the number of integral solutions of $e_1 + e_2 = r$, where $1 \leq e_i \leq 6$.

Then a_r is the coefficient of x^r in the generating function

$$(x + x^2 + x^3 + x^4 + x^5 + x^6)^2.$$

In (*b*) We are looking for the coefficient of x^r in

$(x^2 + x^4 + x^6)(x + x^3 + x^5)$ since $1 \leq e_1 \leq 6$ and e_1 is even while $1 \leq e_2 \leq 6$ and e_2 is odd.

Likewise, the generating function called for in (C) is $(x^2 + x^4 + x^6)^6 (x + x^3 + x^5)^4$.

Problem 3.6. *Find a generating function to count the number of integral solutions to $e_1 + e_2 + e_3 = 10$ if for each i, $0 \leq e_i$.*

Solution. Here we can take two approaches. Of course we are looking for the coefficient of x^{10} in $(1 + x + x^2 + x^3 + ...)^3$.

But since the equation is a model for the distribution of 10 similar balls into 3 boxes we see that each $e_i \leq 10$ for we cannot place more than 10 balls in each box.

Thus we could also interpret the problem as one where we are to find the coefficient of x^{10} in

$$(1 + x + x^2 + ... + x^{10})^3.$$

Problem 3.7. *Use generating functions to find the number of ways to select r objects of n different kinds if we must select atleast one object of each kind.*

Solution. Since we need to select atleast one object of each kind, each of the n kinds of objects contributes the factor $(x + x^2 + x^3 + ...)$ to the generating function G(x) for the sequence $\{a_r\}$, where a_r is the number of ways to select r objects of n different kinds if we need atleast one object of each kind.

Hence, $G(x) = (x + x^2 + x^3 + ...)^n = x^n(1 + x + x^2 + ...)^n = \dfrac{x^n}{(1-x)^n}$

Using the extended Binomial theorem, we have

$$G(x) = \frac{x^n}{(1-x)^n} = x^n . (1-x)^{-n}$$

$$= x^n \sum_{r=0}^{\infty} \binom{-n}{r} (-x)^r = x^n \sum_{r=0}^{\infty} (-1)^r \ C(n+r-1, r)(-1)^r x^r$$

$$= \sum_{r=0}^{\infty} C(n+r-1,\ r)\, x^{n+r}$$

$$= \sum_{t=n}^{\infty} C(t-1, t-n)\, x^t = \sum_{r=n}^{\infty} C(r-1, r-n)\, x^r .$$

We have shifted the summation in the next-to-last equality by setting $t = n + r$ so that $t = n$ when $r = 0$ and $n + r - 1 = t - 1$, and then we replaced t by r as the index of summation in the last equality to our original notation.

Hence, there are $C(r-1, r-n)$ ways to select r objects of n different kinds if we must select at least one object of each kind.

Problem 3.8. *Use generating functions to determine the number of ways to insert tokens worth \$1, \$2, and \$5 into a vending machine to pay for an item that costs r dollars in both the cases wehn the order in which the tokens are inserted does not matter and when the order does matter.*

(For example, there are two ways to pay for an item that costs \$3 when the order in which the tokens are inserted does not matter ; inserting three \$1 token or one \$1 token and a \$2 token. When the order matters, there are three ways ; inserting three \$1 tokens, inserting a \$1 token and then a \$2 token, or inserting a \$2 token and then a \$1 token).

Solution. Consider the case when the order in which the tokens are inserted does not matter.

Here, all we care about is the number of each token used to produce a total of r dollars.

Since we can use any number of \$1 tokens, any number of \$2 tokens, and any number of \$5 tokens, the answer is the coefficient of x^r in the generating function

$$(1 + x + x^2 + x^3 + ...)(1 + x^2 + x^4 + x^6 + ...)\ (1 + x^5 + x^{10} + x^{15} + ...).$$

The first factor in this product represents the \$1 tokens used, the second the \$2 tokens used, and the third the \$5 tokens used.

For example, the number of ways to pay for an item costing \$7 using \$1, \$2 and \$5 tokens is gvien by the coefficient of x^7 in this expansin, which equals 6.

When the order in which the tokens are inserted matters, the number of ways to insert exactly n tokens to produce a total of r dollars is the coefficient of x^r in $(x + x^2 + x^5)^n$.

Since each of the r tokens may be a \$1 token, a \$2 token, or a \$5 token.

Since any number of tokens may be inserted, the number of ways to produce r dollars using \$1, \$2, or \$5 tokens, when the order in which the tokens are inserted matters, is the coefficient of x^r in

$$1 + (x + x^2 + x^5) + (x + x^2 + x^5)^2 + \ldots\ldots$$

$$= \frac{1}{1-(x+x^2+x^5)} = \frac{1}{1-x-x^2-x^5}$$

where we have added the number of ways to insert 0 tokens, 1 token, 2 tokens, 3 tokens, and so on, and where we have used the identity $\frac{1}{(1-x)} = 1 + x + x^2 + \ldots\ldots$, with x replaced with $x + x^2 + x^5$.

For example, the number of ways to pay for an item costing \$7 using \$1, \$2, and \$5 tokens, when the order in which the tokens are used matters, is the coefficient of x^7 in this expansion, which equals 26.

To see that this coefficient equals 26 requires the addition of the coefficients of x^7 in the expansions $(x + x^2 + x^5)^k$ for $2 \le k \le 7$.

This can be done by hand with considerable computation, or a computer algebra system can be used.

Problem 3.9. *In how many different ways can eight identical cookies be distributed among three distinct children to each child receives at least two cookies and no more than four cookies ?*

Solution. Since each child receives at least two but no more than four cookies, for each child there is a factor equal to $(x^2 + x^3 + x^4)$ in the generating function for the sequence $\{C_n\}$, where C_n is the number of ways to distribute n cookies.

Since there are three children, this generating function is

$$(x^2 + x^3 + x^4)^3.$$

We need the coefficient of x^8 in this product. The reason is that the x^8 terms in the expansion correspond to the ways that three terms can be selected, with one from each factor, that have exponents adding upto 8.

Furthermore, the exponents of the term from the first, second and third factors are the numbers of cookies the first, second and third children receive, respectively. Computation shows that this coefficient equals 6.

Hence, there are six ways to distribute the cookies so that each child receives at least two, but no more than four, cookies.

Problem 3.10. *Find the number of solutions of $e_1 + e_2 + e_3 = 17$, where e_1, e_2, and e_3 are non negative integers with $2 \le e \le 5$, $3 \le e_2 \le 6$, and $4 \le e_3 \le 7$.*

Solution. The number of solutions with the indicated constraints is the coefficient of x^{17} in the expansion of

$$(x^2 + x^3 + x^4 + x^5)(x^3 + x^4 + x^5 + x^6)\,(x^4 + x^5 + x^6 + x^7).$$

This follows since we obtain a term equal to x^{17} in the product by picking a term in the first sum x^{e_1}, a term in the second sum x^{e_2}, and a term in the third sum x^{e_3}, where the exponents e_1, e_2 and e_3 satisfy the equation $e_1 + e_2 + e_3 = 17$ and the given constraints. It is not hard to see that the coefficient of x^{17} in this product is 3. Hence there are three solutions.

Problem 3.11. *How many integer solutions are there for the equation $c_1 + c_2 + c_3 + c_4 = 25$ if $0 \le c_i$ for all $1 \le i \le 4$?*

Solution. For each child the possibilities can be described by the polynomial

$$1 + x + x^2 + x^3 + ... + x^{25}$$

Then the answer to this problem is the coefficient of x^{25} in the generating function

$$f(x) = (1 + x + x^2 + ... + x^{25})^4.$$

The answer can also be obtained as the coefficient of x^{25} in the generating function

$$g(x) = (1 + x + x^2 + x^3 + ... + x^{25} + x^{26} + ...)^4$$

if we rephrase the question in terms of distributing, from a large (or unlimited) number of pennies, 25 pennies among four children.

Note that the terms x^k, for all $k \ge 26$.

Problem 3.12. *If there is an unlimited number (or at least 24 of each colour) of red, green white, and black jelly beans, in how many ways can Douglas select 24 of these candies so that he has an even number of white beans and at least six black ones ?*

Solution. The polynomials associated with the jelly bean colours are as follows :

red (green) : $1 + x + x^2 + ... + x^{24}$, where the leading 1 is for $1x^\circ$, because one possibility for the red (and green) jelly beans is that none of that colour is selected.

White : $(1 + x^2 + x^4 + x^6 + ... + x^{24})$

Black : $(x^6 + x^7 + x^8 + ... + x^{24})$.

So the answer to the problem is the coefficient of x^{24} in the generating function

$$f(x) = (1 + x + x^2 + ... + x^{24})^2 \, (1 + x^2 + x^4 + ... + x^{24}) \, (x^6 + x^7 + ... x^{24})$$

One such selection is five red, three green, eight white and eight black jelly beans.

This arises from x^5 in the first factor, x^3 in the second factor, and x^8 in the last two factors.

Problem 3.13. *While shopping one saturday, Mildred buys 12 oranges for her children, Grace, Mary, and Frank. In how many ways can she distribute the oranges so that Grace gets atleast four, and Mary and Frank get at least two, but Frank gets no more than five ?*

Table 1 lists all the possible distributions

G	*M*	*F*	*G*	*M*	*F*
4	*3*	*5*	*6*	*2*	*4*
4	*4*	*4*	*6*	*3*	*3*
4	*5*	*3*	*6*	*4*	*2*
4	*6*	*2*	*7*	*2*	*3*
5	*2*	*5*	*7*	*2*	*3*
5	*3*	*4*	*7*	*3*	*2*
5	*4*	*4*	*8*	*2*	*2*
5	*5*	*2*			

Solution. We see that we have all the integer solutions to the equation $c_1 + c_2 + c_3 = 12$ where $4 \le c_1$, $2 \le c_2$ and $2 \le c_3 \le 5$.

Considering the first two cases in this table, we find the solutions $4 + 3 + 5 = 12$ and $4 + 4 + 4 = 12$. When multiplying polynomials we add the powers of the variable, and here, when we multiply the three polynomials

$$(x^4 + x^5 + x^6 + x^7 + x^8)(x^2 + x^3 + x^4 + x^5 + x^6)(x^2 + x^3 + x^4 + x^5)$$

two of the ways to obtain x^{12} are as follows :

1. From the product $x^4x^3x^5$, where x^4 is taken from $(x^4 + x^5 + x^6 + x^7 + x^8)$, x^3 from $(x^2 + x^3 + x^4 + x^5 + x^6)$, and x^5 from $(x^2 + x^3 + x^4 + x^5)$.

2. From the product $x^4x^4x^4$ where the first x^4 is found in the first polynomial the second x^4 in the second polynomial, and the third x^4 in the third polynomial.

3.3. PARTITIONS OF INTEGERS

In number theory, we are confronted with partitioning a positive integer n into positive summands and seeking the number of such partitions, without regard to order. This number is denoted by P(n).

For example,

$$P(1) = 1 : 1$$
$$P(2) = 2 : 2 = 1 + 1$$
$$P(3) = 3 : 3 = 2 + 1 = 1 + 1 + 1$$
$$P(4) = 5 : 4 = 3 + 1 = 2 + 2 = 2 + 1 + 1 = 1 + 1 + 1 + 1$$
$$P(5) = 7 : 5 = 4 + 1 = 3 + 2 = 3 + 1 + 1 = 2 + 2 + 1$$
$$= 2 + 1 + 1 + 1 = 1 + 1 + 1 + 1 + 1.$$

We should like to obtain P(n) for a given n without having to list all the partitions. We need a tool to keep track of the numbers of 1's, 2's, ... n's that are used to summands for n.

If $n \in Z^+$, the number of 1's we can use is 0 or 1 or 2 or ......

The power series $1 + x + x^2 + x^3 + x^4 + ...$ keeps account of this for us.

In like manner, $1 + x^2 + x^4 + x^6 + ...$ keeps track of the number of 2's in the partition of n, while $1 + x^3 + x^6 + x^9 + ...$ accounts for the number of 3's.

Therefore, in order to determine P(10), for instance, we want the coefficient of x^{10} in $f(x) = (1 + x + x^2 + x^3 + ...) (1 + x^2 + x^4 + ...) (1 + x^3 + x^6 + x^9 + ...) ...... (1 + x^{10} + x^{20} + ...)$ or in

$$g(x) = (1 + x + x^2 + x^3 + ... + x^{10})(1 + x^2 + x^4 + ... + x^{10})$$
$$(1 + x^3 + x^6 + x^9 + ...) ......(1 + x^{10}).$$

$\therefore f(x)$ can be written in the more compact form

$$f(x) = \frac{1}{1-x}\,\frac{1}{(1-x^2)}\,\frac{1}{(1-x^3)}\;......\;\frac{1}{(1-x^{10})} = \prod_{i=1}^{10}\frac{1}{(1-x^i)}$$

If this product is extended beyond $i = 10$, we get $P(x) = \prod_{i=1}^{\infty}\left[\frac{1}{(1-x^i)}\right]$, which generates the sequence P(0), P(1), P(2), P(3) ......, where we define P(0) = 1.

It is impossible to actually calculate the infinite number of terms in the product P(x).

If we consider only $\prod_{i=1}^{r}\left[\frac{1}{(1-x^i)}\right]$ for some fixed r, then the coefficient of x^n here is the number of partitions of n into summands that do not exceed r.

The difficulty in calculating P(n) from P(x) for large values of n, the idea of the generating function will be useful in studying certain kinds of partitions.

Generating functions play an essential role in the theory of partitions. Recall that a partition of a positive integer r is a collection of positive integers with sum r.

$P(r) \equiv$ number of distinct partitions of r

$P_n(r) \equiv$ number of partitions of r into parts at most equal to n

$\equiv$ the number of solutions in non negative integers of $1u_1 + 2u_2 + \ldots\ldots + nu_n = r$

$q_n(r) \equiv$ number of partitions of r into at most n parts

$\equiv$ number of distributions of r identical objects (1 s) among n identical places, empty places being permitted.

3.4. THE EXPONENTIAL GENERATING FUNCTION

For each $n \in Z^+$, $(1+x)^n = \binom{n}{0} + \binom{n}{1}x + \binom{n}{2}x^2 + \ldots\ldots + \binom{n}{n}x^n$.

So $(1+x)^n$ is the (ordinary) generating function for the sequence $\binom{n}{0}, \binom{n}{1}, \binom{n}{2}, \ldots\ldots \binom{n}{n}$, 0, 0, ...

Also $\binom{n}{r}$ = C(n, r) when we wanted to emphasize that $\binom{n}{r}$ represented the number of combinations of n objcets taken r at a time, with $0 \le r \le n$.

Consequently, $(1+x)^n$ generates the sequence

$$C(n, 0), C(n, 1), C(n, 2), \ldots\ldots C(n, n), 0, 0, \ldots\ldots$$

Now for all $0 \le r \le n$,

$$C(n, r) = \frac{n!}{r!(n-r)!} = \left(\frac{1}{r!}\right) P(n, r),$$

where P(n, r) denotes the number of permutations of n objects taken r at a time.

So $(1+x)^n = C(n, 0) + C(n, 1)x + C(n, 2)x^2 + C(n, 3)x^3 + \ldots\ldots + C(n, n)x^n$

$$= P(n, 0) + P(n, 1)x + P(n, 2)\frac{x^2}{2!} + P(n, 3)\frac{x^3}{3!} + \ldots\ldots + P(n, n)\frac{x^n}{n!}.$$

Hence, if in $(1+x)^n$ we consider the coefficient of $\frac{x^r}{r!}$, with $0 \le r \le n$.

For a sequence $a_0, a_1, a_2, a_3, \ldots\ldots$ of real numbers,

$$f(x) = a_0 + a_1 x + a_2 \frac{x^2}{2!} + a_3 \frac{x^3}{3!} + \ldots\ldots$$

$$= \sum_{i=0}^{\infty} a_n \frac{x^n}{n!}$$

is called the **exponential generating** function for the given sequence.

3.5. THE SUMMATION OPERATOR

This section introduces a technique that helps us go from the (ordinary) generating function for the sequence $a_0, a_1, a_2, \ldots\ldots$ to the generating function for the sequence $a_0, a_0 + a_1, a_0 + a_1 + a_2, \ldots\ldots$

For $f(x) = a_0 + a_1 x + a_2 x^2 + \ldots\ldots$

Consider the function $\dfrac{f(x)}{(1-x)}$

$$\frac{f(x)}{(1-x)} = [a_0 + a_1 x + a_2 x^2 + \ldots\ldots][1 + x + x^2 + x^3 + \ldots\ldots]$$

$$= a_0 + (a_0 + a_1)x + (a_0 + a_1 + a_2)x^2 + (a_0 + a_1 + a_2 + a_3)x^3 + \ldots\ldots$$

So $\dfrac{f(x)}{(1-x)}$ generates the sequence of sums $a_0, a_0 + a_1, a_0 + a_1 + a_2, a_0 + a_1 + a_2 + a_3, \ldots\ldots$

This is the convolution of the sequence $a_0, a_1, a_2, \ldots\ldots$ and the sequence $b_0, b_1, b_2, \ldots\ldots$ where $b_n = 1$ for all $n \in \mathbb{N}$.

Problem 3.14. *Find the generating function for the sequence a = 1, 1, 1, 1, ...*

Solution. The given sequence : $a = 1, 1, 1, 1, \ldots\ldots$

Here, the general term $a_n = 1$.

Let $f(x)$ be the binomial generating function for the given sequence, then

$f(x)$ can be written as

$$f(x) = \sum_{n=0}^{\infty} a_n x^n = \sum_{n=0}^{\infty} (1)\, x^n \qquad (\because \ a_n = 1)$$

$$= 1 + x + x^2 + x^3 + \ldots\ldots$$

$$= \frac{1}{1-x} \qquad (\because \text{ the series is geometric with } x \text{ as common ratio})$$

$\therefore \quad f(x) = \dfrac{1}{1-x}$ is the binomial generating function for the sequence $a = 1, 1, 1, 1, \ldots\ldots$

Let $g(x)$ be the exponential generating function for the same sequence, then $g(x)$ can be written as

$$g(x) = \sum_{n=0}^{\infty} a_n \frac{x^n}{n!} = \sum_{n=0}^{\infty} (1) \cdot \frac{x^n}{n!} \qquad (\because \; a_n = 1)$$

$$= 1 + x + \frac{x^2}{2!} + \frac{x^3}{3!} + \ldots\ldots$$

$$= e^x, \qquad (\because \text{ the series is exponential with } x \text{ as base})$$

$\therefore$ $g(x) = e^x$ is the exponential generating function for the sequence $a = 1, 1, 1, 1, \ldots\ldots$.

Problem 3.15. *Find the generating function for the sequence*

$$b = 1, 3, 9, \ldots\ldots, 3^n, \ldots\ldots$$

Solution. The given sequence : $b = 1, 3, 9, \ldots\ldots, 3^n, \ldots\ldots$

Here, the general term $b_n = 3^n$

Let $f(x)$ be the binomial generating function for the given sequence, then $f(x)$ can be written as

$$f(x) = \sum_{n=0}^{\infty} b_n x^n = \sum_{n=0}^{\infty} 3^n \cdot x^n \qquad (\because \; b_n = 3^n)$$

$$= 1 + 3x + (3x)^2 + (3x)^3 + \ldots\ldots$$

$$= \frac{1}{1-3x} \qquad (\because \text{ the series is geometric with } 3x \text{ as common ratio})$$

$\therefore$ $f(x) = \dfrac{1}{1-3x}$ is the binomial generating function for the sequence $b = 1, 3, 3^2, 3^3, \ldots\ldots 3^n, \ldots\ldots$

Let $g(x)$ be the exponential generating function for the same sequence, then $g(x)$ can be written as

$$g(x) = \sum_{n=0}^{\infty} b_n \frac{x^n}{n!} = \sum_{n=0}^{\infty} 3^n \cdot \frac{x^n}{n!} \qquad (\because \; b_n = 3^n)$$

$$= 1 + 3x + \frac{(3x)^2}{2!} + \frac{(3x)^3}{3!} + \ldots\ldots$$

$$= e^{3x} \qquad (\because \text{ the series is exponential with } 3x \text{ as base})$$

$\therefore$ $g(x) = e^{3x}$ is the exponential generating function for the sequence $b = 1, 3, 3^2, \ldots\ldots$

Problem 3.16. *Find the binomial generating function for the sequence*

$$a = 1, 2, 3, \ldots\ldots, r, \ldots\ldots$$

Solution. The given sequence : $a = 1, 2, 3, \ldots\ldots, r, \ldots\ldots$.

Here, the general term $a_r = r$

Let $f(x)$ be the binomial generating function of the given sequence, then

$f(x)$ can be written as

$$f(x) = \sum_{r=0}^{\infty} a_r . x^r = \sum_{r=0}^{\infty} r . x^r \qquad (\because \; a_r = r)$$

$$= 1 + 2x + 3x^2 + 4x^3 + \ldots\ldots \qquad \ldots(1)$$

The equation (1) is arithmetic-geometric in which first factor is following arithmetic progression and second following geometric.

We find the sum of the series by multiplying equation (1) with x and then by subtracting the result from equation (1).

Thus, we get

$$(1 - x) f(x) = 1 + x + x^2 + x^3 + \ldots\ldots$$

$$\Rightarrow \qquad f(x) = \frac{1}{(1-x)^2}$$

this is the required binomial generating function.

Problem 3.17. *Find the discrete numeric function (sequence) whose exponential generating function is given by $2e^x$.*

Solution. The given generating function can be written in series form as

$$2e^x = 2\left[1 + x + \frac{x^2}{2!} + \frac{x^3}{3!} + \frac{x^4}{4!} + \ldots\ldots\right]$$

Clearly, the coefficient of $\frac{x^n}{n!}$ for $n \geq 0$, are 2, 2, 2, ...

$\therefore$ If the discrete numeric function is a, then

$$a = 2, 2, 2, 2, \ldots\ldots .$$

Problem 3.18. *Find the exponential generating function for sequence*

$$t = {}^nP_0, {}^nP_1, {}^nP_2, \ldots\ldots {}^nP_n.$$

Solution. The given sequence is finite. All terms, in this sequence for $m > n$ are zero.

Let $h(x)$ be the exponential generating function for the given sequence, then

$h(x)$ can be written as

$$h(x) = \sum_{i=0}^{\infty} a_i \frac{x^i}{i!} = \sum_{i=0}^{\infty} {}^nP_i \frac{x^i}{i!} \qquad (\because \; a_i = {}^nP_i)$$

$$= {}^nP_0 + {}^nP_1 x + {}^nP_2 \frac{x^2}{2!} + \ldots\ldots + {}^nP_n \frac{x^n}{n!}$$

$$= {}^nC_0 + {}^nC_1 x + {}^nC_2 x^2 + \ldots\ldots + {}^nC_n x^n \qquad (\because \; {}^nP_r = r\,!\,.\,{}^nC_r)$$

$$= (1 + x)^n$$

∴ The exponential generating function for the sequence

$t = {}^nP_0, {}^nP_1, {}^nP_2, \ldots\ldots, {}^nP_n$ is given by

$h(x) = (1 + x)^n$.

Theorem 3.1. *Let f(x) is the generating function for a and g(x) is the generating function for b, then f(x) + g(x) is the generating function for a + b.*

Proof. Let $f(x) = \sum_{n=0}^{\infty} a_n x^n$ and $g(x) = \sum_{n=0}^{\infty} b_n x^n$

be the generating functions for a and b,

then

$$f(x) + g(x) = \sum_{n=0}^{\infty} a_n x^n + \sum_{n=0}^{\infty} b_n x^n$$

$$= \sum_{n=0}^{\infty} (a_n + b_n)\, x^n$$

∴ The sequence generated by $f(x) + g(x)$ is

$(a_0 + b_0), (a_1 + b_1), (a_2 + b_2), \ldots\ldots$

These terms are precisely the sum of the corresponding terms from a and b.

Thus, $f(x) + g(x)$ is generating function for the sum of the sequence $a + b$.

Note 1. If $f(x)$ and $g(x)$ are generating functions for sequences a and b respectively then $f(x) * g(x)$ is not the generating function for $a * b$.

2. Let $f(x)$ and $g(x)$ are generating functions for finite sequences a and b respcetively. The convolution of two sequences a and b is the sequence c defined as

$$C_n = \sum_{r=0}^{n} a_r\, b_{n-r}, \text{ then}$$

$f(x) \,.\, g(x)$ is the generating function for convolution of a and b.

3. For $k > 0$, the function $g(x) = \dfrac{1}{(1-x)^k}$ generates the sequence $a = \{C(n + k - 1, n)/n = 0, 1, 2, 3, \ldots\ldots\}$.

Thus the nth coefficient is the numebr of ways to select n objects of k times.

$$\frac{1}{(1-x)^k} = \left[\frac{1}{(1-x)}\right]^k = (1 + x + x^2 + x^3 + \ldots\ldots)^k$$

If we carry out this multiplication k times, then x^n appear as many times as there are non negative integer solutions to the equation

$$x_1 + x_2 + x_3 + \ldots\ldots + x_k = n$$

We know that, this number is C$(n + k - 1, n)$.

Problem 3.19. *For the following two sequences a and b, whose general terms are given, find a + b and ab.*

$$a_r = \begin{cases} 0 & \text{for } 0 \le r \le 2 \\ 2^{-r} + 5 & \text{for } r \ge 3 \end{cases} \quad \text{and}$$

$$b_n = \begin{cases} 3 - 2^r & \text{for } 0 \le r \le 1 \\ r + 2 & \text{for } r \ge 2 \end{cases}$$

Solution. Let C be the sequence for $a + b$,

at $r = 2$, the general definitions of a and b cannot be simply added.

For $r \ge 3$ and $0 \le r \le 1$, bottom and top definitions will be added, but at $r = 2$ the value will be determined and will be placed in the definition.

Let $d = ab$.

In this case, we shall find the sequence formula for d by multiplying the definitions of a and b and setting the range for sequence position r accordingly.

$$C_r = a_r + b_r = \begin{cases} 3 - 2^r & \text{for } 0 \le r \le 1 \\ 4 & \text{for } r = 2 \\ 2^{-r} + r + 7 & \text{for } r \ge 3 \end{cases}$$

and

$$C_r = a_r * b_r = \begin{cases} 0 & \text{for } 0 \le r \le 2 \\ r2^{-r} + 2^{-r+1} + 5r + 10 & \text{for } r \ge 3 \end{cases}.$$

Problem 3.20. *Find the number of positive integral solution to the equation $x + y + z = 10$.*

Solution. Here x, y and z are positive integers satisfying the given equations.

To find the number of possible solutions to the above equation.

Here x belongs to $\{1, 2, 3, 4, ...\}$ and so are y and z.

If different possible positive integers for x represented by the powers of x then the series for x can be written as

$$x + x^2 + x^3 + ...$$

This problem is reduced to selection of 10 objects of 3 kinds.

The generating function $f(x)$ for this problem is then written as

$$f(x) = (x + x^2 + x^3 + ...)^3$$

$$= \frac{x^3}{(1 - x)^3}$$

The required result is given by the coefficient of x^{10} in $f(x)$.

This is equal to the coefficient of x^7 in $(1 - x)^{-3}$.

$$\Rightarrow \quad {}^{7+3-1}C_7 = {}^9C_7 \qquad (\because \; n = 7, k = 3)$$

$$= \frac{9 * 8}{2} = 36.$$

Problem 3.21. *Find the generating function for the sequence 1, a, a^2, ..., where a is a fixed constant.*

Solution. Let $G(x) = 1 + ax + a^2x^2 + a^3x^3 + \ldots\ldots$

So, $G(x) - 1 = ax + a^2 x^2 + a^3x^3 + \ldots\ldots.$

or $$\frac{G(x) - 1}{ax} = 1 + ax + a^2x^2 + \ldots\ldots$$

or $$\frac{G(x) - 1}{ax} = G(x)$$

$$\Rightarrow \quad G(x) = \frac{1}{1 - ax}$$

The required generating function is therefore, $\frac{1}{1 - ax}$.

Problem 3.22. *Find the generating function of a sequence $\{a_k\}$ if $a_k = 2 + 3k$.*

Solution. The generating function of a sequence whose general term is 2 is $F(x) = \frac{2}{1 - x}$

The generating of a sequence whose general term is $3k$ is $G(x) = \frac{3x}{(1 - x)^2}$.

Hence the required generating function is

$$F(x) + G(x) = \frac{2}{1 - x} + \frac{3x}{(1 - x)^2}.$$

Problem 3.23. *Find the sequences corresponding to the ordinary generating functions (a) $(3 + x)^3$ and (b) $3x^3 + e^{2x}$.*

Solution. (*a*) $(3 + x)^3 = 27 + 27x + 9x^2 + x^3$

The sequence is (27, 27, 9, 1, 0, 0, 0, ...)

$$(b)\; 3x^3 + e^{2x} = 1 + 2x + \frac{2^2}{2!}x^2 + \left(3 + \frac{2^3}{3!}\right)x^3 + \frac{2^4}{4!}x^4 + \frac{2^5}{5!}x^5 + \ldots\ldots.$$

The sequence is $\left(1, 2, \frac{2^2}{2!}, \frac{2^3}{3!} + 3, \frac{2^4}{4!}, \ldots\ldots\right)$.

Problem 3.24. *Find a closed form for the generating function for each of the following sequence*

(*a*) *0, 0, 1, 1, 1, ...*

(*b*) *1, 1, 0, 1, 1, 1, 1, ...*

(*c*) *1, 0, – 1, 0, 1, 0, – 1, 0, 1, ...*

(*d*) *C(8, 0), C(8, 1), C(8, 2),......, C(8, 8), 0, 0, ...*

(*e*) *3, – 3, 3, – 3, 3, – 3, ...*

Solution. (*a*) We know that

$$\frac{1}{1-x} = 1 + x + x^2 + x^3 + \ldots\ldots \infty$$

$$= \sum_{n=0}^{\infty} x^n .$$

So, the generating function of 1, 1, 1, ..., is $\frac{1}{1-x}$.

Now $\frac{x^2}{1-x} = \sum_{n=0}^{\infty} x^{n+2}$.

Hence $\frac{x^2}{1-x}$ is the generating function 0, 0, 1, 1, 1, ...

(*b*) Here $\frac{1}{1-x} - x^2 = 1 + x + x^3 + \ldots\ldots \infty = \sum_{\substack{n=0 \\ n \neq 2}}^{\infty} x^n$

So, the generating function of 1, 1, 0, 1, 1, 1, ... is $\frac{1}{1-x} - x^2$.

(*c*) We know that $\frac{1}{1+x^2} = (1 + x^2)^{-1}$

$$= 1 - x^2 + x^4 - x^6 + x^8 \ldots\ldots \infty$$

$$= 1 + 0.x + (-1)x^2 + 0.x^3 + 1.x^4 + 0.x^5 + (-1) . x^6 + \ldots\ldots$$

So the generating function of 1, 0, – 1, 0, 1, 0, – 1, ... is $\frac{1}{1+x^2}$.

(*d*) We know that

$$(1 + x)^8 = C(8, 0)x^0 + C(8, 1)x + \ldots\ldots + C(8, 8)x^8 + 0 + 0 + \ldots\ldots$$

$$= \sum_{n=0}^{\infty} C(8, n)x^n .$$

So, the generating function of C(8, 0), C(8, 1), ..., C(8, 8), 0, 0, ...

(*e*) We have

$$\frac{3}{1-x} = 3(1 - x)^{-1} = 3(1 - x + x^2 - x^3 + \ldots\ldots)$$

$$= 3 + (-3)x + 3x^2 + (-3)x^3 + \ldots\ldots$$

$$= \sum_{n=0}^{\infty} (-3)^n x^n .$$

Hence, the required generating function is $\dfrac{3}{1+x}$.

Theorem 3.2. Let $f(x) = \sum_{k=0}^{\infty} a_k x^k$ and $g(x) = \sum_{k=0}^{\infty} b_k x^k$. Then

$$f(x) + g(x) = \sum_{k=0}^{\infty} (a_k + b_k) x^k \text{ and}$$

$$f(x)\, g(x) = \sum_{k=0}^{\infty} \left(\sum_{j=0}^{k} a_j\, b_{k-j} \right) x^k .$$

Theorem 3.3. *(The Extended Binomial Theorem)*

Let x be a real number with $|x| < 1$ and let u be a real number. Then $(1+x)^u = \sum_{k=0}^{\infty} \binom{u}{k} x^k$.

Table. Some useful Generating functions :

$G(x)$	a_k
$(1+x)^n = \sum_{k=0}^{n} C(n,k)x^k$ $= 1 + C(n,1)x + C(n,2)x^2 + \ldots\ldots + x^n$	$C(n, k)$
$(1+ax)^n = \sum_{k=0}^{n} C(n,k)a^k x^k$ $= 1 + C(n,1)ax + C(n,2)a^2x^2 + \ldots\ldots + a^n x^n$	$C(n, k)a^k$
$(1+x^r)^n = \sum_{k=0}^{n} C(n,k)x^{rk}$ $= 1 + C(n,1)x^r + C(n,2)x^{2r} + \ldots\ldots + x^{rn}$	$C(n, k/r)$ if r/k, 0, 0 otherwise
$\dfrac{1-x^{n+1}}{1-x} = \sum_{k=0}^{n} x^k = 1 + x + x^2 + \ldots\ldots + x^n$	1 if $k \le n$, 0 otherwise
$\dfrac{1}{1-x} = \sum_{k=0}^{\infty} x^k = 1 + x + x^2 + \ldots\ldots$	1
$\dfrac{1}{1-ax} = \sum_{k=0}^{\infty} a^k x^k = 1 + ax + a^2x^2 + \ldots\ldots$	a^k

(Contd.)

$\frac{1}{1-x^r} = \sum_{k=0}^{\infty} x^{rk} = 1 + x^r + x^{2r} + \ldots\ldots$	1 if r/k, 0 otherwise
$\frac{1}{(1-x)^2} = \sum_{k=0}^{\infty} (k+1)x^k = 1 + 2x + 3x^2 + \ldots\ldots$	$k + 1$
$\frac{1}{(1-x)^n} = \sum_{k=0}^{\infty} C(n+k-1, k)x^k$ $= 1 + C(n, 1)x + C(n+1, 2)x^2 + \ldots\ldots$	$C(n+k-1, k) = C(n+k-1, n-1)$
$\frac{1}{(1+x)^n} = \sum_{k=0}^{\infty} C(n+k-1, k)(-1)^k x^k$ $= 1 - C(n, 1)x + C(n+1, 2)x^2 - \ldots\ldots$	$(-1)^k C(n+k-1, k)$ $= (-1)^k C(n+k-1, n-1)$
$\frac{1}{(1-ax)^n} = \sum_{k=0}^{\infty} C(n+k-1, k)a^k x^k$ $= 1 + C(n, 1)ax + C(n+1, 2)a^2x^2 + \ldots\ldots$	$C(n+k-1, k)a^k$ $= C(n+k-1, n-1)a^k$
$e^x = \sum_{k=0}^{\infty} \frac{x^k}{k!} = 1 + x + \frac{x^2}{2!} + \frac{x^3}{3!} + \ldots\ldots$	$\frac{1}{k!}$
$\ln(1+x) = \sum_{k=0}^{\infty} \frac{(-1)^{k+1}}{k} x^k$ $= x - \frac{x^2}{2} + \frac{x^3}{3} - \frac{x^4}{4} + \ldots\ldots$	$\frac{(-1)^{k+1}}{k}$

Problem 3.25. *What is the generating function for the sequence*

1, 1, 1, 1, 1, 1 ?

Solution. The generating function of 1, 1, 1, 1, 1, 1 is

$$1 + x + x^2 + x^3 + x^4 + x^5.$$

We have $\frac{(x^6 - 1)}{(x - 1)} = 1 + x + x^2 + x^3 + x^4 + x^5.$

Consequently, $G(x) = \frac{(x^6 - 1)}{(x - 1)}$ is the generating function of the sequence 1, 1, 1, 1, 1, 1.

Problem 3.26. *Let m be a positive integer. Let $a_k = C(m, k)$, for k = 0, 1, 2, ..., m. What is the generating function for the sequence $a_0, a_1, \ldots\ldots, a_m$?*

Solution. The generating function for this sequence is

$$G(x) = C(m, 0) + C(m, 1)x + C(m, 2)x^2 + \ldots\ldots + C(m, n)x^m.$$

The binomial theorem shows that $G(x) = (1 + x)^m$.

Problem 3.27. *Let* $f(x) = \dfrac{1}{(1-x)^2}$, *find the coefficients* $a_0, a_1, a_2, \ldots\ldots$ *in the expansion*

$$f(x) = \sum_{k=0}^{\infty} a_k x^k.$$

Solution. We know that

$$\frac{1}{1-x} = 1 + x + x^2 + x^3 + \ldots\ldots$$

Hence, we have

$$\frac{1}{(1-x)^2} = \sum_{k=0}^{\infty}\left(\sum_{j=0}^{k} 1\right)x^k = \sum_{k=0}^{\infty}(k+1)x^k.$$

Problem 3.28. *Find the values of the extended binomial coefficients*

$$\binom{-2}{3} \text{ and } \binom{1/2}{3}.$$

Solution. Taking $u = -2$ and $k = 3$ in

$$\binom{u}{k} = \begin{cases} u(u-1)\ldots\ldots(u-k+1)/k! & \text{if } k > 0 \\ 1 & \text{if } k = 0 \end{cases}$$

which gives, $\displaystyle\binom{-2}{3} = \frac{(-2)(-3)(-4)}{3!} = -4$

Similarly, taking $u = \dfrac{1}{2}$ and $k = 3$ gives us

$$\binom{1/2}{3} = \frac{(1/2)(1/2-1)(1/2-2)}{3!}$$

$$= \left(\frac{1}{2}\right)\left(-\frac{1}{2}\right)\left(-\frac{3}{2}\right)\Big/6 = \frac{1}{16}.$$

Problem 3.29. *Find the generating functions for $(1 + x)^{-n}$ and $(1 - x)^{-n}$, where n is a positive integer, using the extended Binomial theorem.*

Solution. By the extended Binomial theorem, it follows that

$$(1+x)^{-n} = \sum_{k=0}^{\infty}\binom{-n}{k}x^k$$

Using $\binom{-n}{r} = (-1)^r\, C(n + r - 1, r)$, which provides a simple formula for $\binom{-n}{k}$, we obtain

$$(1 + x)^{-n} = \sum_{k=0}^{\infty} (-1)^k \;\; C(n + k - 1, k)x^k.$$

Replacing x by $-x$, we find that

$$(1 - x)^{-n} = \sum_{k=0}^{\infty} C(n + k - 1, k)x^k.$$

Problem 3.30. *Use generating functions to find the number of k-combinations of a set with n elements. (Assume that the Binomial theorem has already been established).*

Solution. Each of the n elements in the set contributes the term $(1 + x)$ to the generating function

$$f(x) = \sum_{k=0}^{n} a_k x^k .$$

Here $f(x)$ is the generating function for $\{a_k\}$, where a_k represents the number of k-combinations of a set with n elements.

Hence, $f(x) = (1 + x)^n$.

But by Binomial Theorem, we have

$$f(x) = \sum_{k=0}^{n} \binom{n}{k} x^k,$$

where $$\binom{n}{k} = \frac{n!}{k!(n-k)!}.$$

Hence, C(n, k), the number of k-combinations of a set with n elements, is $\dfrac{n!}{k!(n-k)!}$.

Problem 3.31. *Use generating functions to find the number of r-combinations from a set with n elements when repetition of elements is allowed.*

Solution. Let G(x) be the generating function for the sequence $\{a_r\}$, where a_r equals the number of r-combinations of a set with n elements with repetitions allowed.

That is, $$G(x) = \sum_{r=0}^{\infty} a_r\, x^r.$$

Since we can select any number of a particular member of the set with n elements when we form an r-combinations with repetition allowed, each of the n elements contributes $(1 + x + x^2 + x^3 + \ldots\ldots)$ to a product expansion for G(x).

Each element contributes this factor since it may be selected zero times, one time, two times, three times, and so on, when an r-combinations is formed (with a total of r elements selected).

Since there are n elements in the set and each contributes this same factor to G(x), we have

$$G(x) = (1 + x + x^2 + ...)^n$$

As long as $| x | < 1$, we have $1 + x + x^2 + ... = \dfrac{1}{1-x}$

So, $G(x) = \dfrac{1}{(1-x)^n} = (1-x)^{-n}$

Applying the extended Binomial Theorem, it follows that

$$(1-x)^{-n} = (1 + (-x))^{-n} = \sum_{r=0}^{\infty} \binom{-n}{r} (-x)^r$$

The number of r-combinations of a set with n elements with repetitions allowed, when r is a positive integer, is the coefficient a_r of x^r in this sum.

Consequently, we find that a_r equals

$$\binom{-n}{r}(-1)^r = (-1)^r\, C(n+r-1, r)\,.\,(-1)^r$$

$$= C(n+r-1, r).$$

Problem 3.32. *A box contains many identical red, blue, white, and green marbles. Find the ordinary generating function corresponding to the problem of finding the number of ways of choosing r marble from the box such that the sample does not have more than 2 red, more than 3 blue, more than 4 white, and more than 5 green.*

Solution. The generating function is

$$(1 + x + x^2)(1 + x + x^2 + x^3)(1 + x + ... + x^4)(1 + x + ... + x^5)$$
$$= (1 - x^3)\,(1 - x^4)\,(1 - x^5)\,(1 - x^6)\,(1 - x)^{-4}.$$

Problem 3.33. (Uniqueness of Base-b Representation)

If b is an integer greater than 1, prove by means of a generating function that an arbitrary positive integer r can be written as

$$r = r_0 b^0 + r_1 b^1 + r_2 b^2 + ...,\quad 0 \le r_i \le b - 1 \qquad ...(1)$$

in 1 and only 1 way.

Solution. The generating function for the number of solution vectors $(r_0, r_1, r_2, ...)$ of (1) is obviously

$$[1 + x^{1(1)} + x^{2(1)} + ... + x^{(b-1)(1)}]\,[1 + x^{1(b)} + x^{2(b)} + ... + x^{(b-1)(b)}]$$
$$\times\,[1 + x^{1(b^2)} + x^{2(b^2)} + ... + x^{(b-1)(b^2)}]\, ...$$

$$= \frac{1-x^b}{1-x}\;\frac{1-x^{b^2}}{1-x^b}\;\frac{1-x^{b^3}}{1-x^{b^2}}\;......$$

$$= \frac{1}{1-x} = 1 + x + x^2 + x^3 + \ldots\ldots$$

Each coefficient in the generating function is 1,

i.e., any r has a unique base-b representation.

Problem 3.34. *Given a positive integer k, use the generating-function method to find the number of solutions in non negative integers of* $u_1 + u_2 + \ldots\ldots + u_k = n$,

(*a*) *the first 2 variables are atmost 2, and*

(*b*) *the sum of the first two variables is atmost 2.*

Solution. (*a*) $f(x) = (1 + x + x^2)^2 (1 + x + x^2 + \ldots\ldots)^{k-2}$

$$= (1 - x^3)^2 (1 - x)^{-k}.$$

(*b*) **Case 1.** Both variables are 0, and the sum is 0.

Case 2. One of them is 1 and the order is 0, and the sum is 1.

Case 3. Both are 1 or one of them is 2 and the other is 0, and the sum is 2.

These cases occur in 1, 2, and 3 ways, respectively.

So the generating function is

$$(1 + 2x + 3x^2)(1 + x + x^2 + \ldots\ldots)^{k-2} = (1 + 2x + 3x^2)(1 - x)^{-k+2}.$$

Problem 3.35. *Find an ordinary generating function that solves the problem of finding the number of positive 5-digit integers with digit sum r.*

Solution. The leading digit is atleast 1 and most 9 ; the other 4 digits are non negative and atmost 9.

Hence, the generating function is

$$(x + x^2 + \ldots\ldots + x^9)(1 + x + x^2 + \ldots\ldots + x^9)^4 = x(1 - x^9)(1 - x^{10})(1 - x)^{-2}.$$

Problem 3.36. *Let* $f(x) = (1 + x + \ldots\ldots + x^n)^3$ *and* $g(x) = (1 + x + \ldots\ldots + x^{n-1})^3$. *Use a combinatorial argument to show that the coefficient of* x^{2n+1} *in f(x) is equal to the coefficient of* x^{2n-2} *in g(x).*

Solution. Consider the equation $a + b + c = 2n + 1$, where the 3 variables are non negative integers at most equal to n.

The number of solutions is the coefficient of x^{2n+1} in $f(x)$.

But no variable in this equation is 0, for then 1 of the remaining 2 variables would have to exceed n.

So the number of solutions is also equal to the coefficient of x^{2n+1} in

$$(x + x^2 + \ldots\ldots + x^n)^3 = x^3 g(x)$$

which is the coefficient of x^{2n-2} in $g(x)$.

Problem 3.37. *Find the coefficient of* x^{27} *in*

(*a*) $(x^4 + x^5 + x^6 + \ldots\ldots)^5$ *and* (*b*) $(x^4 + 2x^5 + 3x^6 + \ldots\ldots)^5$.

Solution. (*a*) Since $(x^4 + x^5 + \ldots\ldots)^5 = x^{20} (1 - x)^{-5}$, what is required is the coefficient of x^7 in $(1 - x)^{-5}$.

This is C(11, 4).

(*b*) Since $(x^4 + 2x^5 + 3x^6 + ...)^5 = x^{20}\,[(1 - x)^{-2}]^5 = x^{20}(1 - x)^{-10}$ we require the coefficient of x^7 in $(1 - x)^{-10}$, which is C(16, 9).

Problem 3.38. *Find the sequences corresponding to the ordinary generating functions (a) $(3 + x)^3$ (b) $3x^3 + e^{2x}$ and (c) $2x^2(1 - x)^{-1}$.*

Solution. (*a*) $(3 + x)^3 = 27 + 27x + 9x^2 + x^3$

The sequence is $<27, 27, 9, 1, 0, 0, 0, ...>$.

(*b*) $$3x^3 + e^{2x} = 1 + 2x + \frac{2^2}{2!}x^2 + \left(3 + \frac{2^3}{3!}\right)x^3 + \frac{2^4}{4!}x^4 + \frac{2^5}{5!}x^5 + ...$$

The sequence is $$\left\langle 1, 2, \frac{2^2}{2!}, \frac{2^3}{3!} + 3, \frac{2^4}{4!} \right\rangle.$$

(*c*) $2x^2(1 - x)^{-1} = 2x^2\,(1 + x + x^2 + x^3 + ...)$

The sequence is $<0, 0, 2, 2, 2, ...>$.

Problem 3.39. *Using the generating function of $\sum_{k=0}^{n} C(n, k)x^k = (1 + x)^n$, establish Pascal's identity $C(n + 1, r) = C(n, r) + C(n, r - 1)$.*

Solution. The coefficient of x^r in $(1 + x)^{n+1}$ is $C(n + 1, r)$.

But $(1 + x)^{n+1} = (1 + x)^n$, and the coefficient of x^r in the right-hand side is $C(n, r) + C(n, r - 1)$.

Problem 3.40. *Find the ordinary generating function of the sequence*

$$<C(r + n - 1, n - 1)>_{r \geq 0}.$$

(a) by a combinatorial argument, and

(b) by differentiation of the infinite geometric series.

Solution. (*a*) From $x_1 + x_2 + ... + x_m = n$, it is known that $C(r + n - 1, n - 1)$ counts the non negative integral solutions of $u_1 + u_2 + ... + u_n = r$.

Therefore, one can write

$$\frac{1}{(1-x)^n} = \left(\frac{1}{1-x}\right)\left(\frac{1}{1-x}\right) ...... \left(\frac{1}{1-x}\right)$$

$$= (1 + x + x^2 + ... + x^{u_1} + ...)(1 + x + x^2 + ... + x^{u_2} + ...)$$
$$...... (1 + x + x^2 + ... + x^{u_n} + ...)$$

$$= \sum_{r=0}^{\infty}\left(\sum_{\substack{u_1 + u_2 + ... + u_n = r \\ u_i \geq 0}} 1\right)x^r$$

$$= \sum_{r=0}^{\infty} C(r + n - 1, n - 1)x^r$$

(*b*) Differentiate $\dfrac{1}{1-x} = \sum\limits_{k=0}^{\infty} x^k$

$n - 1$ times to obtain.

$$\frac{(n-1)!}{(1-x)^n} = \sum_{k=n-1}^{\infty} k(k-1) \;......\; (k-n+2)x^{k-n+1}$$

$$= \sum_{r=0}^{\infty} (r+n-1)(r+n-2) \;.......\; (r+1)x^r$$

$$= \sum_{r=0}^{\infty} \frac{(r+n-1)!}{r} x^r$$

Division of both sides by $(n-1)$! reproduces the result of (*a*).

Problem 3.41. *Use the generating function method*

(*a*) *to count the distinct binary solutions of* $u_1 + u_2 + \;......\; + u_n = r$

(*b*) *to establish the pigeonhole principle.*

Solution. (*a*) The generating function (on *r*) is

$$(1+x)^n = \sum_{r=0}^{n} C(n, r)x^r$$

Thus these $C(n, r)$ solutions ($= 0$ for $r > n$)

(*b*) The function $(1+x)^n$ is also the generating function corresponding to the problem of distributing *r* identical pigeons among *n* distinct pigeonholes so that each hole receives fewer than 2 pigeons.

The coefficient of x^{n+1} in the generating function is zero.

Hence, when $n + 1$ pigeons are distributed, some hole receives at least 2 pigeons.

Problem 3.42. *If throwing a die 5 times constitutes a trial, with the 5 throws considered distinguishable, find the number of trials that produce a total of 12 or fewer dots.*

Solution. Let $a_r \equiv$ number of trials that produce *r* dots

$A_r \equiv$ number of trials that produce atmost *r* dots

Clearly, $\langle a_r \rangle$, has the generating function

$$(x^1 + x^2 + \;......\; + x^6)^5 = \left(\frac{x - x^7}{1-x}\right)^5$$

Hence $\langle A_r \rangle$ has the generating function

$$(1-x)^{-1}\left(\frac{x-x^7}{1-x}\right)^5 = (x - x^7)^5 (1-x)^{-6}$$

$$= (x^5 - 5x^{11} + 10x^{17} - \;......)\sum_{r=0}^{\infty} C(r+5, 5)x^r$$

The coefficient of x^{12} on the right is

$$A_{12} = 1.C(12, 5) - 5 . C(6, 5) = 762.$$

Problem 3.43. *Prove, for all r and n, $P_n(r) = q_n(r)$, without drawing the star diagram.*

Solution. The system $1u_1 + 2u_2 + 3u_3 + ... + nu_n = r$ (u_i a non negative integer)

Which by definition has precisely $P_n(r)$ solutions, is taken by the substitution

$$\begin{aligned} u_n &= w_1 \\ u_{n-1} &= w_2 - w_1 \\ u_{n-2} &= w_3 - w_2 \\ &\ldots\ldots\ldots\ldots \\ u_1 &= w_n - w_{n-1} \end{aligned} \qquad ...(1)$$

into the system

$$w_1 + w_2 + w_3 + ... + w_n = r$$

$$w_1 \le w_2 \le w_3 \le ...... \le w_n \qquad (w_i \text{ a non negative integer})$$

This later system has precisely $q_n(r)$ solutions.

But the mapping (1) is obviously bijective, so that

$$P_n(r) = q_n(r).$$

Problem 3.44. (Euler's Theorem)

Let P(r, ODD) be the number of ways of partitioning r into (possibly repeated) odd parts. Show that, for every r, P(r, ODD) = $P^{\#}(r)$.

Solution. The ordinary generating function of $< P(r, ODD) >$ is given by

$$\frac{1}{(1-x^2)(1-x^3)(1-x^5)\ldots\ldots} = \left(\frac{1-x^2}{1-x^1}\right)\left(\frac{1-x^4}{1-x^2}\right)\left(\frac{1-x^6}{1-x^3}\right)\left(\frac{1-x^8}{1-x^4}\right)\ldots\ldots$$

$$= (1+x^1)(1+x^2)(1+x^3)(1+x^4)\ldots\ldots$$

$= \text{generating function of} < P^{\#}(r) >$

$$= \sum_{r=0}^{\infty} P^{\#}(r)\, x^r.$$

Problem 3.45. *Find the exponential generating function of $<a_r>$, where a_r is the number of r sequences of the set E = $\{e_1, e_2, ..., e_n\}$.*

Solution. Looking at the product

$$G(x) \equiv \left(1 + x + \frac{x^2}{2!} + \ldots\ldots + \frac{x^{i_1}}{i_1!} + \ldots\ldots\right)\left(1 + x + \frac{x^2}{2!} + \ldots\ldots + \frac{x^{i_2}}{i_2!} + \ldots\ldots\right)$$

$$\ldots\ldots \left(1 + x + \frac{x^2}{2!} + \ldots\ldots + \frac{x^{i_n}}{i_n!} + \ldots\ldots\right)$$

One sees that the r sample of E consisting of i_1e_1's, i_2e_2's, ..., i_ne_n's where $i_1, i_2, ..., i_n$ are non negative integers with sum r-contributes

$$\frac{r!}{i_1!\,i_2!\ldots\ldots i_n!} = \mathrm{P}(r\,;\,i_1, i_2, \ldots\ldots i_n)$$

to the coefficeint of $\frac{x^r}{r!}$ in G(x).

The number of r sequences of E generated by permutation of the given r sample.

Hence, the total coefficient of $\frac{x^r}{r!}$ is just a_r.

This is, G(x) = $(e^x)^n$ is the desired exponential generating function.

Problem 3.46. *If a leading digit of O is permitted, find the numbers of r-digit binary numbers that can be formed using*

(*a*) *an even number of 0^s and an even number 1 s,*

(*b*) *an odd number of 0^s and an odd number of 1 s.*

Solution. Here we are counting r sequences of the set {0, 1} that obey certain restrictions.

(*a*) The exponential generating function is

$$\mathrm{F}_e(x) = \left(1 + \frac{x^2}{2!} + \frac{x^4}{4!} + \ldots\ldots\right)^2 = \left(\frac{e^x + e^{-x}}{2}\right)^2$$

$$= \frac{1}{4}(e^{2x} + e^{-2x} + 2)$$

The coefficient of $\frac{x^r}{r!}$ in $\mathrm{F}_e(x)$ is 2^{r-1} if r is even, and 0 if r is odd.

$$(b)\ \mathrm{F}_0(x) = \left(x + \frac{x^3}{3!} + \frac{x^5}{5!} + \ldots\ldots\right)^2 = \left(\frac{e^x - e^{-x}}{2}\right)^2$$

$$= \mathrm{F}_e(x) - 1$$

Thus the answer is the same as in (*a*).

Problem 3.47. *The exponential generating function for the Bell numbers was found in*

$$B_n = n!\sum_{{}^1P_1 + {}^2P_2 + \ldots\ldots + {}^nP_n = n} \frac{1}{[P_1!\,(1!)P_1]\,[P_2!\,(2!)^{P_2}]\ldots\ldots[P_n!\,(n!)^{P_n}]}$$

and $B_n = \left.\frac{d^n}{dx^n}e^{e^x-1}\right|_{x=0}$ *to be* e^{e^x-1}.

Check this result by use of the exponential generating function for the stirling numbers of the second kind.

Solution. By the definition of the two kinds of numbers

$$\mathrm{B}_n = \sum_{m=0}^{\infty} \mathrm{S}(n, m),\ \mathrm{B}_0 = \mathrm{S}(0, 0) = 1\,;$$

If $n \geq 1$, only n of the summands are non zero.

Because the exponential generating of a sum is the sum of the generating functions and

$$(e^x - 1)^m = \left(x + \frac{x^2}{2!} + \ldots\ldots + \frac{x^{n_1}}{n_1!} + \ldots\ldots\right)\left(x + \frac{x^2}{2!} + \ldots\ldots + \frac{x^{n_2}}{n_2!} + \ldots\ldots\right)$$

$$x + \frac{x^2}{2!} + \ldots\ldots + \frac{x^{nm}}{n_m!} + \ldots\ldots$$

gives :

Exponential generating function of

$$<B_n> = \sum_{m=0}^{\infty} \frac{1}{m!}(e^x - 1)^m = e^{e^x - 1}.$$

Problem 3.48. *Find the number of r-letter sequences that can be formed using the letters P, Q, R and S such that in each sequence there are an odd number of P's and an even number of Q's.*

Solution. The answer is the coefficient of $\frac{x^r}{r!}$ in

$$\left(x + \frac{x^3}{3!} + \frac{x^5}{5!} + \ldots\ldots\right)\left(1 + \frac{x^2}{2!} + \frac{x^4}{4!} + \ldots\ldots\right)(e^x)(e^x)$$

$$= \left(\frac{e^x - e^{-x}}{2}\right)\left(\frac{e^x + e^{-x}}{2}\right)e^{2x} = \frac{1}{4}(e^{4x} - 1)$$

This coefficient is 4^{r-1}.

Problem 3.49. *The sequence of Bernoulli numbers, $<b_n>_{n \geq 0}$, has the exponential generating function $\frac{x}{e^x - 1}$.*

Show that (a) $b_3 = b_5 = b_7 = \ldots\ldots = 0$ and

(b) $b_n = \sum_{m=0}^{\infty} \frac{(-1)^m}{m+1} m!\, S(n, m)$.

Solution. (*a*) $b_1 \frac{x^1}{1!} + b_3 \frac{x^3}{3!} + b_5 \frac{x^5}{5!} + \ldots\ldots = \frac{1}{2}\left(\frac{x}{e^x - 1} - \frac{-x}{e^{-x} - 1}\right) = -\frac{1}{2}x$

whence $b_1 = -\frac{1}{2}$ and $b_3 = b_5 = \ldots\ldots = 0$.

(*b*) Two sequences are identical if and only if they have the same exponential generating function.

The exponential generating function of the numbers

$$\sum_{m=0}^{\infty} \frac{(-1)^m}{m+1} m!\,S(n, m)$$

is given by

$$\sum_{m=0}^{\infty} \frac{(-1)^m}{m+1}(e^x - 1)^m = \frac{1}{1-e^x} \sum_{m=0}^{\infty} \frac{(1-e^x)^{m+1}}{m+1}$$

$$= \frac{1}{1-e^x} \int_0^{1-e^x} \frac{1}{1-u}\,du$$

$$= \frac{1}{1-e^x}(-x) = \frac{x}{e^x - 1}.$$

Problem 3.50. *Find the number of ways of distributing 10 distinguishable books among 4 distinguishable shelves so that each. Self gets atleast 2 and atmost 7 books.*

Solution. This is a problem in restricted sequences.

Here we want to count the 10-sequences of a 4 set (the ith term of a sequence is the shelf to which the ith book is assigned) that can exist under the given limitations.

The appropriate exponential generating function is

$$G(x) = \left(\frac{x^2}{2!} + \frac{x^3}{3!} + \ldots\ldots + \frac{x^7}{7!}\right)^4$$

$$= x^8 \left(\frac{1}{2!} + \frac{x}{3!} + \frac{x^2}{4!} + \ldots\ldots\right)^4$$

and the answer is the coefficient of $\frac{x^{10}}{10!}$ in G(x), which is $\frac{10!}{16} = 226,800.$

Problem 3.51. (Dobinski's Equality)

$$\textit{Prove that } B_n = \begin{cases} 1 & n = 0 \\ e^{-1} \sum_{k=1}^{\infty} \frac{k^n}{k!} & n = 1, 2, 3, \ldots\ldots \end{cases}$$

Solution. The exponential generating function of the number on the right of the asserted equality is

$$1 + e^{-1} \sum_{n=1}^{\infty}\left(\sum_{k=1}^{\infty} \frac{k^n}{k\,!}\right)\frac{x^n}{n\,!} = 1 + e^{-1} \sum_{k=1}^{\infty} \frac{1}{k\,!} \sum_{n=1}^{\infty} \frac{(kx)^n}{n\,!}$$

$$= 1 + e^{-1} \sum_{k=1}^{\infty} \frac{1}{k\,!} [(e^x)^k - 1]$$

$$= 1 + e^{-1} (e^{e^x} - e) = e^{e^x} - 1.$$

which is just the exponential generating function of $<B_n>$.

Problem 3.52. *Derive the linear recursion relation for the Bell numbers from the exponential generating function*

$$\left(B_n = \sum_{k=0}^{n-1} C(n-1, k) B_k\right).$$

Solution. Differentiation of $e^{e^x - 1} = \sum_{r=0}^{\infty} B_r \frac{x^r}{r\,!}$

gives $(e^{e^x - 1})(e^x) = \sum_{r=0}^{\infty} B_r \frac{x^{r-1}}{(r-1)\,!}$.

On the right the coefficient of $\frac{x^r}{r\,!}$ is B_{r+1}, on the left it is

$$\sum_{i=0}^{r} C(r, i) B_i$$

Therefore, $B_{r+1} = \sum_{i=0}^{r} C(r, i) B_i$ or $B_r = \sum_{i=0}^{r-1} C(r-1, i) B_i$.

Problem 3.53. *Obtain a linear recurrence relation for the Bernoulli numbers.*

Solution. We have $x = (e^x - 1) \sum_{n=0}^{\infty} b_n \frac{x^n}{n\,!}$

The left-hand side is the exponential generating function of < 0, 1, 0, 0, 0, ...>. Since $e^x - 1$ generates <0, 1, 1, 1, ...>.

The right hand side generates $\left\langle \sum_{i=1}^{n} C(n, i) b_{n-i} \right\rangle$

Consequently, $b_0 = 1$ and for $n \geq 2$

$$\sum_{i=1}^{n} C(n, i) b_{n-i} = 0 \text{ or } b_{n-1} = -\frac{1}{n} \sum_{i=2}^{n} C(n, i) b_{n-i} .$$

Problem 3.54. *For every positive integer n let* $S_0(n) = n$ *and*

$$S_m(n) = 1^m + 2^m + 3^m + \ldots\ldots + (n-1)^m$$

where m is a positive integer. Obtain the exponential generating function of the sequence

$$<S_m(n)>_{m \geq 0}.$$

Solution. By linearity,

generating function of $<S_m(k+1)>_{m \geq 0}$ – generating function of $< S_m(k) >_{m \geq 0}$ = generating function of $<S_m(k+1) - S_m(k)>_{m \geq 0}$

$$= \text{generating function of } <k^m>_{m \geq 0} = e^{kx} = (e^x)$$

This simple recurrence relation is solved by summation

$$[S_m(0) \equiv 0]\ ;$$

$$\text{generating function of } <S_m(n)>_{m \geq 0} = \sum_{k=0}^{n-1} (e^x)^k = \frac{e^{nx} - 1}{e^x - 1}.$$

Problem 3.55. *The Bernoulli polynomial of degree m is defined by*

$$B_m(t) = \sum_{i=0}^{m} C(m, i) b_i\, t^{m-i}$$

(*a*) *Show that* $S_{m-1}(n) = [B_m(n) - B_m(0)]/m$.

(*b*) *Obtain the exponential generating function (in the variable x) of the sequence* $<B_m(t)>_{m \geq 0}$.

Solution. (*a*) We have $S_m(n) = \dfrac{1}{m+1} \displaystyle\sum_{i=0}^{m} C(m+1, i) b_i n^{m+1-i}$

$$S_{m-1}(n) = \frac{1}{m}\left[\sum_{i=0}^{m} C(m, i) b_i n^{m-1} - C(m, m) b_m\, n^{m-m}\right]$$

$$= \frac{1}{m}[B_m(n) - B_m(0)].$$

(*b*) The sequences $<b_m>_{m \geq 0}$ and $<t^m>_{m \geq 0}$ have respective exponential generating functions

$$\frac{x}{e^x - 1} \text{ and } e^{tx}.$$

Then by the binomial convolution

$$\left\langle \sum_{i=0}^{m} C(m, i) b_i t^{m-i} \right\rangle_{m \geq 0} = <B_m(t)>_{m \geq 0}.$$

must have the product $\dfrac{xe^{tx}}{(e^x - 1)}$ as its exponential generating function.

Problem 3.56. *Give the ordinary generating function of the sequence* $<n(3 + 5n)>$.

Solution. We have $n(3 + 5n) = 3n + 5n^2$

By $g(x) = \sum_{n=0}^{\infty} a_n x^n$, the respective generating functions of $<n>$ and $<n^2>$ are

$$\frac{x}{(1-x)^2} \text{ and } \frac{x(1+x)}{(1-x)^3}.$$

Hence, the answer is

$$\frac{3x}{(1-x)^2} + \frac{5x(1+x)}{(1-x)^3} = \frac{8x+2x^2}{(1-x)^3}.$$

Problem 3.57. (Restricted Partitions)

Given a collection k of n distinct positive integers $\alpha_1 < \alpha_2 < \ldots\ldots < \alpha_n$, *and an arbitrary positive integer r, let* $f_n(r) \equiv$ *number of partitions of r into parts selected (with replacement) from k.*

Determine the ordinary generating function of $<f_n(0) \equiv 1, f_n(1), \ldots>$.

Solution. Here we want to count the solutions in non negative integers of

$$\alpha_1 u_1 + \alpha_2 u_2 + \ldots\ldots + \alpha_n u_n = r$$

and so we write

$$\prod_{i=1}^{n} \frac{1}{1-x^{\alpha_i}} = [1 + x^{\alpha_1} + (x^{\alpha_1})^2 + \ldots\ldots + (x^{\alpha_1})^{u_1} + \ldots\ldots]$$

$$\times [1 + x^{\alpha_2} + (x^{\alpha_2})^2 + \ldots\ldots + (x^{\alpha_2})^{u_2} + \ldots\ldots] \times \ldots\ldots$$

$$\ldots\ldots \times [1 + x^{\alpha_n} + (x^{\alpha_n})^2 + \ldots\ldots + (x^{\alpha_n})^{u_n} + \ldots\ldots]$$

$$= \sum_{r=0}^{\infty} f_n(r)\, x^r.$$

Problem 3.58. *In an experiment, 4 differently coloured dice are thrown simultaneously, and the numbers are added. Find the numbers of distinct experiments such that (a) the total is 18 and (b) the total is 18 and the green die shows an even number.*

Solution. (*a*) The answer is the coefficient of x^{18} in the generating function

$$(x^1 + x^2 + \ldots\ldots + x^6)^4 = (x - x^7)^4\,(1 - x^{-4})$$

$$= (x^4 - 4x^{10} + 6x^{16} - \ldots\ldots) \sum_{r=0}^{\infty} C(r+3, 3)x^r$$

which is seen to be

$$1 \,.\, C(17, 3) - 4 \,.\, C(11, 3) + 6 \,.\, C(5, 3) = 80.$$

(*b*) Now the generating function is

$$(x^2 + x^4 + x^6)(x^1 + x^2 + \ldots\ldots + x^6)^3$$

$$= (x^2 + x^4 + x^6)(x^3 - 3x^9 + 3x^{15} - x^{21}) \sum_{r=0}^{\infty} C(r + 2, 2)x$$

$$= (x^5 + x^7 + x^9 - 3x^{11} - 3x^{13} - 3x^{15} + 3x^{17} + ...) \sum_{r=0}^{\infty} C(r + 2, 2)x$$

in which the coefficient of x^{18} is

$$1.C(15, 2) + 1.C(13, 2) + 1.C(11, 2) - 3.C(9, 2) - 3.C(7, 2) - 3.C(5, 2) + 3.C(3, 2) \text{ or } 46.$$

Problem 3.59. *Find the number of ways of forming a committee of 9 people drawn from 3 different parties so that no party has an absolute majority in the committee.*

Solution. If any party is excluded, one of the other parties will have an absolute majority. So there must be atleast 1 person from each party. And no party can have more than 4 repersentatives in the committee.

Thus, the generating function is

$$f(x) = (x + x^2 + x^3 + x^4)^3$$

$$= (x^3 - 3x^7 + 3x^{11} - x^{15}) \sum_{r=0}^{\infty} C(r + 2, 2)x^r$$

The answer is the coefficient of x^9 in $f(x)$, which is

$$1.C(8, 2) - 3.C(4, 2) = 10.$$

Problem 3.60. *Establish the upper bound $P(r) < \exp. (3\sqrt{r})$.*

Solution. If $g(x)$ is the generating function of $<P(r)>$ then for any r and all $0 < x < 1$,

where $$g(x) = \frac{1}{(1-x)(1-x^2)(1-x^3)......}$$

$$P(r)x^r < g(x) \text{ or } \log P(r) < \log g(x) - r \log x$$

From the well-known expansion

$$\log \frac{1}{1-u} = u + \frac{u^2}{2} + \frac{u^3}{2} + ... \quad (0 \le u < 1)$$

it follows that

$$\log g(x) = \left(x + \frac{x^2}{2} + \frac{x^3}{3} + ...\right) + \left(x^2 + \frac{x^4}{2} + \frac{x^6}{3} + ...\right)$$

$$+ \left(x^3 + \frac{x^6}{2} + \frac{x^9}{3} + ...\right) + ...$$

$$= (x + x^2 + x^3 + \ldots\ldots) + \frac{1}{2}(x^2 + x^4 + x^6 + \ldots\ldots) + \frac{1}{3}(x^3 + x^6 + x^9 + \ldots\ldots)$$

$$= \frac{x}{1-x} + \frac{1}{2}\frac{x^2}{1-x^2} + \frac{1}{3}\frac{x^3}{1-x^3} + \ldots\ldots$$

Now, for $0 < x < 1$ and $k = 1, 2, 3, \ldots\ldots$

$$\frac{x^k}{1-x^k} = \frac{x}{1-x}\frac{x^{k-1}}{1 + x + x^2 + \ldots\ldots + x^{k-1}} < \frac{x}{1-x}\frac{x^{k-1}}{x^{k-1} + x^{k-1} + \ldots\ldots + x^{k-1}}$$

$$= \frac{x}{1-x}\frac{1}{k}$$

Whence $\log g(x) < \frac{x}{1-x}\left[1 + \left(\frac{1}{2}\right)^2 + \left(\frac{1}{3}\right)^2 + \ldots\ldots\right]$

$$= \frac{x}{1-x}\frac{\pi^2}{6}$$

Further, $\quad -\log x = \log\frac{1}{x} = \int_1^{1/x}\frac{dt}{t} < \int_1^{1/x} dt = \frac{1-x}{x}.$

Therefore we have

$$\log P(r) < \frac{\pi^2}{6}\frac{x}{1-x} + r\frac{1-x}{x} \qquad \ldots(1)$$

Using standard calculus to minimize the right-hand side of (1) over $0 < x < 1$, one obtains

$$\log P(r) < \frac{2\pi}{\sqrt{6}}\sqrt{r} < 3\sqrt{r}$$

and the proof is complete.

Problem 3.61. *Establish the lower bound $P(r) \geq 2^q$ for $r \geq 2$, where $q \equiv \lfloor\sqrt{r}\rfloor$.*

Solution. The bound may be established by inspection for $q = 1, 2$; so assume $q \geq 3$. It is asserted that each non empty subset S of

$X = \{1, 2, 3, \ldots\ldots, q\}$ generates a partition of r.

In fact, if $\sigma(S) \equiv$ Sum of the elements of S

then $$\sigma(S) \leq \sigma(X) = \frac{q^2 + q}{2} \leq \frac{r + \sqrt{r}}{2} < \frac{r + r}{2} = r.$$

So that $S \cup \{r - \sigma(S)\}$ is the desired partition of r.

Furthermore, distinct subsets generate distinct partitions.

To see that this is so, let

$$S_1 \cup \{r - \sigma(S_1)\} \text{ and } S_2 \cup \{r - \sigma(S_2)\} \qquad \text{...(1)}$$

be the partitions generated by the distinct k subsets $(1 \le k \le q - 1)S_1$ and S_2.

For $i = 1, 2$,

we have $$r - \sigma(S_i) \ge q^2 - [\sigma(X) - 1] = q^2 - \left[\frac{q^2 + q - 2}{2}\right]$$

$$= \frac{q^2 - q + 2}{2} = q + \frac{(q-2)(q-1)}{2}$$

Consequently, for $q \ge 3$, $r - \sigma(S_i) > q$ ...(2)

If the 2 partitions (1) coincided, and if $\sigma(S_1) = \sigma(S_2)$, then S_1 must coincide with S_2, which is contrary to the hypothesis.

On the otherhand, if the 2 partitions (1) coincided, and if $\sigma(S_1) \ne \sigma(S_2)$, then $r - \sigma(S_1)$ would have to be an element of S_2, which is ruled out by (2).

The conclusion is that the number of partitions of r must exceed the number of non null subsets of X :

$$P(r) > 2^q - 1 \ge 2^q.$$

Problem 3.62. *Define*

$q^{\#}(r, E)$ = number of partitions of r into an even number of unequal parts

$q^{\#}(r, 0)$ = number of partitions of r into an odd number of unequal parts

Prove that $(1 - x)(1 - x^2)(1 - x^3) \ldots\ldots = \sum_{r=0}^{\infty} [q^{\#}(r, E) - q^{\#}(r, 0)]\, x^r$.

Solution. Because $(1 - x)(1 - x^2)(1 - x^3) \ldots\ldots$

$$= [1 + (-1)x][1 + (-1)x^2][1 + (-1)x^3] \ldots\ldots$$

any partition of r into an even number, e, of unequal parts will contribute $(-1)^e = +1$ to the coefficient of x^r in the infinite product.

Analogously, any partition of r into an odd number, 0, of unequal parts will contribute $(-1)^0 = -1$.

Therefore, the coefficient of x^r is

$$q^{\#}(r, E)(+1) + q^{\#}(r, 0)(-1) = q^{\#}(r, E) - q^{\#}(r, 0) \text{ as asserted.}$$

Problem 3.63. (Euler's First Identity)

Derive $(1 + x^1)(1 + x^3)(1 + x^5) \ldots\ldots = \sum_{k=0}^{\infty} \frac{x^{k^2}}{(1 - x^2)(1 - x^4)(1 - x^6) \ldots\ldots (1 - x^{2k})}$

(an empty product equals unity).

Solution. The left hand side is the ordinary generating function of $< P^{\#}(r, \text{ODD})>$.

In view of $P^{\#}(r, \text{ODD}) = \sum_{k=1}^{\lfloor\sqrt{r}\rfloor} P_{2k}(r-k^2, \text{EVEN})$

$$= \sum_{k=1}^{\lfloor\sqrt{r}\rfloor} q_k(r-k^2, \text{EVEN}).$$

it is enough to prove that the kth summand ($k = 0, 1, 2, \ldots\ldots$) on the right is the generating function of $< P_{2k}(r-k^2, \text{EVEN})>_{r \geq 0}$. But this is obvious, for

$$\frac{x^{k^2}}{(1-x^2)(1-x^4)\ldots\ldots(1-x^{2k})} = x^{k^2}(1+x^2+x^4+\ldots\ldots)(1+x^4+x^8+\ldots\ldots)$$

$$\ldots\ldots(1+x^k+x^{4k}+\ldots\ldots)$$

$$= x^{k^2}\sum_{s=0}^{\infty} P_{2k}(S, \text{EVEN})\, x^s = \sum_{s=0}^{\infty} P_{2k}(S, \text{EVEN})\, x^{s+k^2}$$

$$= \sum_{r=0}^{\infty} P_{2k}(r-k^2, \text{EVEN})\, x^r$$

Problem 3.64. (Euler's Second Identity)

Show that $(1+x^2)(1+x^4)(1+x^6)\ldots\ldots = \sum_{k=0}^{\infty} \dfrac{x^{k(k+1)}}{(1-x^2)(1-x^4)(1-x^6)\ldots\ldots(1-x^{2k})}$.

Solution. Suppose that we are given a partition of r into k distinct even parts.
Then subtraction of 1 from each part yields a unique partition of $r-k$ into k distinct odd parts.
Conversely, addition of 1 ...... yields a unique ......
By this one-to-one correspondence, and by the result

$$P^{\#}(r, \text{ODD}) = \sum_{k=1}^{\lfloor\sqrt{r}\rfloor} P_{2k}(r-k^2, \text{EVEN}) = \sum_{k=1}^{\lfloor\sqrt{r}\rfloor} q_k(r-k^2, \text{EVEN})$$

$$\begin{pmatrix}\text{No. of partitions of } r \text{ into} \\ k \text{ distinct even parts}\end{pmatrix} = \begin{pmatrix}\text{No. of partitions of } r-k \text{ into} \\ k \text{ distinct odd parts}\end{pmatrix}$$

$$= P_{2k}((r-k)-k^2, \text{EVEN})$$
$$= P_{2k}(r-k(k+1), \text{EVEN})$$

Therefore, the proof of Euler's second identity reduces to establishing that the ordinary generating function of

$$<P_{2k}(r - k(k + 1), \text{EVEN})>_{r \geq 0} \text{ is just}$$

$$\frac{x^{k(k+1)}}{(1 - x^2)(1 - x^4)(1 - x^6) \ldots\ldots (1 - x^{2k})}$$

This may be carried out by inspection.

Problem 3.65. *The number of partitions of r into n distinct (unequal) parts is denoted by $q^{\#}(r, n)$ Prove that*

$$q^{\#}(r, n) = q(r - C(n, 2), n)$$

Solution. By definition the system

$$\begin{aligned} &u_1 + u_2 + \ldots\ldots + u_n = r \\ &0 < u_1 < u_2 < \ldots\ldots < u_n \end{aligned} \qquad \ldots(1)$$

has precisely $q^{\#}(r, n)$ solutions in integers u_i. Under the bijective transformation

$$\begin{aligned} u_1 &= w_1 \\ u_2 &= w_2 + 1 \\ u_3 &= w_3 + 2 \\ &\ldots\ldots\ldots\ldots\ldots\ldots \\ u_n &= w_n + (n - 1) \end{aligned}$$

(1) goes over into

$$\begin{aligned} &w_1 + w_2 + \ldots\ldots + w_n = r - C(n, 2) \\ &0 < w_1 \leq w_2 \leq w_3 \leq \ldots\ldots \leq w_n \end{aligned} \qquad \ldots(2)$$

But, again by definitions, (2) has exactly

$q(r - C(n, 2), n)$ solutions in integers w_i.

Problem 3.66. *Show that*

$$P^{\#}(r, ODD) = \sum_{k=1}^{\lfloor \sqrt{r} \rfloor} P_{2k}(r - k^2, EVEN) = \sum_{k=1}^{\lfloor \sqrt{r} \rfloor} q_k(r - k^2, EVEN).$$

Solution. As in Fig. 3.1, represent a partition of r into distinct odd parts by nested elbows.

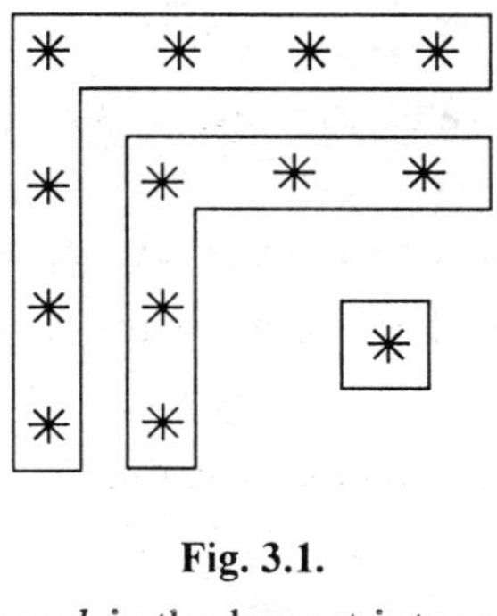

Fig. 3.1.

Let k be the number of parts. Then k is the largest integer such that the diagram contains a $k \times k$ square (called the DURFEE square) having as one corner the asterisk in the first row and first column.

Clearly, $1 \le k \le \lfloor \sqrt{r} \rfloor$, in Fig. 3.1, which diagrams

$$23 = 11 + 9 + 3, \quad \text{one has } k = 3.$$

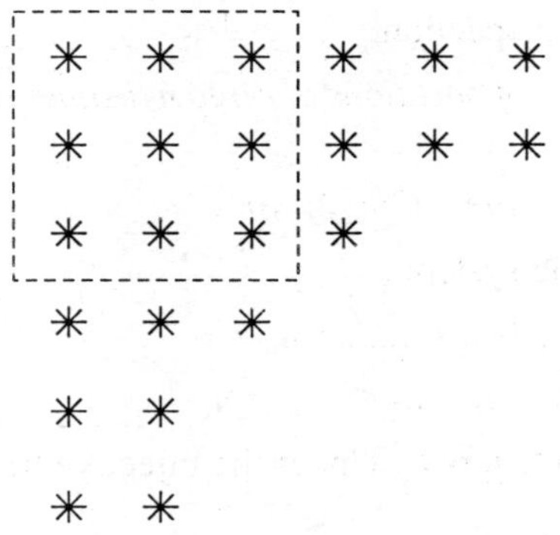

Fig. 3.2.

The remaining $r - k^2$ asterisks can be assembled into a partition of $r - k^2$ with all parts even, in two different ways :

(*i*) If the ith part in the partition is the total number of asterisks in the $(k + i)^{\text{th}}$ row and the $(k + i)^{\text{th}}$ column, then all parts are less than or equal to $2k$.

(*ii*) If the jth part in the partition is the number of asterisks in the jth elbow which lie outside the square, then there are at most k parts.

Since the geometric argument is reversible, we conclude that the number of partitions of r into k distinct odd parts is given by either $P_{2k}(r - k^2, \text{EVEN})$ or $q_k(r - k^2, \text{EVEN})$.

Hence, a summation on k yields the required results. Note that, as usual, the sums may be extended from $k = 0$ to $k = \infty$, since only null summands are thereby introduced.

Problem 3.67. *Let $P^{\#}(r)$ be the number of partitions of r into unequal parts. Obtain the ordinary generating function of $\langle P^{\#}(r) \rangle$.*

Solution. A given positive integer i appears either 0 times or 1 time among the parts of r ;

$$\text{Hence, } (1 + x^1)(1 + x^2)(1 + x^3) \ldots\ldots (1 + x^i) = \sum_{r=0}^{\infty} P^{\#}(r)x^r$$

It is evident that for a particular value of r say, $r = s$ on the first s factors of the infinite product need be retained.

Problem 3.68. *Let $P(r, n) \equiv$ number of partitions of r with largest part n*

$q(r, n)$ = number of partitions of r into exactly n parts.

(a) Prove that, for all r and n, $P(r, n) = q(r, n)$

(b) Determine the ordinary generating function (on r) of either sequence.

Solution. (*a*) $P^*(r, n) = P_n(r) - P_{n-1}(r)$

$$= q_n(r) - q_{n-1}(r) = q(r, n).$$

(*b*) The ordinary generating function of $<P_n(r)> = <q_n(r)>$ is

$$g_n(x) = \frac{1}{(1-x)(1-x^2)(1-x^3)\ldots\ldots(1-x^n)}$$

$$f_n(x) \equiv \sum_{r=0}^{\infty} P(r, n)x^r = \sum_{r=0}^{\infty} P_n(r)x^r - \sum_{r=0}^{\infty} P_{n-1}(r)x^r$$

$$= g_n(x) - g_{n-1}(x) = x^n g_n(x).$$

Problem 3.69. *Establish the recurrence relation*

$$q_n(r) = q_{n-1}(r) + q_n(r-n)$$

(also satisfied by $P_n(r)$)

(a) by solving a distribution problem, and

(b) by use of theorem ; the ordinary generating function of

$$<P_n(r)> = <q_n(r)> \text{ is } g_n(x) = \frac{1}{(1-x)(1-x^2)(1-x^3)\ldots\ldots(1-x^n)} \quad \ldots(1)$$

Solution. (*a*) Imagine you are given a heap of *r* identical 1 s and a row of *n* identical boxes.

Partitioning *r* into exactly *n* parts, which, by definition, can be accomplished in $q_n(r) - q_{n-1}(r)$ ways is tantamount to first putting *a*1 into each box (1 way) and then arbitrarily distributing the remaining $(r-n)$ 1 s among the *n* boxes [$q_n(r-n)$ ways].

Thus, by the product rule

$$q_n(r) - q_{n-1}(r) = 1 \,.\, q_n(r-n).$$

(*b*) From (1), $(1-x^n)\, g_n(x) = g_{n-1}(x)$

Equating coefficients of x^r,

$$q_n(r) - q_n(r-n) = q_{n-1}(r).$$

Problem 3.70. *Find the generating function for $P_d(n)$, the number of partitions of a positive integer n into distinct summands.*

Solution. Let us consider the 11 partitions of 6 :

1. 1 + 1 + 1 + 1 + 1 + 1
2. 1 + 1 + 1 + 1 + 2
3. 1 + 1 + 1 + 3
4. 1 + 1 + 4
5. 1 + 1 + 2 + 2
6. 1 + 5
7. 1 + 2 + 3
8. 2 + 2 + 2
9. 2 + 4
10. 3 + 3
11. 6

Partitions (6), (7), (9), and (11) have distinct summands so $P_d(6) = 4$.

In calculating $P_d(n)$, for each $k \in Z^+$ there are two choices : either *k* is not used as one of the summands of *n*, or it is. This can be accounted for by the polynomial $1 + x^k$, and consequently, the generating function for these partitions is

$$P_d(x) = (1+x)(1+x^2)(1+x^3) \ldots\ldots = \prod_{i=1}^{\infty} (1+x^i)$$

For each $n \in Z^+$, $P_d(n)$ is the coefficient of x^n in $(1+x)(1+x^2) \ldots\ldots (1+x^n)$.

When $n = 6$, the coefficient of x^6 in $(1+x)(1+x^2) \ldots\ldots (1+x^6)$ is 4.

Problem 3.71. *Find the generating function for the number of ways an advertising agent can purchase n minutes ($n \in Z^+$) of air time if time slots for commercials come in block of 30, 60 or 120 seconds.*

Solution. Let 30 seconds represent one time unit. Then the answer is the number of integer solutions to the equation $a + 2b + 4c = 2n$ with $0 \leq a, b, c$.

The associated generating function is

$$f(x) = (1 + x + x^2 + \ldots\ldots)(1 + x^2 + x^4 + \ldots\ldots)(1 + x^4 + x^8 + \ldots\ldots)$$

$$= \frac{1}{1-x}\,\frac{1}{1-x^2}\,\frac{1}{1-x^4}.$$

and the coefficient of x^{2n} is the number of partitions of $2n$ into 1's, 2's and 4's.

Problem 3.72. *A ship carries 48 flags, 12 each of the colors red, white, blue, and black. Twelve of these flags are placed on a vertical pole in order to communicate a signal to other ships.*

(*a*) *How many of these signals use an even number of blue flags and an odd number of black flags ?*

(*b*) *How many of the signals have atleast three white flags or no white flags at all ?*

Solution. (*a*) The exponential generating function

$$f(x) = \left(1 + x + \frac{x^2}{2!} + \frac{x^3}{3!} + \ldots\ldots\right)^2 \left(1 + \frac{x^2}{2!} + \frac{x^4}{4!} + \ldots\ldots\right)\left(x + \frac{x^3}{3!} + \frac{x^5}{5!} + \ldots\ldots\right)$$

Consider all such signals made up of n flags, where $n \geq 1$.

The last two factors in $f(x)$ restrict the signals to an even number of blue flags and an odd number of black flags, respectively.

Since $f(x) = (e^x)^2 \left(\frac{e^x + e^{-x}}{2}\right)\left(\frac{e^x - e^{-x}}{2}\right)$

$$= \left(\frac{1}{4}\right)(e^{2x})(e^{2x} - e^{-2x})$$

$$= \frac{1}{4}(e^{4x} - 1)$$

$$= \frac{1}{4}\left(\sum_{i=0}^{\infty} \frac{(4x)^i}{i!} - 1\right) = \left(\frac{1}{4}\right)\sum_{i=1}^{\infty} \frac{(4x)^i}{i!}$$

The co-efficient of $\frac{x^{12}}{12!}$ in $f(x)$ yields $\left(\frac{1}{4}\right)(4^{12}) = 4^{11}$ signals made up of 12 flags with an even number of blue flags and an odd number of black flags.

(*b*) The exponential generating function

$$g(x) = \left(1 + x + \frac{x^2}{2!} + \frac{x^3}{3!} + ...\right)\left(1 + \frac{x^3}{3!} + \frac{x^4}{4!} + ...\right)\left(1 + x + \frac{x^2}{2!} + \frac{x^3}{3!} + ...\right)^2$$

$$= e^x\left(e^x - x - \frac{x^2}{2!}\right)(e^x)^2$$

$$= e^{3x}\left(e^x - x - \frac{x^2}{2!}\right)$$

$$= e^{4x} - xe^{3x} - \left(\frac{1}{2}\right)x^2e^{3x}$$

$$= \sum_{i=0}^{\infty} \frac{(4x)^i}{i!} - x\sum_{i=0}^{\infty} \frac{(3x)^i}{i!} - \left(\frac{x^2}{2}\right)\left(\sum_{i=0}^{\infty} \frac{(3x)^i}{i!}\right)$$

Here the factor $\left(1 + \frac{x^3}{3!} + \frac{x^4}{4!} + ...\right) = e^x - x - \frac{x^2}{2!}$ in $g(x)$ restricts the signals to those that contain three or more of the 12 white flags, or none at all.

The answer for the number of signals sought here is the coefficient of $\frac{x^{12}}{12!}$ in $g(x)$.

We consider each summand, we find :

(*i*) $\sum_{i=0}^{\infty} \frac{(4x)^i}{i!}$, here we have the term $\frac{(4x)^{12}}{12!} = 4^{12}\left(\frac{x^{12}}{12!}\right)$. So the coefficient of $\frac{x^{12}}{12!}$ is 4^{12}.

(*ii*) $x\left(\sum_{i=0}^{\infty} \frac{(3x)^i}{i!}\right)$. Now we see that in order to get $\frac{x^{12}}{12!}$ we need to consider the term

$x\left[\frac{(3x)^{11}}{11!}\right] = 3^{11}\left(\frac{x^{12}}{11!}\right) = (12)(3^{11})\left(\frac{x^{12}}{12!}\right)$ and here the coefficient of $\frac{x^{12}}{12!}$ is $(12)(3^{11})$ and

(*iii*) $\left(\frac{x^2}{2}\right)\left(\sum_{i=0}^{\infty} \frac{(3x)^i}{i}\right)$, for this summand we observe that $\left(\frac{x^2}{2}\right)\left[\frac{(3x)^{10}}{10!}\right] = \left(\frac{1}{2}\right)(3^{10})\left(\frac{x^{12}}{10!}\right)$

$= \left(\frac{1}{2}\right)(12)(11)(3^{10})\left(\frac{x^{12}}{12!}\right)$, where the coefficient of $\frac{x^{12}}{12!}$ is $\left(\frac{1}{2}\right)(12)(11)(3^{10})$.

Consequently, the number of 12 flags signals with atleast three white flags, or none at all is

$$4^{12} - 12(3^{11}) - \left(\frac{1}{2}\right)(12)(11)(3^{10}) = 10,754,218.$$

Problem 3.73. *Find a formula to express $0^2 + 1^2 + 2^2 + \ldots\ldots + n^2$ as a function of n.*

Solution. We start with $g(x) = \dfrac{1}{(1-x)} = 1 + x + x^2 + \ldots\ldots$

Then $(-1)(1-x)^{-2}(-1) = \dfrac{1}{(1-x)^2} = \dfrac{dg(x)}{de} = 1 + 2x + 3x^2 + 4x^3 + \ldots\ldots$

So $\dfrac{x}{(1-x)^2}$ is the generating function for 0, 1, 2, 3, 4, ...

Repeating this technique, we find that

$$x\frac{d}{dx}\left[x\left(\frac{dg(x)}{dx}\right)\right] = \frac{x(1+x)}{(1-x)^3} = x + 2^2x^2 + 3^2x^3 + \ldots\ldots$$

So $\dfrac{x(1+x)}{(1-x)^3}$ generates $0^2, 1^2, 2^2, 3^2, \ldots\ldots$

$$\frac{x(1+x)}{(1-x)^3} \cdot \frac{1}{(1-x)} = \frac{x(1+x)}{(1-x)^4}$$

is the generating function for $0^2, 0^2 + 1^2, 0^2 + 1^2 + 2^2, 0^2 + 1^2 + 2^2 + 3^2, \ldots\ldots$

Hence the coefficient of x^n in $\dfrac{x(1+x)}{(1-x)^4}$ is $\sum_{i=0}^{n} i^2$. But the coefficient of x^n in $\dfrac{x(1+x)}{(1-x)^4}$ can also be calculated as follows :

$$\frac{x(1+x)}{(1-x)^4} = (x + x^2)(1-x)^{-4} = (x + x^2)\left[\binom{-4}{0} + \binom{-4}{1}(-x) + \binom{-4}{2}(-x)^2 + \ldots\ldots\right]$$

So the coefficient of x^n is

$$\binom{-4}{n-1}(-1)^{n-1} + \binom{-4}{n-2}(-1)^{n-2}$$

$$= (-1)^{n-1}\binom{4+(n-1)-1}{n-1}(-1)^{n-1} + (-1)^{n-2}\binom{4+(n-2)-1}{n-2}(-1)^{n-2}$$

$$= \binom{n+2}{n-1} + \binom{n+1}{n-2} = \frac{(n+2)!}{3!(n-1)!} + \frac{(n+1)!}{3!(n-2)!}$$

$$= \frac{1}{6}[(n+2)(n+1)(n) + (n+1)(n)(n-1)]$$

$$= \frac{1}{6}(n)(n+1)[(n+2) + (n-1)]$$

$$= (n)(n+1)(2n+1)/6.$$

Problem 3.74. *A company hires 11 new employees, each of whom is to be assigned to one of four sub divisions. Each sub-division will get atleast one new employee. In how many ways can these assignments be made ?*

Solution. Calling the subdivisions A, B, C and D, we can equivalently count the number of 11-letter sequences in which there is atleast one occurrence of aech of the letters A, B, C and D.

The exponential generating function for these arrangements is

$$f(x) = \left(x + \frac{x^2}{2!} + \frac{x^3}{3!} + \frac{x^4}{4!} + \ldots\ldots\right)^4$$

$$= (e^x - 1)^4 = e^{4x} - 4e^{3x} + 6e^{2x} - 4e^x + 1$$

The answer then is the coefficient of $\frac{x^{11}}{11!}$ in $f(x)$:

$$4^{11} - 4(3^{11}) + 6(2^{11}) - 4(1^{11})$$

$$= \sum_{i=0}^{4} (-1)^i \binom{4}{i} (4-i)^{11}.$$

3.6. CALCULATIONAL TECHNIQUES

Definition and examples

The most important concept we introduce is that of division of formal **power series.** First let us discuss the meaning of $\frac{1}{A(x)}$. If $A(x) = \sum_{n=0}^{\infty} a_n x^n$ is a formal power series then A(x) is said to have a multiplicative inverse if there is a formal power series $B(x) = \sum_{k=0}^{\infty} b_k kx^k$ such that A(x) B(x) = 1.

In particular, if A(x) has a **multiplicative inverse**, then we see that $a_0 b_0 = 1$, so that a_0 must be non zero.

The converse is also true. Infact, if $a_0 \neq 0$, then we can determine the coefficients of B(x) by writing down the coefficients of successive powers of x in A(x) B(x) from the definition of product of 2 power series, and then equating these to the coefficients of like powers of x in the powe series 1. Therefore we have

$$a_0b_0 = 1$$
$$a_0b_1 + a_1b_0 = 0$$
$$a_0b_2 + a_1b_1 + a_2b_0 = 0$$
$$a_0b_3 + a_1b_2 + a_2b_1 + a_3b_0 = 0$$
$$\ldots\ldots\ldots\ldots\ldots\ldots\ldots\ldots$$
$$a_0b_n + a_1b_{n-1} + \ldots\ldots + a_nb_0 = 0 \text{ and so on.}$$

From the first equation, we can solve for $b_0 = \frac{1}{a_0}$, from the second, we find $b_1 = \frac{-a_1b_0}{a_0} = \frac{-a_1}{a_0^2}$

in the third equation, we get

$$b_2 = \frac{-a_1b_1 - a_2b_0}{a_0} = \frac{a_1^2 - a_2a_0}{a_0^3}$$

from the fourth, we find

$$b_3 = \frac{-a_1b_2 - a_2b_1 - a_3b_0}{a_0}$$

We can substitute into this expression for b_0, b_1 and b_2 to obtain an expression for b_3 involving only the coefficients of A(x). Continuing in this manner, we can solve for each coefficient of B(x).

Thus, we established that a formal power series

$A(x) = \sum_{n=0}^{\infty} a_nx^n$ has a multiplicative inverse iff the constant term a_0 is different from zero.

If A(x) and C(x) are power series, we say that A(x) divides C(x) if there is a formal power series D(x) such that C(x) = A(x) D(x), and we write $D(x) = \frac{C(x)}{A(x)}$.

Of course, for arbitrary formal power series, A(x) and C(x), it need not be the case that A(x) divides C(x).

However, if $A(x) = \sum_{n=0}^{\infty} a_nx^n$ is such that $a_0 \neq 0$, then A(x) has a multiplicative inverse B(x) $= \frac{1}{A(x)}$ and then A(x) divides any C(x), let $D(x) = C(x)B(x) = \left(C(x)\frac{1}{A(x)}\right)$.

If $A(x) = \sum_{n=0}^{\infty} a_nx^n$ and $a_0 = 0$, but some coefficient of A(x) is not zero, then let a_k be the first non zero coefficient of A(x), and $A(x) = x^k A_1(x)$, where a_k, the constant term of $A_1(x)$ is non zero.

Then in order for A(x) to divide C(x) it must be true that x^k is also a factor of C(x), that is, $C(x) = x^k C_1(x)$ where $C_1(x)$ is a formal power series. If this is the case, then cancel the common powers of x from both A(x) and C(x) and then we can find $\dfrac{C(x)}{A(x)} = \dfrac{C_1(x)}{A_1(x)}$ by using the multiplicative inverse of $A_1(x)$.

3.6.1. Geometric Series

Let us use the multiplicative inverse for $A(x) = 1 - x$.

Let $B(x) = \dfrac{1}{A(x)} = \sum_{n=0}^{\infty} b_n x^n$.

Solving successively for b_0, b_1, ..., as above, we see that

$$b_0 = \frac{1}{a_0} = 1$$

$$b_1 = \frac{-a_1 b_0}{a_0} = \frac{-(-1)(1)}{(1)} = 1$$

$$b_2 = \frac{-a_1 b_1 - a_2 b_0}{a_0} = \frac{-(-1)(1)-(0)(1)}{1} = 1$$

$$b_3 = \frac{-a_1 b_2 - a_2 b_1 - a_3 b_0}{a_0} = 1, \text{ and so on.}$$

We see that each $b_i = 1$ so that we have an expression for the geometric series

$$\frac{1}{1-x} = \sum_{r=0}^{\infty} x^r \quad ...(1)$$

If we replace in the above expression x by ax where a is a real number, then we see that

$$\frac{1}{1-ax} = \sum_{r=0}^{\infty} a^r x^r \quad ...(2)$$

the so called geometric series (with common ratio a).

In particular, let $a = -1$, then we get

$$\frac{1}{1+x} = \sum_{r=0}^{\infty} (-1)^r x^r = 1 - x + x^2 - x^3 \quad ...(3)$$

the so called alternating geometric series.

Likewise, $$\frac{1}{1+ax} = \sum_{r=0}^{\infty} (-1)^r a^r x^r \quad ...(4)$$

Suppose that n is a positive integer. If $B_1(x)$, $B_2(x)$, ... $B_n(x)$ are the multiplicative inverse of $A_1(x)$, $A_2(x)$ ...... and $A_n(x)$, respectively, then $B_1(x)\,B_2(x)$ ...... $B_n(x)$ is the multiplicative inverse of $A_1(x)A_2(x)$ ...... $A_n(x)$, just multiply

$A_1(x)A_2(x)$ ...... $A_n(x)$ by $B_1(x)\,B_2(x)$ ...... $B_n(x)$ and use the facts that $A_i(x)B_i(x) = 1$ for each i.

In particular, if $B(x)$ is the multiplicative inverse of $A(x)$, then $(B(x))^n$ is the multiplicative inverse of $(A(x))^n$. Let us apply this observation to $A(x) = 1 - x$.

For n a positive integer,

$$\frac{1}{(1-x)^n} = \left(\sum_{k=0}^{\infty} x^k\right) = \sum_{r=0}^{\infty} C(n-1+r, r)x^r \qquad ...(5)$$

The fact that $\sum_{k=0}^{\infty} x^k$ is the multiplicative inverse of $1 - x$.

The equality $\frac{1}{(1-x)^n} = \sum_{r=0}^{\infty} C(n-1+r, r)x^r$ could also be proved by mathematical induction and use of the identity $C(n-1, 0) + C(n, 1) + C(n+1, 2) + \ldots\ldots + C(n+r-1, r) = C(n+r, r)$

By replacing x by $-x$ in the above we get the following identity :

For n a positive integer,

$$\frac{1}{(1+x)^n} = \sum_{r=0}^{\infty} C(n-1+r, r)(-1)^r x^r \qquad ...(6)$$

Following this pattern, replace x by ax in (5) and (6) to obtain

$$\frac{1}{(1-ax)^n} = \sum_{r=0}^{\infty} C(n-1+r, r)a^r x^r \qquad ...(7)$$

$$\frac{1}{(1+ax)^n} = \sum_{r=0}^{\infty} C(n-1+r, r)(-a)^r x^r \qquad ...(8)$$

Likewise, replace x by x^k in (1) to get for k a positive integer,

$$\frac{1}{1-x^k} = \sum_{r=0}^{\infty} x^{kr} = 1 + x^k + x^{2k} + \ldots\ldots \qquad ...(9)$$

and

$$\frac{1}{1+x^k} = \sum_{r=0}^{\infty} (-1)^r x^{kr} \qquad ...(10)$$

If a is a non zero real number,

$$\frac{1}{a-x} = \frac{1}{a}\left(\frac{1}{1-\frac{x}{a}}\right) = \frac{1}{a}\sum_{r=0}^{\infty}\frac{x^r}{a^r} \quad ...(11)$$

and

$$\frac{1}{x-a} = -\frac{1}{a-x} = -\frac{1}{a}\sum_{r=0}^{\infty}\frac{x^r}{a^r} \quad ...(12)$$

If n is a positive integer,

$$1 + x + x^2 + ... + x^n = \frac{1-x^{n+1}}{1-x} \quad ...(13)$$

If n is a positive integer,

$$(1+x)^n = 1 + \binom{n}{1}x + \binom{n}{2}x^2 + ... + \binom{n}{n}x^n$$

$$(1+x^k)^n = 1 + \binom{n}{1}x^k + \binom{n}{2}x^{2k} + ... + \binom{n}{n}x^{nk}$$

$$(1-x)^n = 1 - \binom{n}{1}x + \binom{n}{2}x^2 + ... + (-1)^n\binom{n}{n}x^n$$

$$(1-x^k)^n = 1 - \binom{n}{1}x^k + \binom{n}{2}x^{2k} + ... + (-1)^n\binom{n}{n}x^{nk}.$$

3.6.2. Use of Partial Fraction Decomposition

If A(x) and C(x) are polynomials, we compute $\frac{C(x)}{A(x)}$ by using the above identities and partial fractions.

If A(x) is a product of linear factors,

$A(x) = a_n(x-\alpha_1)^{r_1}(x-\alpha_2)^{r_2} ...... (x-\alpha_k)^{r_k}$ and if C(x) is any polynomial of degree less than the degree of A(x), then $\frac{C(x)}{A(x)}$ can be written as the sum of elementary fractions as follows :

$$\frac{C(x)}{A(x)} = \frac{A_{11}}{(x-\alpha_1)^{r_1}} + \frac{A_{12}}{(x-\alpha_1)^{r_1-1}} + ... + \frac{A_{1r_1}}{(x-\alpha_1)}$$

$$+ \frac{A_{21}}{(x-\alpha_2)^{r_2}} + \frac{A_{22}}{(x-\alpha_2)^{r_2-1}} + ... + \frac{A_{2r_2}}{(x-\alpha_2)} + ...$$

$$+\frac{Ak_1}{(x-\alpha_k)^{r_k}}+\frac{Ak_2}{(x-\alpha_k)^{r_k-1}}+\ldots\ldots\frac{Ak_{rk}}{(x-\alpha_k)}.$$

To find the numbers A_{11}, ..., A_{krk}, we multiply both sides of the last equation by $(x-\alpha_1)^{r_1}(x-\alpha_2)^{r_2}$ ...... $(x-\alpha_k)^{r_k}$ to clear of denominators and then we equate coefficients of the same powers of x. Then the required coefficients can be solved from the resulting system of equations.

Problem 3.75. *Calculate* $B(x)=\sum_{r=0}^{\infty} b_r x^r = \frac{1}{(x^2-5x+6)}$.

Solution. Since $x^2-5x+6=(x-3)(x-2)$.

We see that $\frac{1}{x^2-5x+6}=\frac{A}{x-3}+\frac{B}{x-2}$

Thus, $A(x-2)+B(x-3)=1$.

Let $x=2$ and we find $B=-1$

Let $x=3$ and we see that $A=1$.

Thus $\frac{1}{x^2-5x+6}=\frac{1}{x-3}-\frac{1}{x-2}$

Then we use

$$\frac{1}{a-x}=\frac{1}{a}\left(\frac{1}{1-\frac{x}{a}}\right)=\frac{1}{a}\sum_{r=0}^{\infty}\frac{x^r}{a^r}$$

$$\frac{1}{x-a}=-\frac{1}{a-x}=-\frac{1}{a}\sum_{r=0}^{\infty}\frac{x^r}{a^r}$$

to see that

$$\frac{1}{x^2-5x+6}=-\frac{1}{3-x}+\frac{1}{2-x}$$

$$=-\frac{1}{3\left(1-\frac{x}{3}\right)}+\frac{1}{2\left(1-\frac{x}{2}\right)}$$

$$=-\frac{1}{3}\sum_{r=0}^{\infty}\left(\frac{1}{3}\right)^r x^r+\frac{1}{2}\sum_{r=0}^{\infty}\left(\frac{1}{2}\right)^r x^r$$

$$= \sum_{r=0}^{\infty} \left(-\frac{1}{3^{r+1}} + \frac{1}{2^{r+1}} \right) x^r$$

$$= \text{B}(x)$$

Therefore, for each r, $b_r = -\dfrac{1}{3^{r+1}} + \dfrac{1}{2^{r+1}}$

$$\text{Thus } \frac{x^5}{x^2 - 5x + 6} = x^5 \sum_{r=0}^{\infty} \left(-\frac{1}{3^{r+1}} + \frac{1}{2^{r+1}} \right) x^r$$

$$= \sum_{r=0}^{\infty} \left(-\frac{1}{3^{r+1}} + \frac{1}{2^{r+1}} \right) x^{r+5}$$

and if we make the substitution $k = r + 5$ we see that

$$\frac{x^5}{x^2 - 5x + 6} = \sum_{k=5}^{\infty} \left(-\frac{1}{3^{k-4}} + \frac{1}{2^{k-4}} \right) x^k = \sum_{k=0}^{\infty} d_k x^k$$

and what this final equality says is that

$$d_0 = d_1 = d_2 = d_3 = d_4 = 0$$

$$d_5 = -\frac{1}{3} + \frac{1}{2}$$

$$d_6 = -\frac{1}{3^2} + \frac{1}{2^2}$$

$$d_k = -\frac{1}{3^{k-4}} + \frac{1}{2^{k-4}} \text{ if } k \geq 5 \text{ and so on.}$$

Problem 3.76. *Compute the coefficients of*

$$\sum_{r=0}^{\infty} d_r x^r = \frac{x^2 - 5x + 3}{x^4 - 5x^2 + 4}.$$

Solution. Since $x^4 - 5x^2 + 4 = (x^2 - 1)(x^2 - 4)$

$$= (x - 1)(x + 1)(x - 2)(x + 2)$$

We can write

$$\frac{x^2 - 5x + 3}{x^4 - 5x^2 + 4} = \frac{\text{A}}{x - 1} + \frac{\text{B}}{x + 1} + \frac{\text{C}}{x - 2} + \frac{\text{D}}{x + 2}$$

Multiplication by $x^4 - 5x^2 + 4$ gives

$$x^2 - 5x + 3 = \text{A}(x + 1)(x - 2)(x + 2) + \text{B}(x - 1)(x - 2)(x + 2)$$
$$+ \text{C}(x - 1)(x + 1)(x + 2) + \text{D}(x - 1)(x + 1)(x - 2)$$

Let $x = 1$, then all terms of the right hand side that involve the factor $x - 1$ vanish, and we have

$-1 = -6$ A or A $= \frac{1}{6}$.

Similarly putting $x = -1$, $x = 2$, and $x = -2$, we find B $= \frac{3}{2}$, C $= -\frac{1}{4}$, and D $= -\frac{17}{12}$.

$$\text{Thus, } \frac{x^2 - 5x + 3}{x^4 - 5x^2 + 4} = \frac{1}{6(x-1)} + \frac{3}{2(x+1)} - \frac{1}{4(x-2)} - \frac{17}{12(x+2)}$$

$$= \frac{1}{2}\left[-\frac{1}{3(1-x)} + \frac{3}{1+x} + \frac{1}{4\left(1-\frac{x}{2}\right)} - \frac{17}{12\left(1+\frac{x}{2}\right)}\right]$$

$$= \frac{1}{2}\left[-\frac{1}{3}\sum_{r=0}^{\infty} x^r + 3\sum_{r=0}^{\infty}(-1)^r x^r + \frac{1}{4}\sum_{r=0}^{\infty}\left(\frac{1}{2^r}\right)x^r - \frac{17}{12}\sum_{r=0}^{\infty}\left(\frac{1}{2}\right)^r x^r\right]$$

$$= \frac{1}{2}\sum_{r=0}^{\infty}\left[\left(-\frac{1}{3}\right) + 3(-1)^r + \frac{1}{4}\frac{1}{2^r} - \frac{17}{12}\left(-\frac{1}{2}\right)^r\right]x^r$$

$$\text{Therefore, } d_r = \frac{1}{2}\left[-\frac{1}{3} + 3(-1)^r + \frac{1}{2^{r+2}} - \frac{17}{3}(-1)^r\frac{1}{2^{r+2}}\right]$$

which can be simplified to

$$d_r = \begin{cases} \frac{1}{2}\left[-\frac{1}{3} + 3 + \frac{1}{2^{r+2}}\left(1 - \frac{17}{3}\right)\right] = \frac{1}{3}\left(4 - \frac{14}{2^{r+3}}\right) \text{ if } r \text{ is even} \\ \frac{1}{2}\left[-\frac{1}{3} - 3 + \frac{1}{2^{r+2}}\left(1 + \frac{17}{3}\right)\right] = \frac{1}{3}\left(-5 + \frac{5}{2^{r+1}}\right) \text{ if } r \text{ is odd} \end{cases}$$

After doing these examples we see that it is desirable to write $\frac{C(x)}{A(x)}$ in the form

$$\frac{B_{11}}{\left[1-\left(\frac{x}{\alpha_1}\right)\right]^{r_1}} + \frac{B_{12}}{\left[1-\left(\frac{x}{\alpha_1}\right)\right]^{r_1-1}} + \ldots\ldots + \frac{B_1 r_1}{\left[1-\left(\frac{x}{\alpha_1}\right)\right]} + \ldots\ldots + \frac{Bk_1}{\left[1-\left(\frac{x}{\alpha_k}\right)\right]^{r_k}}$$

$$+ \frac{Bk_2}{\left[1-\left(\frac{x}{\alpha_k}\right)\right]^{r_k-1}} + \ldots\ldots + \frac{Bk_{rk}}{\left[1-\left(\frac{x}{\alpha_k}\right)\right]}$$

where $A(x) = a_n(x - \alpha_1)^{r_1}(x - \alpha_2)^{r_2} \ldots\ldots (x - \alpha_k)^{r_k}$ and then solve for the constants $B_{11}, \ldots\ldots, B_1 r_1, \ldots\ldots, B_{k_1} \ldots\ldots, B_{kr_k}$, by algebraic techniques. This is desirable because in this form we can readily apply the formulas

$$\frac{1}{1-x} = \sum_{r=0}^{\infty} x^r \text{ through } \frac{1}{(1+ax)^n} = \sum_{r=0}^{\infty} C(n-1+r, r)(-a)^r x^r.$$

Without having to resort to the intermediate step of applying $\frac{1}{a-x} = \frac{1}{a}\left(\frac{1}{1-\frac{x}{a}}\right) = \frac{1}{a}\sum_{r=0}^{\infty}\frac{x^r}{a^r}$

and $$\frac{1}{x-a} = -\frac{1}{a-x} = -\frac{1}{a}\sum_{r=0}^{\infty}\frac{x^r}{a^r}$$

Problem 3.77. *Find the coefficient of x^{20} in $(x^3 + x^4 + x^5 \ldots\ldots)^5$.*

Solution. Simplify the expression by extracting x^3 from each factor.

Thus, $(x^3 + x^4 + x^5 + \ldots\ldots)^5 = [x^3(1 + x + \ldots\ldots)]^5$

$$= x^{15}\left(\sum_{r=0}^{\infty} x^r\right)^5 = \frac{x^{15}}{(1-x)^5}$$

$$= x^{15}\sum_{r=0}^{\infty} C(5-1+r, r)x^r$$

The coefficient of x^{20} in the original expression becomes the coefficient of x^5 in $\sum_{r=0}^{\infty} C(4+r, r)x^r$.

Thus, the coefficient we seek is when $r = 5$ in the last power series, that is, the coefficient is

$$C(4 + 5, 5) = C(9, 5).$$

Problem 3.78. *Calculate the coefficient of x^{15} in*

$$A(x) = (x^2 + x^3 + x^4 + x^5)(x + x^2 + x^3 + x^4 + x^5 + x^6 + x^7)(1 + x + \ldots\ldots + x^{15}).$$

Soltuion. Note that we can rewrite the expression for $A(x)$ as

$$x^2(1 + x + x^2 + x^3)(x)(1 + x + \ldots\ldots + x^6)(1 + x + \ldots\ldots + x^{15})$$

$$= x^3\frac{(1-x^4)}{(1-x)}\frac{(1-x^7)}{1-x}\frac{(1-x^{16})}{1-x}$$

$$= x^3\frac{(1-x^4)(1-x^7)(1-x^{16})}{(1-x)^3}$$

The coefficient of x^{15} in A(x) is the same as the coefficient of x^{12} in $\dfrac{(1-x^4)(1-x^7)(1-x^{16})}{(1-x)^3}$

$$= (1-x^4)(1-x^7)(1-x^{16})\left(\sum_{r=0}^{\infty} C(r+2, r)x^r\right)$$

Since the coefficient of x^{12} in a product of several factors can be obtained by taking one term from each factor so that the sum of their exponents equals 12, we see that the term x^{16} in the third factor and all terms of degree greater than 12 in the last factor need not be considered.

Hence we look for the coefficient of x^{12} in

$$\cdot(1-x^4)(1-x^7)\sum_{r=0}^{12} C(r+2, r)x^r$$

$$= (1-x^4-x^7+x^{11})\sum_{r=0}^{\infty} C(r+2, r)x^r.$$

Problem 3.79. *Find the number of ways of placing 20 similar balls into 6 numbered boxes so that the first box contains any number of balls between and 5 inclusive and the other 5 boxes must contain 2 or more balls each.*

Solution. The integer solution of an equation model is : count the number of integral solutions to $e_1 + e_2 + e_3 + e_4 + e_5 + e_6 = 20$ where $1 \le e_1 \le 5$ and $2 \le e_2, e_3, e_4, e_5, e_6$.

First, we will count the solutions where $1 \le e_1$ and $2 \le e_i$ for $i = 2, 3, 4, 5, 6$.

We do this by placing 1 ball in box number one, 2 balls each in the other 5 boxes, and then counting the number of ways to distribute the remaining 9 balls into 6 boxes with unlimited repetition. There are C(14, 9) ways to do this.

But then we wish to discard the number of solutions for which $6 \le e_1$ and $2 \le e_i$ for $i = 2, 3, 4, 5, 6$. There are C(9, 4) of these.

Hence the total number of solutions subject to the constraints is C(14, 9) – C(9, 4).

Now let us solve this problem with generating function.

The generating function we consider is

$$(x+x^2+x^3+x^4+x^5)\,(x^2+x^3+\ldots\ldots)^5$$
$$= x(1+x+x^2+x^3+x^4)\,[x^2(1+x+x^2+\ldots\ldots)]^5$$
$$= x(1+x+x^2+x^3+x^4)(x^{10})(1+x+x^2+\ldots\ldots)^5$$
$$= x^{11}(1+x+x^2+x^3+x^4)(1+x+x^2+\ldots\ldots)^5$$

We desire to compute the coefficient of x^{20} in this last product but we need only compute the coefficient of x^9 in

$(1+x+x^2+x^3+x^4)\,(1+x+x^2+\ldots\ldots)^5$, which can be rewritten as

$$\left(\frac{1-x^5}{1-x}\right)\left(\frac{1}{1-x}\right)^5 = (1-x^5)\left(\frac{1}{1-x}\right)^6$$

$$= (1-x^5)\left(\sum_{r=0}^{\infty} C(r+5, r)x^r\right)$$

Thus, the coefficient of x^9 in this last product is C(14, 9) – C(9, 4).

Problem 3.80. *Determine the coefficient of* x^8 *in* $\dfrac{1}{(x-3)(x-2)^2}$.

Solution. Since $\dfrac{1}{x-a} = \left(-\dfrac{1}{a}\right)\left(\dfrac{1}{\left(1-\left(\dfrac{x}{a}\right)\right)}\right)$

$$= \left(-\frac{1}{a}\right)\left[1+\left(\frac{x}{a}\right)+\left(\frac{x}{a}\right)^2+......\right] \text{ for any } a \neq 0.$$

We could solve this problem by finding the coefficient of x^8 in $\dfrac{1}{[(x-3)(x-2)^2]}$ expressed as

$$\left(-\frac{1}{3}\right)\left[1+\left(\frac{x}{3}\right)+\left(\frac{x}{3}\right)^2+......\right]\left(\frac{1}{4}\right)\left[\binom{-2}{0}+\binom{-2}{1}\left(\frac{-x}{2}\right)\right.$$

$$\left.+\binom{-2}{2}\cdot\left(\frac{-x}{2}\right)^2+......\right]$$

An alternative techniques uses the partial fraction decomposition

$$\frac{1}{(x-3)(x-2)^2} = \frac{A}{x-3}+\frac{B}{x-2}+\frac{C}{(x-2)^2}$$

This decomposition implies that

$$1 = A(x-2)^2 + B(x-2)(x-3) + C(x-3)$$

or $$0.x^2 + 0.x + 1 = 1 = (A+B)x^2 + (-4A + -5B + C)x + (4A + 6B - 3C)$$

By comparing coefficients, we find that A + B = 0,

$$-4A - 5B + C = 0 \text{ and } 4A + 6B - 3C = 1,$$

Solving these equations yields A = 1, B = – 1, C = – 1

Hence, $$\frac{1}{(x-3)(x-2)^2} = \frac{1}{x-3} + \frac{1}{x-2} - \frac{1}{(x-2)^2}$$

$$= \left(\frac{-1}{3}\right)\frac{1}{1-\left(\frac{x}{3}\right)} + \left(\frac{1}{2}\right)\frac{1}{\left(1-\left(\frac{x}{2}\right)\right)} + \left(\frac{-1}{4}\right)\frac{1}{\left(1-\left(\frac{x}{2}\right)\right)^2}$$

$$= \left(\frac{-1}{3}\right)\sum_{i=0}^{\infty}\left(\frac{x}{3}\right)^i + \left(\frac{1}{2}\right)\sum_{i=0}^{\infty}\left(\frac{x}{2}\right)^i$$

$$+ \left(\frac{-1}{4}\right)\left[\binom{-2}{0} + \binom{-2}{1}\left(\frac{-x}{2}\right) + \binom{-2}{2}\left(\frac{-x}{2}\right)^2 + \ldots\ldots\right]$$

The coefficient of x^8 is $\left(\frac{-1}{3}\right)\left(\frac{1}{3}\right)^8 + \left(\frac{1}{2}\right)\left(\frac{1}{2}\right)^8 + \left(\frac{-1}{4}\right)\binom{-2}{8}\left(\frac{-1}{2}\right)^8$

$$= -\left[\left(\frac{1}{3}\right)^9 + 7\left(\frac{1}{2}\right)^{10}\right].$$

Problem 3.81. *In how many ways can a police contain distribute 24 rifle shells to four police officer so that each officer gets at least three shells, but not more than eight ?*

Solution. The choices for the number of shells each officer receives are given by

$$x^3 + x^4 + \ldots\ldots + x^8.$$

There are four officers, so the resulting generating function is

$$f(x) = (x^3 + x^4 + \ldots\ldots + x^8)^4$$

We seek the coefficient of x^{24} in $f(x)$, with

$$(x^3 + x^4 + \ldots\ldots + x^8)^4 = x^{12}(1 + x + x^2 + \ldots\ldots + x^5)^4$$

$$= x^{12}\left(\frac{(1-x^6)}{(1-x)}\right)^4.$$

The answer is the coefficient of x^{12} in

$$(1-x^6)^4(1-x)^{-4} = \left[1 - \binom{4}{1}x^6 + \binom{4}{2}x^{12} - \binom{4}{3}x^{18} + x^{24}\right]$$

$$\left[\binom{-4}{0} + \binom{-4}{1}(-x) + \binom{-4}{2}(-x)^2 + \ldots\ldots\right]$$

which is $$\left[\binom{-4}{12}(-1)^{12}-\binom{4}{1}\binom{-4}{6}(-1)^{6}+\binom{4}{2}\binom{-4}{0}\right]$$

$$=\left[\binom{15}{12}-\binom{4}{1}\binom{9}{6}+\binom{4}{2}\right]=125.$$

Problem 3.82. *Verify that for all* $n \in Z^+$, $\binom{2n}{n}=\sum_{i=0}^{n}\binom{n}{i}^2$.

Solution. Since $(1+x)^{2n}=[(1+x)^n]^2$, by comparison of coefficients the coefficients of x^n in $(1+x)^{2n}$, which is $\binom{2n}{n}$, must equal the coefficient of x^n in

$$\left[\binom{n}{0}+\binom{n}{1}x+\binom{n}{2}x^2+\ldots\ldots+\binom{n}{n}x^n\right]^2, \text{ and this is}$$

$$\binom{n}{0}\binom{n}{n}+\binom{n}{1}\binom{n}{n-1}+\binom{n}{2}\binom{n}{n-2}+\ldots\ldots+\binom{n}{n}\binom{n}{0}$$

with $\binom{n}{r}=\binom{n}{n-r}$, for all $0 \le r \le n$.

Problem Set 3.1

1. Find a genarating function for the sequence $A=\{a_r\}_{r=0}^{\infty}$ where

$$a_r=\begin{cases}1 & \text{if } 0\le r\le 2\\ 3 & \text{if } 3\le r\le 5\\ 0 & \text{if } r\ge 6\end{cases}$$

2. Find the generating function for the number of r-combinations of $\{3.a, 5.b, 2.c\}$
3. Write a generating function for a_n, the number of ways of obtaining the sum n when tossing q distinguishable dice. Then find a_{25}.
4. How many ways are there to paint 20 identical rooms in a hotel with 5 colours if there is only enough blue, pink and green paint to paint 3 rooms ?
5. Write the generating function for the sequence $\left\{a_r\right\}_{r=0}^{\infty}$ defined by

(i) $a_r=(-1)^r$ (ii) $a_r=(-1)^r 3^r$ (iii) $a_r=5^r$

(iv) $a_r=r+1$ (v) $a_r=6(r+1)$ (vi) $a_r=C(r+3, r)$

(*vii*) $a_r = (r + 3)(r + 2)(r + 1)$

(*viii*) $a_r = \dfrac{(-1)^r (r+2)(r+1)}{2!}$ (*ix*) $a_r = 5^r + (-1)^r 3^r + 8C(r + 3, r)$

(*x*) $a_r = (r + 1) 3^r$ (*xi*) $a_r = (r + 3)(r + 1)3^r$.

6. Build a generating function for a_r = the number of integral solution to the equation $e_1 + e_2 + e_3 = r$ if

(*i*) $0 \le e_i \le 3$ for each i

(*ii*) $2 \le e_i \le 5$ for each i

(*iii*) $0 \le e_i$ for each i

(*iv*) $0 \le e_1 \le 6$ and e_1 is even, $2 < e_2 \le 7$ and e_2 is odd, $5 \le e_3 \le 7$.

7. Write a generating function for a_r when a_r is

(*i*) the number of ways of selecting r balls from 3 red balls, 5 blue balls, 7 white balls.

(*ii*) the number of ways of selecting r coins from an unlimited supply of pennies, nickels, dimes and quarters.

(*iii*) the number of r-combinations formed from n letters where the first letter can appear an even number of times up to 12, the second letter can appear an odd number of times upto 7, the remaining letters can occur an unlimited number of times.

(*iv*) the number of ways of obtaining a total of r upon tossing 50 distinguishable dice.

(*v*) the number of integers between 0 and 999 whose sum of digits is r.

8. Find a generating function for a_r = the number of ways of distributing r similar balls into 7 numbered boxes where the second, third, fourth and fifth boxes are nonempty.

9. (*a*) Find a generating function for the number of ways to distribute 30 balls into 5 numbered boxes where each box contains at least 3 balls and at most 7 balls.

(*b*) Factor out x^{15} from the above functions and interpret this revised generating function combinatorially.

10. Build a gererating function for a_r = the number of ways to distribute r similar balls into 5 numbered boxes with

(*i*) at most 3 balls in each box.

(*ii*) 3, 6, or 8 balls in each box.

(*iii*) at least 1 ball in each of the first 3 boxes and at least 3 balls in each of the last 2 boxes.

(*iv*) at most 5 balls in box 1, at most 7 balls in the last 4 boxes.

(*v*) a multiple of 5 balls in box 1, a multiple of 10 balls in box 2, a multiple of 25 balls in box 3, a multiple of multiple of 50 balls in box 4, and a multiple of 100 balls in box 5.

11. (*a*) Find a generating function for the number of ways to select 6 non consecutive integers from 1, 2, ... , n.

(*b*) Which coefficient do we want to find in case $n = 20$?

(*c*) Which coefficient do we want for general n ?

12. In $(1 + x^5 + x^9)^{10}$ find

(*i*) the coefficient of x^{23} (*ii*) the coefficient of x^{32}.

13. Find the coefficient of x^{12} in $\dfrac{1 - x^4 - x^7 + x^{11}}{(1-x)^5}$.

14. Find the coefficient of x^{10} in

(*a*) $(1 + x + x^2 + ...)^2$ (*b*) $\dfrac{1}{(1-x)^3}$ (*c*) $\dfrac{1}{(1-x)^5}$

(*d*) $\dfrac{1}{(1+x)^5}$ (*e*) $(x^3 + x^4 + ...)^2$

(*f*) $x^4(1 + x + x^2 + x^3)(1 + x + x^2 + x^3 + x^4)(1 + x + x^2 + ... + x^{12})$.

15. Find the coeficient of x^{12} in

(*a*) $\dfrac{x^2}{(1-x)^{10}}$ (*b*) $\dfrac{x^5}{(1-x)^{10}}$ (*c*) $(1 - x)^{20}$ (*d*) $(1 + x)^{20}$

(*e*) $(1 + x)^{-20}$ (*f*) $(1 - 4x)^{-5}$ (*g*) $(1 - 4x)^{15}$ (*h*) $(1 + x^3)^{-4}$

(*i*) $\dfrac{x^2 - 3x}{(1-x)^4}$ (*j*) $(1 - 2x)^{19}$.

16. Find a_r be the number of non negative integral solutions to $x_1 + x_2 + x_3 = r$.

(*a*) Find a_{10} if $0 \le x_i \le 4$ for each i

(*b*) Find a_{50} where $2 \le x_1 \le 50$, $0 \le x_2 \le 50$, $5 \le x_3 \le 25$.

17. Let a_r be the number of ways the sum r can be obtained by tossing 50 distinguishable dice. Write a generating function for the sequence $\left\{a_r\right\}_{r=0}^{\infty}$. Then find the number of ways to obtain the sum of 100, that is, find a_{100}.

18. (*a*) Find the coefficient of x^{50} in $(x^{10} + x^{11} + ... + x^{25})\,(x + x^2 + ... + x^{15})\,(x^{20} + x^{21} + ... + x^{45})$

(*b*) Find the coefficient of x^{25} in $(x^2 + x^3 + x^4 + x^5 + x^6)^7$.

19. Find the coefficient of x^{20} in $(x + x^2 + x^3 + x^4 + x^5)(x^2 + x^3 + x^4 + ...)^5$.

20. Find the coefficient of x^{14} in

(*a*) $(1 + x + x^2 + x^3)^{10}$

(*b*) $(1 + x + x^2 + x^3 + x^4 + ... + x^8)^{10}$

(*c*) $(x^2 + x^3 + x^4 + x^5 + x^6 + x^7)^4$.

21. How many ways are these to place an order for 12 chocolate sundaes if there are 5 types of sundaes, and at most 4 sundaes of one type are allowed ?

22. Use generating functions to find the number of ways to select 10 balls from a large pile of red, white and blue balls if

(*a*) the selection has at least 2 balls of each colour

(*b*) the selection has at most 2 red balls, and

(*c*) the selection has an even number of blue balls.

23. (*a*) Find the generating function for the number of ways to select 10 candy bars from large supplies of six different kinds.

(*b*) Find the generating function for the number of ways to select, with repetitions allowed, r objects from a collection of n distinct objects.

24. Find the generating function for the number of integer solutions to the equation $c_1 + c_2 + c_3 + c_4 = 20$ where $-3 \le c_1$, $-3 \le c_2$, $-5 \le c_3 \le 5$, and $0 \le c_4$.

25. Determine the generating function for the number of integer solutions for the following equation.

(*a*) $c_1 + c_2 + c_3 + c_4 = 20$, $0 \le c_i \le 7$ for all $1 \le i \le 4$

(*b*) $c_1 + c_2 + c_3 + c_4 = 20$, $0 \le c_i$ for all $1 \le i \le 4$, with c_2 and c_3 even

(*c*) $c_1 + c_2 + c_3 + c_4 + c_5 = 30$, $2 \le c_1 \le 4$ and $3 \le c_i \le 8$ for all $2 \le i \le 5$

(*d*) $c_1 + c_2 + c_3 + c_4 + c_5 = 30$, $0 \le c_i$ for all $1 \le i \le 5$, with c_2 even and c_3 odd.

Problem Set 3.2

1. Find the closed form of the gererating function for each of the following sequences :

(*a*) 0, 0, 0, 1, 1, 1, ...

(*b*) 0, 0, 1, 2, 3, 4, ...

(*c*) 3, – 3, 3, – 3, 3, – 3, ...

(*d*) 3, 9, 27, 81, ...

(*e*) 1, – 2, 3, – 4, ...

2. Find the generating function for each of the following sequences if G(x) is the generating function for the sequence $\{a_n\}$:

(*a*) $0, 0, 0, a_0, a_1, a_2, \ldots\ldots$

(*b*) $0, 0, a_2, a_3, \ldots\ldots$

(*c*) $a_0, 0, a_1, 0, a_2, 0, \ldots\ldots$

(*d*) $a_0, 3a_1, 9a_2, 27a_3, \ldots\ldots$.

3. Find the closed form of the generating function for each of the following sequence $\{a_n\}$ where :

(*a*) $a_n = 3$ (*b*) $a_n = n + 3$ (*c*) $a_n = 3^n$ (*d*) $a_n = n(3 + 5n)$

(*e*) $a_n = n(n - 1)$.

4. Find the coefficients of

(*a*) x^{10} in $1/(1 - x)^6$ (*b*) x^{10} in $1/(1 - 3x)$

(*c*) x^{12} in $x^2/(1 - x)^{10}$ (*d*) x^{10} in $(x^3 + x^4 + x^5 + \ldots\ldots)^3$.

5. Find the generating function for the sequence A = $\{a_n\}$

where $a_n = \begin{cases} 1 & \text{if} \quad 0 \le n \le 3 \\ 5 & \text{if} \quad 4 \le n \le 7. \\ 0 & \text{if} \quad n \ge 8 \end{cases}$

6. Use generating functions to determine eight identical balls can be distributed among three children if each child receives at least two balls and atmost four balls.

7. Find the number of ways in which 9 balls can be distributed among three distinct boxes so that no box will contain more than 4 balls.

8. Find the generating function for each of the following discrete numeric functions :

(*i*) 1, – 2, 3, – 4, 5, 6, ... (*ii*) 1, 2/3, 3/9, 4/27, ...

(*iii*) 1, 1, 2, 2, 3, 3, 4, 4, ... (*iv*) $0 * 1, 1 * 2, 2 * 3, 3 * 4, ...$

(*v*) $0 * 5^0, 1 * 5^1, 2 * 5^2, 3 * 5^3, ...$

9. If x, y and z are digits, then find the number of possible solutions to the following equations

(*i*) $x + y + z = 10$ (*ii*) $x - y + z = 17$

(*iii*) $x + 2y + 3z = 8$ (*iv*) $x - 2y + 2z = 37$.

10. If x, y and z are positive integers, then find the number of possible solutions to the following equations

(*i*) $x + y + z = 16$ (*ii*) $x - y + z = -7$

(*iii*) $x + 2y + 3z = 30$ (*iv*) $x - 2y + 2z = 0$.

11. If x, y and z are non-negative integers, then find the number of possible solutions to the following

(*i*) $x + y + z = 100$ (*ii*) $x - y + z = 27$

(*iii*) $x + 2y + 3z = 10$ (*iv*) $x - 2y + 2z = 71$.

12. In how many ways can we select seven non-consecutive integers from { 1, 2, 3, ... 50} ?

13. Determine the sequence generated by each of the following generating functions.

(*a*) $f(x) = (2x - 3)^3$ (*b*) $f(x) = x^4/(1 - x)$

(*b*) $f(x) = x^3/(1 - x^2)$ (*d*) $f(x) = 1/(1 + 3x)$

(*e*) $f(x) = 1/(3 - x)$ (*f*) $f(x) = 1/(1 - x) + 3x^2 - 11$.

14. (*a*) Find the coefficient of x^7 in $(1 + x + x^2 + x^3 + ...)^{15}$

(*b*) Find the coefficient of x^7 in $(1 + x + x^2 + x^3 + ...)^n$ for $n \in Z^+$.

15. Find the coefficient of x^{20} in $(x^2 + x^3 + x^4 + x^5 + x^6)^5$.

16. Find the coefficient of x^{15} in each of the following

(*a*) $x^3(1 - 2x)^{10}$ (*b*) $(x^3 - 5x)/(1 - x)^3$ (*c*) $(1 + x)^4/(1 - x)^4$.

17. Determine the constant (that is, the coefficient of x^0) in $(3x^2 - (2/x))^{15}$.

18. Find the coefficient of x^{50} in $(x^7 + x^8 + x^9 + ...)^6$.

19. Find the generating functions for the following sequences

(*a*) $\binom{8}{0}, \binom{8}{1}, \binom{8}{2}, \ldots \binom{8}{8}$

(*b*) $\binom{8}{1}, 2\binom{8}{2}, 3\binom{8}{3}, \ldots .8\binom{8}{8}$

(*c*) $1, -1, 1, -1, 1, -1, \ldots$

(*d*) $0, 0, 0, 6, -6, 6, -6, 6, \ldots$

(*e*) $1, 0, 1, 0, 1, 0, 1, \ldots$

(*f*) $0, 0, 1, a, a^2, a^3, \ldots, a \neq 0$.

20. Find the partitions of 7.

21. In $f(x) = [1/(1-x)]\,[1/(1-x^2)]\,[1/(1-x^3)]$, the coefficient of x^6 is 7. Interpret this result in terms of partitions of 6.

22. Find the generating function for the number of integer solutions of

(*a*) $2w + 3x + 5y + 7z = n, 0 \leq w, x, y, z$

(*b*) $2w + 3x + 5y + 7z = n, 0 \leq w, 4 \leq x, y\,; 5 \leq z$.

23. Find the generating function for the number of partitions of the non negative integer n into summands where

(*a*) each summand must appear an even number of times ;

(*b*) each summand must be even.

24. Find the exponential generating function for the sequence 0! 1! 2! 3! ......

25. Find the exponential generating function for each of the following sequences.

(*a*) $1, -1, 1, -1, 1, -1, \ldots$

(*b*) $1, 2, 2^2, 2^3, 2^4, \ldots$

(*c*) $1, -a, a^2, -a^3, a^4, \ldots\ a \in R$.

(*d*) $1, a^2, a^4, a^6, \ldots\ a \in R$

(*e*) $a, a^3, a^5, a^7, \ldots\ a \in R$

(*f*) $0, 1, 2(2), 3(2^2), 4(2^3), \ldots$.

26. Find the generating function for the sequence $a_0, a_1, a_2, \ldots\ldots$, where $a_n = \sum_{i=0}^{n}\left(\frac{1}{i!}\right), n \in N$.

27. Let $f(x)$ be the generating function for the sequence $a_0, a_1, a_2, \ldots\ldots$. For what sequence is $(1-x)\,f(x)$ the generating function ?

28. Find the generating function for each of the following sequences

(*a*) $7, 8, 9, 10, \ldots\ldots$

(*b*) $1, a, a^2, a^3, a^4, \ldots\ldots\ a \in R$

(*c*) $1, (1+a), (1+a)^2, (1+a)^3, \ldots\ldots\ a \in R$

(*d*) $2, 1+a, 1+a^2, 1+a^3, \ldots\ldots\ a \in R$.

29. Find the coefficient of x^{83} in

$$f(x) = (x^5 + x^8 + x^{11} + x^{14} + x^{17})^{10}.$$

30. (*a*) For what sequence of numbers is

$g(x) = (1 - 2x)^{-5/2}$ the exponential generating function ?

(*b*) Find a and b so that $(1 - ax)^b$ is the exponential genearting function for the sequence 1, 7, 7.11, 7.11.15, ...

Answers 3.1

1. $1 + x + x^2 + 3x^3 + 3x^4 + 3x^5$ **2.** $(1 + x + x^2 + x^3)(1 + x + ... + x^5)(1 + x + x^2)$.

3. $A(x) = (x + x^2 + x^3 + x^4 + x^5 + x^6)^9$

$= x^9(1 + x + ... + x^5)^9 = x^9(1 - x^6)^9(1 - x)^{-9}$

the coefficient of x^{16} in $(1 - x^6)^9 (1 - x)^{-9}$,

$a_{25} = C(24, 16) - 9C(18, 10) + C(9, 2)\, C(12, 4)$.

4. The co-efficient of x^{20} in $(1 + x + x^2 + x^3)^3 \left(\sum_{r=0}^{\infty} x^r\right)^2$ is C(24, 20) – 3C(20, 16) + 3C(16, 12) – C(12, 8).

6. (*i*) $(1 + x + x^2 + x^3)^3$ (*ii*) $(x^2 + x^3 + x^4 + x^5)^3$

(*iii*) $(x + x^2 + ...)^3$ (*iv*) $(1 + x^2 + x^4 + x^6)(x^3 + x^5 + x^7)(x^5 + x^6 + x^7)$.

7. (*i*) $(1 + x + x^2 + x^3)(1 + x + ... + x^5)(1 + x + ... + x^7)$

(*ii*) $(1 + x + ... + x^n + ...)^4$

(*iii*) $(1 + x^2 + ... + x^{12})(x + x^3 + x^5 + x^7)(1 + x + ... + x^n ......)^{n-2}$

(*iv*) $(x + ... + x^6)^{50}$, a_{100} (*v*) $(1 + x + ... + x^9)^3$.

8. $(1 + x + ...)^3 (x + x^2 + ...)^4$.

11. (*a*) $(1 + x + ...)^2 (x + x^2 + ...)^5$, think of the 6 integers chosen as dividers for 7 boxes where the first and last box can be empty and the other 5 boxes are non empty.

(*b*) Coefficient of x^{14}

(*c*) Coefficient of x^{n-6}.

12. (*i*) To solve $e_1 + e_2 + ... + e_{10} = 23$, where $e_i = 0, 5, 9$. This can be done only with one 5, two 9's and seven 0's. Hence the coefficient is $\dfrac{10!}{1!2!7!}$.

(*ii*) 32 can be obtained only with three 9's, one 5, and 60's. Thus the coefficient of x^{32} is $\dfrac{10!}{3!1!6!}$.

14. (*a*) $\dfrac{1}{(1-x)^2} = \sum_{r=0}^{\infty} C(r+1, r)x^r = \sum_{r=0}^{\infty} (r+1)x^r$, coefficient of x^{10} is 11.

(*b*) $\frac{1}{(1-x)^3} = \sum_{r=0}^{\infty} C(r+2, r)x^r = \sum_{r=0}^{\infty} \frac{(r+2)(r+1)}{2} x^r$, coefficient of x^{10} is (12)(11)/2.

(*c*) C(14, 10).

(*d*) $(-1)^{10}$ C(14, 10) = C(14, 10).

(*e*) $[x^3(1+x+x^2+ ......)]^2 = x^6 \left[\frac{1}{(1-x)^2}\right]$, coefficient of x^{10} is the coefficient of x^4 in $\frac{1}{(1-x)^2}$

$= \sum_{r=0}^{\infty} C(r+1, r)x^r$, coefficient = 5.

(*f*) C(8, 6) – C(4, 2) – C(3, 1).

15. (*a*) C(19, 10) (*b*) C(16, 7) (*c*) C(20, 12) (*d*) C(20, 12)
(*e*) C(31, 12) (*g*) 4^{12} C(15, 12) (*j*) $(-2)^{12}$ C(19, 12).

16. (*a*) C(12, 10) – 3C(7, 5) + 3 (*b*) C(45, 43) – C(24, 22).

17. $(x+x^2+x^3+x^4+x^5+x^6)^{50} = x^{50}(1-x^6)^{50} \frac{1}{(1-x)^{50}}$, coefficeint of x^{100} is C(99, 50)

– C(44 + 49, 44) C(50, 1) + C(49 + 38, 38) C(50, 2) – C(49 + 32, 32) C(50, 3) ....

18. (*a*) C(21, 19) – C(6, 4) – C(5, 3) (*b*) C(17, 11) – 7C(12, 6) + C(7, 2) C(7, 1).

19. C(14, 9) – C(9, 4).

20. (*a*) $(1+x+x^2+x^3)^{10} = \left(\frac{1-x^4}{1-x}\right)^{10} = \frac{(1-x^4)^{10}}{(1-x)^{10}} = (1-x^4)^{10} \sum_{r=0}^{\infty} C(r+9, r)x^r$

$= [1 - C(10, 1)x^4 + C(10, 2)x^8 - C(10, 3)x^{12} + ... x^{40}]$

$\sum_{r=0}^{\infty} C(r+9, r)x^r$,

Coefficient of x^{14} is C(23, 9) – C(10, 1) C(19, 10) + C(10, 2) C(15, 6) – C(10, 3) C(11, 2).

(*b*) C(14, 9) – C(9, 4).

21. C(16, 12) – C(5, 1) C(11, 7) + C(5, 2) C(6, 2).

22. (*a*) C(6, 4) (*b*) C(12, 10) – C(9, 7).

23. (*a*) The coefficient of x^{10} in $(1+x+x^2+x^3+ ......)^6$.

(*b*) The coefficient of x^r in $(1+x+x^2+x^3+ ......)^n$.

24. The answer is the coefficient of x^{31} in the generating function

$(1+x+x^2+x^3+ ......)^3(1+x+x^2+ ...... +x^{10})$.

25. (*a*) The coefficient of x^{20} in $(1+x+x^2+ ...... +x^7)^4$.

(*b*) The coefficient of x^{20} in $(1+x+x^2+ ...... +x^{20})^2(1+x^2+x^4+ ...... +x^{20})^2$

or $(1 + x + x^2 + ...)^2 (1 + x^2 + x^4 + ...)^2$.

(*c*) The coefficient of x^{30} in $(x^2 + x^3 + x^4)(x^3 + x^4 + ... + x^8)^4$.

(*d*) The coefficient of x^{30} in $(1 + x + x^2 + ... + x^{30})^3 (1 + x^2 + x^4 + ... + x^{30})$
$(x + x^3 + x^5 + ... + x^{29})$

or $(1 + x + x^2 + ...)^3 (1 + x^2 + x^4 + ...)(x + x^3 + x^5 + ...)$.

Answers 3.2

1. (*a*) $\dfrac{x^3}{1-x}$ (*b*) $\dfrac{x^2}{(1-x)^2}$ (*c*) $\dfrac{3}{1+x}$ (*d*) $\dfrac{3}{1-3x}$ (*e*) $\dfrac{1}{(1-x)^2}$.

2. (*a*) $x^3\,G(x)$ (*b*) $G(x) - a_0 - a_1x$ (*c*) $G(x^2)$ (*d*) $G(3x)$.

3. (*a*) $\dfrac{3}{1-x}$ (*b*) $\dfrac{x}{(1-x)^2} + \dfrac{3}{1-x}$ (*c*) $\dfrac{1}{1-3x}$ (*d*) $\dfrac{3x}{(1-x)^2} + \dfrac{5x(1+x)}{(1-x)^3}$ (*e*) $\dfrac{2x^2}{(1-x)^3}$.

4. (*a*) C(15, 10) (*b*) 3^{10} (*c*) C(19, 10) (*d*) 3.

5. $1 + x + x^2 + x^3 + 5x^4 + 5x^5 + 5x^6 + 5x^7$.

6. 6 **7.** 10 **14.** (*a*) $\dbinom{21}{7}$ (*b*) $\dbinom{n+6}{7}$

15. $\dbinom{14}{10} - 5\dbinom{9}{5} + \dbinom{5}{2}$

16. (*a*) 0 (*b*) $\dbinom{14}{12} - 5\dbinom{16}{14}$ (*c*) $\dbinom{18}{15} + 4\dbinom{17}{14} + 6\dbinom{16}{13} + 4\dbinom{15}{12} + \dbinom{14}{11}$.

19. (*a*) $(1 + x)^8$ (*b*) $8(1 + x)^7$ (*c*) $(1 + x)^{-1}$ (*d*) $\dfrac{6x^3}{(1+x)}$ (*e*) $(1 - x^2)^{-1}$ (*f*) $\dfrac{x^2}{(1-ax)}$.

20. 7 : 6 + 1 ; 5 + 2 ; 5 + 1 + 1 ; 4 + 3 ; 4 + 2 + 1 ; 4 + 1 + 1 + 1 ; 3 + 3 + 1 ; 3 + 2 + 2 ; 3 + 2 + 1 + 1 ; 3 + 1 + 1 + 1 + 1 ; 2 + 2 + 2 + 1 ; 2 + 2 + 1 + 1 + 1 ; 2 + 1 + 1 + 1 + 1 + 1 ; 1 + 1 + 1 + 1 + 1 + 1 + 1.

21. The number of partitions of 6 into 1's, 2's and 3's is 7.

23. (*a*) and (*b*)

$$(1 + x^2 + x^4 + x^6 + ...)(1 + x^4 + x^8 + ...)(1 + x^6 + x^{12} + ...)\ ...... = \prod_{i=1}^{\infty} \frac{1}{1-x^{2i}}.$$

24. $\dfrac{1}{1-x} = 1 + x + x^2 + x^3 + ... = (0\,!)\dfrac{x^0}{0!} + (1\,!)\dfrac{x^1}{1!} + (2\,!)\dfrac{x^2}{2!} + (3\,!)\dfrac{x^3}{3!} + ...$

25. (*a*) e^{-x} (*b*) e^{2x} (*c*) e^{-ax} (*d*) e^{a^2x} (*e*) ae^{a^2x} (*f*) xe^{2x}. **26.** $f(x) = \left[\dfrac{e^x}{(1-x)}\right]$.

27. $a_0, a_1, -a_0, a_2, -a_1, a_3, -a_2, \ldots\ldots$

28. (*a*) $\dfrac{6}{(1-x)} + \dfrac{1}{(1-x)^2}$ (*b*) $\dfrac{1}{(1-ax)}$

(*c*) $\dfrac{1}{[1-(1+a)x]}$ (*d*) $\dfrac{1}{\left[(1-x) + \dfrac{1}{(1-ax)}\right]}$.

30. (*a*) 1, 5, (5)(7), (5)(7)(9), (5)(7)(9)(11), ... (*b*) $a = 4, b = -\dfrac{7}{4}$.

CHAPTER

4

Recurrence Relations

INTRODUCTION

A wide variety of recurrence relations occur in models. Some of these recurrence relations can be solved using iteration or some other adhoc technique. However, one important class of recurrence relations can be explicitly solved in a systematic way. These are recurrence relations that express the terms of a sequence as linear combinations of previous terms.

4.1. The First-order Linear Recurrence Relation

Suppose n is a natural number, we define 2^n as

$$2^n = \underbrace{2.2.2.......2}_{n2\text{'s}}$$

or $\qquad 2' = 2$, and for $k \geq 1$, $2^{k+1} = 2.2^k$

We write $\quad 0! = 1$ and for $k \geq 0$, $(k+1)! = (k+1)\,k!$

* A sequence is a function whose domain is some infinite set of integers (often N) and whose range is a set of real numbers.

* The sequence which is the function $f: \mathrm{N} \rightarrow \mathrm{R}$ defined by $f(n) = n^2 = 1, 4, 9, 16, ...$

The numbers in the list are called the **terms of the sequence,** the terms are denoted $a_0, a_1, a_2, ...$.

The sequence 2, 4, 8, 16, ... can be defined recursively like : $a_1 = 2$ and for $f \geq 1$, $a_{k+1} = 2a_k$ (*) setting $k = 1, 2, 3, ...$ and $a_1 = 2$ in (*) gives 2, 4, 8, ...

* The equation $a_{k+1} = 2a_k$ in (*), which defines one member of the sequence in terms of a previous one, is called a **recurrence relation.** The equation $a_1 = 2$ is called an **initial condition.**

For example, we write

$a_0 = 2$ and for $k \geq 0$, $a_{k+1} = 2a_k$ or we say $a_1 = 2$ and for $k \geq 2$, $a_k = 2a_{k-1}$

In (*), for instance, $a_n = 2^n$, we say that $a_n = 2^n$ **is the solution** to the recurrence relation.

* A sequence of numbers like 50, 64, 78, 92, ... where each term is determined by adding the same fixed number to the previous one, is called an **arithmetic sequence.** The fixed number is called **the common difference** of the sequence.

* The arithmetic sequence with first term a and common difference d is the sequence defined by

$$a_1 = a \text{ and } k \geq 1,\ a_{k+1} = a_k + d$$

The general arithmetic sequence, takes the form

$$a, a+d, a+2d, ...$$

and for $n \geq 1$, the nth term of the sequence is $a_n = a + (n-1)\,d$.

The sum of n terms of the arithmetic sequence with first term a and common difference d is

$$\mathrm{S} = \frac{n}{2}\,[2a + (n-1)\,d]$$

* The geometric sequence with first term a and common ratio r is the sequence defined by

$$a_1 = a \text{ and for } k \geq 1,\ a_{k+1} = ra_k$$

The general geometric sequence, this has the form

$$a,\ ar,\ ar^2,\ ar^3,\ ....$$

the nth term being $a_n = ar^{n-1}$, the sum S of n terms $(r \neq 1)$, $S = a(1 - r^n)/(1 - r)$.

* The Fibonacci sequence,

$$f_1 = 1, f_2 = 1 \text{ and for } k \geq 2, f_{k+1} = f_k + f_{k-1}$$

the nth term of the Fibonacci sequence is the closed integer to the number

$$\frac{1}{\sqrt{5}}\left(\frac{1+\sqrt{5}}{2}\right)^n$$

For example, $a_1 = 1$ and for $k > 1$,

$$a_k = \begin{cases} 1 + a_{k/2} & \text{if } k \text{ is even} \\ 1 + a_{3k-1} & \text{if } k \text{ is odd.} \end{cases}$$

A linear homogeneous recurrence relation of degree k **with constant coefficients** is a recurrence relation of the form

$$a_n = c_1a_{n-1} + c_2a_{n-2} + ... + c_ka_{n-k}$$

where $c_1, c_2, ..., c_k$ are real numbes, and $c_k \neq 0$.

The recurrence relation in the definition is **linear** since the right-hand side is a sum of multiples of the previous terms of the sequence. The recurrence relation is **homogeneous** since no terms occur that are not multiples of the a_js. The coefficients of the terms of the sequence are all **constants,** rather than functions that depend on n. The degree is k because a_n is expressed in terms of the previous k terms of the sequene. A consequence of the principle of Mathematical induction is that a sequence satisfying the recurrence relation in the definition is uniquely determined by this recurrence relation and k initial conditions

$$a_0 = c_0,\ a_1 = c_1,\ ....\ a_{k-1} = c_{k-1}.$$

Problem 4.1. (Compound Interest)

A person invests Rs. 10,000/- @ 12% interest compounded annually. How much will be there at the end of 15 years.

Solution. Let A_n represents the amount at the end of n years.

So at the end of $n - 1$ years, the amount is A_{n-1}.

Since the amount after n years equals the amount after $n - 1$ years plus interest for the nth year.

Thus the sequence $\{A_n\}$ satisfies the recurrence relation

$$A_n = A_{n-1} + (0.12)\ A_{n-1} = (1.12)\ A_{n-1},\ n \geq 1.$$

With initial condition $A_0 = 10,000$.

The recurrence relation with the initial condition allow us to compute the value of A_n for any n.

For example, $A_1 = (1.12)\ A_0$

$$A_2 = (1.12)\ A_1 = (1.12)^2\ A_0$$

$$A_3 = (1.12)\, A_2 = (1.12)^3\, A_0$$
$$\vdots$$
$$A_n = (1.12)^n\, A_0$$

which is an explicit formula and the required amount can be derived from the formula by putting $n = 15$.

So, $A_{15} = (1.12)^{15}\,(10000)$.

Problem 4.2. (Tower of Hanoi)

The Tower of Hanoi is a puzzle consisting of three pegs mounted on a board with disks of different diameters. Initially, these disks are placed on the first peg in order of size, with the largest on the bottom (as shown in Fig. 4.1). These disks are to be transfered on the second peg with their relative positions unchanged. The rules of the puzzle allow disk to be moved one at a time from one peg to another as long as a disk is never placed on the top of smaller one.

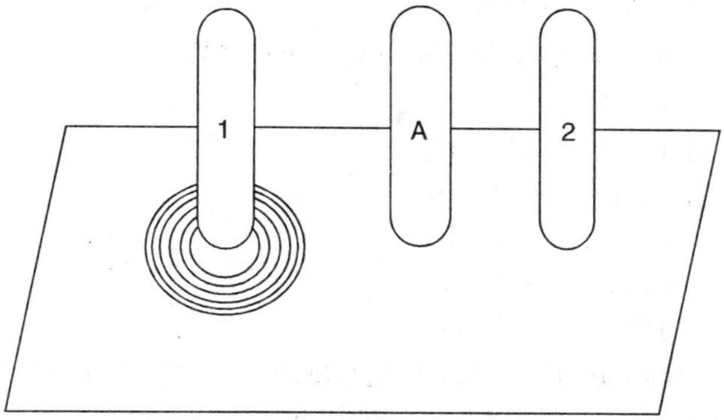

(*a*) : Initial position

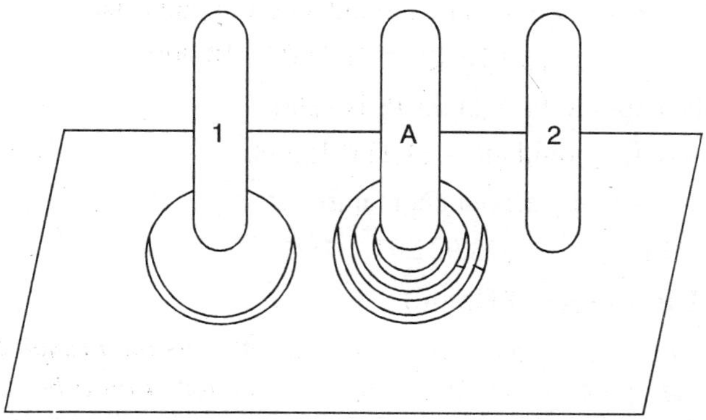

(*b*) : Position after moving $n - 1$ disks from 1 to A.

Fig. 4.1.

Solution. Let C_n denotes the number of moves needed to solve the n-disk puzzle.

If there is only one disk, we simply move it to the second peg.

If we have $n > 1$, we cab transfer $n - 1$ disks to peg A using C_{n-1} moves.

During the moves, the largest disk at the bottom of peg 1 stays fixed.

There we use one moev to transfer the largest disk to the second peg.

We can now move again $n - 1$ disks on peg A to peg 2 using C_{n-1} moves.

Therefore $C_n = 2C_{n-1} + 1, n > 1$ with initial condition $C_1 = 1$.

Problem 4.3. (Compound Interest)

Suppose that a person deposits \$10,000 in a savings account at a bank yielding 11% per year with interest compounded annually. How must will be in the account after 30 years ?

Solution. To solve this problem. Let P_n denote the amount in the account after n years.

Since the amount in the account after n years equals the amount in the account after $n - 1$ years plus interest for the nth year, we see that sequence $\{P_n\}$ satisfies the recurrence relation

$$P_n = P_{n-1} + 0.11\ P_{n-1} = (1.11)\ P_{n-1}$$

This initial condition is $P_0 = 10{,}000$.

We can use an interative approach to find a formula for P_n.

Note that $P_1 = (1.11)\ P_0$

$$P_2 = (1.11)\ P_1 = (1.11)^2\ P_0$$

$$P_3 = (1.11)\ P_2 = (1.11)^3\ P_0$$

$$\dots\dots\dots\dots\dots\dots$$

$$\dots\dots\dots\dots\dots\dots$$

$$P_n = (1.11)\ P_{n-1} = (1.11)^n\ P_0$$

when we insert the initial condition $P_0 = 10{,}000$, the formula $P_n = (1.11)^n\ 10{,}000$ is obtained.

We can use mathetmatical induction to establish its validity. That the formula is valid for $n = 0$ is a consequence of the initial condition.

Now assume that $P_n = (1.11)^n\ 10{,}000$.

Then, from the recurrence relation and the induction hypothesis.

$P_{n+1} = (1.11)\ P_n = (1.11)\ (1.11)^n\ 10{,}000 = (1.11)^{n+1}\ 10{,}000$.

This shows that the explicit formula for P_n is valid.

Inserting $n = 30$ into the formula $P_n = (1.11)^n\ 10{,}000$

Shows that after 30 years the account contains

$$P_{30} = (1.11)^{30}\ 10{,}000 = \$228{,}922.97.$$

Problem 4.4. (The Towers of Hanoi)

A popular puzzle of the late nineteenth century invented by the French Mathematician Édouard Lucas, called the Tower of Hanoi, consists of three pegs mounted on a board together with disks of different sizes. Initially these disks are placed on the first peg in order of size with the largest on the bottom (as shown in figure). The rules of the puzzle allow disks to be moved one at a time from one peg to another as long as a disk is never placed on top of a smaller disk. The goal of the puzzle is to have all the disks on the second peg in order of size, with the largest on the bottom.

Let H_n denote the number of moves needed to solve the Tower of Hanoi problem with n disks. Set up a recurrence relation for the sequence $\{H_n\}$.

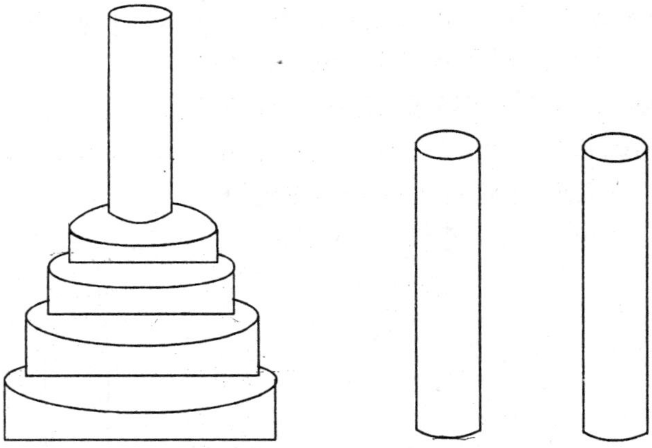

Fig. 4.2. The initial position in the Tower of Hanoi.

Solution. Begin with n disks on peg 1. We can transfer the top $n-1$ disks, following the rules of the puzzle, to peg 3 using H_{n-1} moves (as shown in Fig. 4.2).

We keep the largest disk fixed during these moves. Then, we use one move to transfer the largest disk to the second peg.

We can transfer the $n-1$ disk on peg 3 to peg 2 using H_{n-1} additional moves, placing them on top the largest disk, which always stays fixed on the bottom of peg 2. Moreover, it is easy to see that the puzzle cannot be solved using fewer steps.

This shows that $H_n = 2H_{2n-1} + 1$

This initial condition is $H_1 = 1$, since one disk can be transferred from peg 1 to peg 2, according to the rules of the puzzle, in one move.

We can use an iterative approach to solve this recurrence relation.

Note that
$$\begin{aligned}
H_n &= 2H_{n-1} + 1 \\
&= 2(2H_{n-2} + 1) + 1 = 2^2 H_{n-2} + 2 + 1 \\
&= 2^2 (2H_{n-3} + 1) + 2 + 1 = 2^3 H_{n-3} + 2^2 + 2 + 1 \\
&\ldots\ldots\ldots\ldots\ldots\ldots \\
&\ldots\ldots\ldots\ldots\ldots\ldots \\
&= 2^{n-1}H_1 + 2^{n-2} + 2^{n-3} + \ldots\ldots + 2 + 1 \\
&= 2^{n-1} + 2^{n-2} + \ldots\ldots + 2 + 1 \\
&= 2^n - 1.
\end{aligned}$$

We have to used the recurrence relation repeatedly to express H_n in terms of previous terms of the sequence. In the next to last equality, the initial condition $H_1 = 1$ has been used.

The last equality is based on the formula for the sum of the terms of a geometric series,

The iterative approach has produced the solution to the recurrence relation $H_n = 2H_{n-1} + 1$ with the initial condition $H_1 = 1$.

This formula can be proved using mathematical induction. A myth created to accompany the puzzle tells of a tower in Hanoi where monks are transferring 64 gold disks from one peg to another, according to the rules of the puzzle. The myth says that the world will end when they finish the puzzle. How long after the monks started will the world end if the monks take one second to move a disk ?

From the explicit formula, the monks require

$$2^{64} - 1 = 18,446,744,073,709,551,615.$$

Moves to transfer the disks.

Making one move per second, it will take them more than 500 billion years to solve the puzzle, so the world should survive a while longer than already has.

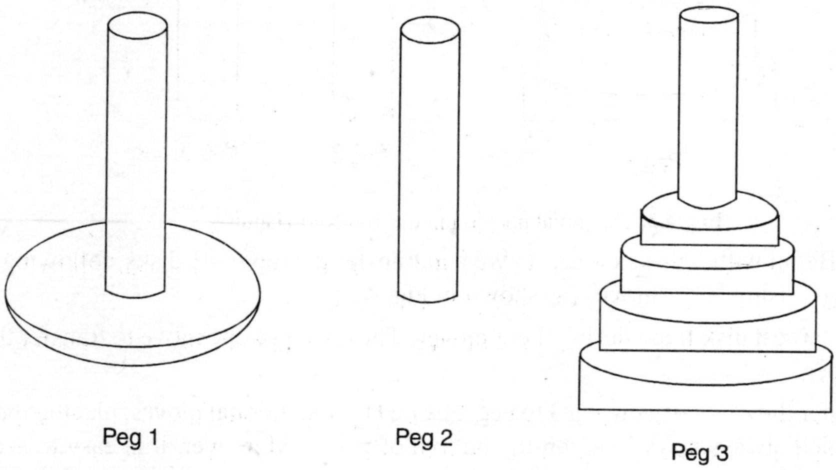

Fig. 4.3. An intermediate position in the Tower of Hanoi

Problem 4.5. (Rabbits and the Fibonacci Numbers)

Consider this problem, which was originally posed by Leonardo di pisa, also known as Fibonacci, in the thirteenth century in this book Liber abaci. A young pair of rabbits (one of each sex) is placed on an island. A pair of rabbits does not breed until they are 2 months old. After they are 2 months old, each pair of rabbits produces another pair each month, as shown in Fig.. Find a recurrence relation for the number of pairs of rabbits on the island after n months, assuming that no rabbits ever die.

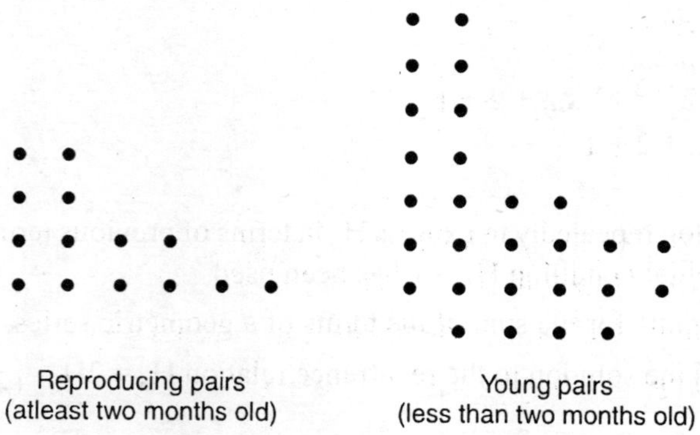

Month	Reproduction pairs	Young pairs	Total pairs
1	0	1	1
2	0	1	1
3	1	1	2
4	1	2	3
5	2	3	5
6	3	5	8

Fig. 4.4. Rabbits on an Island.

Solution. Denote by f_n the number of pairs of rabbits after n months. We will show that f_n, $n = 1, 2, 3, ...$ are the terms of the Fibonacci sequence.

The rabbit population can be modeled using a recurrence relation. At the end of the first month, the number of pairs of rabbits on the island is $f_1 = 1$. Since this pair does not breed during the second month, $f_2 = 1$ also. To find the number of pairs after n months, add the number on the island the previous month, f_{n-1}, and the number of new born pairs, which equals f_{n-2}, since each new born pair comes from a pair of least 2 months old.

Consequently, the sequence $\{f_n\}$ satisfies the recurrence relation $f_n = f_{n-1} + f_{n-2}$ for $n \geq 3$, together with the initial conditions $f_1 = 1$ and $f_2 = 1$. Since this recurrence relation and the initial conditions uniquely determine this sequence, the number of pairs of rabbits on the island after n months is given by the nth Fibonacci number.

Problem 4.6. (Codeword Enumeration)

A computer system considers a string of decimal digit a valid codeword if it contains an even number of 0 digits. For instance, 1230407869 is valid, whereas 120987045608 is not valid. Let a_n the number of valid n-digit codewords. Find a recurrence relation for a_n.

Solution. Note that $a_1 = 9$ since there are 10 one-digit strings, and only one, namely, the string 0, is not valid.

A recurrence relation can be derived for this sequence by considering how a valid n-digit string can be obtained from strings of $n - 1$ digits. There are two ways to form a valid string with n digits from a string with one fewer digit.

First, a valid string of n digits can obtained by appending a valid string of $n - 1$ digits with a digit other than 0. This appending can be done in nine ways. Hence, a valid string with n digits can be formed in this manner in $9a_{n-1}$ ways.

Second, a valid string of n digits can be obtained by appending a 0 to a string of length $n - 1$ that is not valid.

This produces a string with an even number of 10 digits since the invalid string of length $n - 1$ has an odd number of 0 digits.

The number of ways that this can be done equals the number of invalid $(n - 1)$-digit strings.

Since these are 10^{n-1} strings of length $n - 1$, and a_{n-1} are valid, these are $10^{n-1} - a_{n-1}$ valid n-digit strings obtained by appending an invalid string of length $n - 1$ with $a0$.

Since all valid strings of length n are produced in one of these two ways, it follows that these are

$$\begin{aligned} a_n &= 9a_{n-1} + (10^{n-1} - a_{n-1}) \\ &= 8a_{n-1} + 10^{n-1} \end{aligned}$$

valid strings of length n.

Problem 4.7. *Let $\{a_n\}$ be a sequence that satisfies the recurrence relation $a_n = a_{n-1} - a_{n-2}$ for $n = 2, 3, 4, ...$ and suppose that $a_0 = 3$ and $a_1 = 5$. What are a_2 and a_3 ?*

Solution. We see from the recurrence relation that

$$a_2 = a_1 - a_0 = 5 - 3 = 2 \quad \text{and} \quad a_3 = a_2 - a_1 = 2 - 5 = -3$$

We can find a_4, a_5 and each successive term in a similar way.

Problem 4.8. *Determine whether the sequence $\{a_n\}$ is a solution of the recurrence relation $a_n = 2a_{n-1} - a_{n-2}$ for $n = 2, 3, 4, \ldots\ldots$ where $a_n = 3n$ for every non negative integer n. Answer the same question where $a_n = 2^n$ and where $a_n = 5$.*

Solution. Suppose that $a_n = 3n$ for every non negative integer n. Then, for $n \geq 2$, we see that $2a_{n-1} - a_{n-2}$

$$= 2[3(n-1)] - 3(n-2) = 3n = a_n.$$

Therefore, $\{a_n\}$, where $a_n = 3n$, is a solution of the recurrence relation.

Suppose that $a_n = 2^n$ for every non-negative integer n.

Note that $a_0 = 1$, $a_1 = 2$ and $a_2 = 4$.

Since $2a_1 - a_0 = 2.2 - 1 = 3 \neq a_2$, we see that $\{a_n\}$, where $a_n = 2^n$, is not a solution of the recurrence relation.

Suppose that $a_n = 5$ for every non-negative integer n. Then for $n \geq 2$, we see that $a_n = 2a_{n-1} - a_{n-2} = 2.5 - 5 = 5 = a_n$.

Therefore, $\{a_n\}$, where $a_n = 5$, is a solution of the recurrence relation.

Problem 4.9. *Find a recurrence relation and give initial conditions for the number of bit strings of length n that do not have two consecutive 0s. How many such bit strings are there of length five ?*

Solution. Let a_n denote the number of bit strings of length n that do not have two consecutive 0s.

To obtain a recurrence relation for $\{a_n\}$, note that by the sum rule, the number of bit strings of length n that do not have two consecutive 0s equals the number of such bit strings ending with *a* 0 plus the number of such bit strings ending what *a* 1.

We will assume that $n \geq 3$, so that the bit string has at least three bits.

The bit strings of length n ending with 1 that do not have two consecutive 0s are precisely the bit strings of length $n - 1$ with no two consecutive 0s with *a* 1 added at the end.

Consequently, there are a_{n-1} such bit strings.

Bit strings of length n ending with *a* 0 that do not have two consecutive 0s must have 1 as their $(n-1)$st bit, otherwise they would end with a pair of 0s.

It follows that the bit strings of length n ending with *a* 0 that have no two consecutive 0s are precisely the bit strings of length $n-2$ with no two consecutive 0s with 10 added at the end. Consequently, there are a_{n-2} such bit strings.

We conclude that, $a_n = a_{n-1} + a_{n-2}$ for $n \geq 3$.

The initial conditions are $a_1 = 2$, since both it strings of length one, 0 and 1 do not have consecutive 0s, and $a_2 = 3$, since the valid bit strings of length two are 01, 10 and 11.

To obtain a_5, we use the recurrence relation three times to find that

$$a_3 = a_2 + a_1 = 3 + 2 = 5,$$

$$a_4 = a_3 + a_2 = 5 + 3 = 8,$$

$$a_5 = a_4 + a_3 = 8 + 5 = 13.$$

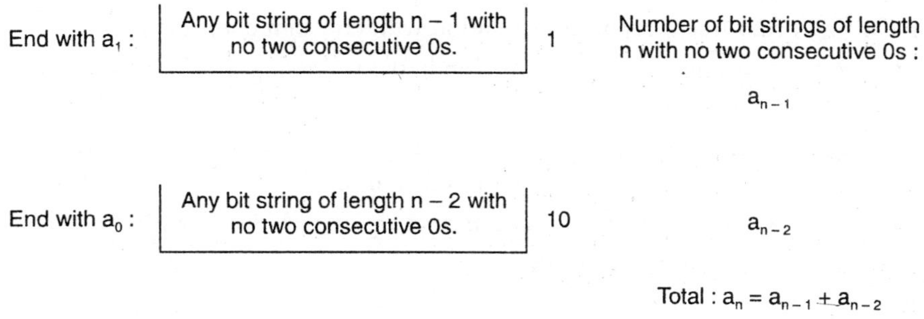

Fig. 4.5. Counting Bit Strings of length n with no two consecutive 0s.

Problem 4.10. *Find a recurrence relation for C_n, the number of ways to parenthesize the product of $n + 1$ numbers, $x_0 . x_1 . x_2 ...... x_n$ to specify the order of multiplication. For example, $C_3 = 5$ since there are five ways to parenthesize $x_0 . x_1 . x_2 . x_3$ to determine the order of multiplication.*

$((x_0 . x_1) . x_2) . x_3 \quad (x_0 . (x_1 . x_2)) . x_3 \quad (x_0 . x_1) . (x_2 . x_3) \quad x_0 . ((x_1 . x_2) . x_3) \quad x_0 . (x_1 . (x_2 . x_3)).$

Solution. To develop a recurrence relation for C_n, we note that however we insert parentheses in the product $x_0 . x_1 . x_2 .... x_n$, one "." operator remains outside all parenthese, namely, the operator for the final multiplication to be performed.

For example, in $(x_0 . (x_1 . x_2)) . x_3$ it is the final ".", while in $(x_0 . x_1) . (x_2 . x_3)$ it is the second "."

Thie final operator appears between two of the $n + 1$ numbers say, x_k and x_{k+1}.

There are C_kC_{n-k-1} ways to inset parentheses to determine the order of the $n + 1$ numbers to be multiplied when the final operator appears between x_k and x_{k+1}, since there are C_k ways to insert parentheses in the product $x_0 . x_1 ...... x_k$ to determine the order in which these $k + 1$ numbers are to be multiplied and C_{n-k-1} ways to inset parentheses in the product $x_{k+1} x_{k+2} ...... x_n$ to determine the order in which these $n - k$ numbers are to be multiplied.

Since this final operator can appear between any two of the $n + 1$ numbers, it follows that

$$C_n = C_0C_{n-1} + C_1C_{n-2} + ... C_{n-2}C_1 + C_{n-1}C_0$$

$$= \sum_{k=0}^{n-1} C_kC_{n-k-1} .$$

Note that the initial conditions are $C_0 = 1$ and $C_1 = 1$.

4.1.1. Back Tracking Method

In this method, we shall start from a_n and move backward towards a_1 to find a pattern, if any, to solve the problem.

To backtrack, we keep on substituting the definition of a_n, a_{n-1}, a_{n-2} and so on. Until a recognizable pattern appears.

4.1.2. Forward Chaining Method

In this method, we begin from initial (terminating) condition and keep on moving towards the nth term until we get a clear pattern.

4.1.3. Summation Method

To solve a first order linear recurrence relation with constant coefficient.

In this method, we arrange the given equation in the following form : $a_n - ka_{n-1} = f(n)$ and then backtrack till terminating condition.

In the process, we get a number of equations. Add these equations in such a way that all intermediate terms gets cancelled. Finally, we get the required solution.

Problem 4.11. *Solve the recurrence equation* $a_n = a_{n-1} + 3$ *with* $a_1 = 2$.

Solution. (*i*) **Backtracking Method :**

We have $a_n = a_{n-1} + 3$ with $a_1 = 2$

$$a_n = a_{n-2} + 3 + 3 \quad \text{(since } a_{n-1} = a_{n-2} + 3\text{)}$$
$$= a_{n-2} + 2 \times 3$$
$$= a_{n-3} + 3 + 2 \times 3 \quad \text{(since } a_{n-2} = a_{n-3} + 3\text{)}$$
$$= a_{n-3} + 3 \times 3$$
$$= a_{n-4} + 3 + 3 \times 3 \quad \text{(since } a_{n-3} = a_{n-4} + 3\text{)}$$
$$= a_{n-4} + 4 \times 3$$

- - - - - - - - - - - -

- - - - - - - - - - - -

$$= a_{n-(n-1)} + (n-1) \times 3$$
$$= a_1 + 3(n-1)$$
$$= 2 + 3(n-1) \quad \text{(since } a_1 = 2 \text{ is the terminating condition)}$$

$\therefore \quad a_n = 2 + 3(n-1)$

(*ii*) **Forward Chaining Method :**

Given, initial condition : $a_1 = 2$

Now
$$a_1 = 2$$
$$a_2 = a_1 + 3$$
$$a_3 = a_2 + 3$$
$$= a_1 + 2 \times 3$$
$$a_4 = a_3 + 3$$
$$= a_1 + 2 \times 3 + 3$$
$$= a_1 + 3 \times 3$$
$$= a_1 + (4-1) \times 3$$
$$a_5 = a_1 + (5-1) \times 3$$

- - - - - - - - - - - -

- - - - - - - - - - - -

$$a_n = a_1 + (n-1)\,3$$

$\therefore \quad a_n = 2 + 3(n-1)$

(*iii*) **Summation Method :**

The given equation can be rearranged as

$$a_n - a_{n-1} = 3$$
$$a_{n-1} - a_{n-2} = 3$$
$$a_{n-2} - a_{n-3} = 3$$

- - - - - - - - - - - -

- - - - - - - - - - - -

$$a_3 - a_2 = 3$$
$$a_2 - a_1 = 3$$

We stop here, since $a_1 = 2$ is given.

Adding all, we get

$$a_n - a_1 = 3 + 3 + 3 + \ldots\ldots + (n-1) \text{ times.}$$
$$= 3(n-1)$$

$\Rightarrow \quad a_n = a_1 + 3(n-1).$

Problem 4.12. *Solve the recurrence equation* $t_n = 2t_{n/2} + n$ *with* $t_1 = 1$.

Solution. ***Backtracking Method :***

We have $\quad t_n = 2t_{n/2} + n \qquad \ldots(1)$

By repeated substitution of the definition of $t_{n/2}, t_{n/4}, t_{n/8}$ etc. we get the following results :

$$t_n = 2[2t_{n/4} + n/2] + n$$
$$= 2^2\, t_{n/4} + 2 * n/2 + n$$
$$= 2^2\, t_{n/4} + n + n \qquad \ldots(2)$$
$$t_n = 2^2\, [2\, t_{n/8} + n/4] + n + n$$
$$= 2^3\, t_{n/8} + 2^2 * n/4 + n + n$$
$$= 2^3\, t_{n/8} + n + n + n \qquad \ldots(3)$$

Equations (1), (2) and (3) can be written as

$$t_n = 2^{\log_2^2}\, t_{n/2} + n \log_2^2, \qquad \text{from (1)}$$
$$t_n = 2^{\log_2^4}\, t_{n/4} + n \log_2^4, \qquad \text{from (2)}$$
$$t_n = 2^{\log_2^8}\, t_{n/8} + n \log_2^8, \qquad \text{from (3)}$$

Similarly, we can write,

$$t_n = 2^{\log_2^n}\, t_{n/n} + n \log_2^n$$
$$t_n = 2^{\log_2^n}\, t_1 + n \log_2^n$$
$$t_n = 2^{\log_2^n} + n \log_2^n \qquad \because \quad t_1 = 1$$

$\therefore \quad t_n = 2^{\log_2^n} + n \log_2^n$

Summation Method :

Arranging the given recurrence equation, we get

$$t_n - 2t_{n/2} = n$$
$$t_{n/2} - 2t_{n/4} = \frac{n}{2}$$

$$t_{n/4} - 2t_{n/8} = \frac{n}{4}$$

- - - - - - - - - -
- - - - - - - - - -

$$t_4 - 2t_2 = 4$$
$$t_2 - 2t_1 = 2$$

The number equations in this set of equations is $\log_2^n$.

Now, if we multiply the above equations by

$1, 2, 3, 4, \ldots\ldots, 2^{\log_2^n}$ respectively from top to bottom, and then by adding them all together, we get

$$t_n - 2^{\log_2^n} = n + n + n + \ldots\ldots \log_2^n \text{ times.}$$

$$t_n - 2^{\log_2^n} = n \log_2^n$$

$$\Rightarrow \qquad t_n = 2^{\log_2^n} + n \log_2^n.$$

Problem 4.13. *Solve the recurrence equation*

$$t(n) = t(\sqrt{n}) + C \log_2{}^n \quad \text{with} \quad t(1) = 1.$$

Solution. We have $t(n) = t(\sqrt{n}) + \text{C} \log_2{}^n$

$$t(2) = t(2^{1/2}) + \text{C} \log_2{}^2$$
$$= t(2^{1/4}) + \text{C} \log_2^{2^{1/2}} + \text{C} \log_2{}^2$$
$$= t(2^{1/8}) + \text{C} \log_2^{2^{1/4}} + \text{C} \log_2^{2^{1/2}} + \text{C} \log_2{}^2$$
$$= t(2^{1/2n}) + \text{C} \log_2^{2^{1/n}} + \ldots\ldots + \text{C} \log_2^{2^{1/4}} + \text{C} \log_2^{2^{1/2}} + \text{C} \log_2{}^2$$

Since t is defined for $n = 1$ only, we have to let

$t(2^{1/2n}) \longrightarrow t(1)$. This is possible when $n \to \infty$

i.e., when $n \to \infty \quad \Rightarrow \quad \dfrac{1}{2n} \to 0 \quad \Rightarrow \quad 2^{1/2n} \to 1$

Therefore, $t(2) = t(1) + \left[\text{C} \log_2^2 + \text{C} \log_2^{2^{1/2}} + \text{C} \log_2^{2^{1/4}} + \ldots\ldots\right]$

$$= t(1) + \text{C} \log_2{}^2 \left[1 + \frac{1}{2} + \frac{1}{4} + \frac{1}{8} + \ldots\ldots\right]$$
$$= t(1) + 2\,\text{C} \log_2{}^2$$

i.e., $t(2) = t(1) + 2\,\text{C} \log_2{}^2$

Similarly, $t(3) = t(1) = 2\,\text{C} \log_2{}^3$

$t(4) = t(1) + 2\,\text{C} \log_2{}^4$

- - - - - - - - - -
- - - - - - - - - -

$$t(n) = t(1) + 2\,C \log_2{}^n$$

Therefore $\quad t(n) = t(1) + 2\,C \log_2{}^n$.

Problem 4.14. *Solve the recurrence relation $a_n = a_{n-1} + 3$ with $a_1 = 2$ defines the sequence 2, 5, 8, ...*

Solution. We backtrack the value of a_n by substituting the definition of a_{n-1}, a_{n-2}, and so on until a pattern is clear.

$$\begin{aligned} a_n &= a_{n-1} + 3 \\ &= (a_{n-2} + 3) + 3 \\ &= ((a_{n-3} + 3) + 3) + 3 \end{aligned} \qquad \text{or} \qquad \begin{aligned} a_n &= a_{n-1} + 3 \\ &= a_{n-2} + 2.3 \\ &= a_{n-3} + 3.3 \end{aligned}$$

Eventually this process will produce

$$\begin{aligned} a_n &= a_{n-(n-1)} + (n-1)\,.\,3 \\ &= a_1 + (n-1)\,.\,3 \\ &= 2 + (n-1)\,.\,3 \end{aligned}$$

An explicit formula for the sequence is $\quad a_n = 2 + (n-1)\,3$.

Problem 4.15. *Backtrack to find an explicit formula for the sequence defined by the recurrence relation $b_n = 2b_{n-1} + 1$ with initial condition $b_1 = 7$.*

Solution. We begin by substituting the definition of the previous term in the defining formula.

$$\begin{aligned} b_n &= 2b_{n-1} + 1 \\ &= 2\,(2b_{n-2} + 1) + 1 \\ &= 2[2\,(2b_{n-3} + 1) + 1] + 1 \\ &= 2^3\, b_{n-3} + 4 + 2 + 1 \\ &= 2^3\, b_{n-3} + 2^2 + 2^1 + 1. \end{aligned}$$

A pattern is emerging with these rewritting of b_n.

(**Note :** There are no set rules for how to rewrite these expressions and a certain amount of experimentation may be necessary.)

The backtracking will end at

$$\begin{aligned} b_n &= 2^{n-1} b_{n-(n-1)} + 2^{n-2} + 2^{n-3} + \ldots\ldots + 2^2 + 2^1 + 1 \\ &= 2^{n-1} b_1 + 2^{n-1} - 1 \\ &= 7\,.\,2^{n-1} + 2^{n-1} - 1 && \text{(using } b_1 = 7) \\ &= 8\,.\,2^{n-1} - 1 \quad \text{or} \quad 2^{n+2} - 1. \end{aligned}$$

Problem 4.16. *Write down the first six terms of the sequence defined by $a_1 = 1$, $a_{k+1} = 3a_k + 1$ for $k \geq 1$. Guess a formula for a_n and prove that your formula is correct.*

Solution. The first six terms are

$$\begin{aligned} a_1 &= 1 \\ a_2 &= 3a_1 + 1 = 3(1) + 1 = 4 \\ a_3 &= 3a_2 + 1 = 3(4) + 1 = 13 \\ a_4 &= 40,\ a_5 = 121,\ a_6 = 364. \end{aligned}$$

Since there is multiplication by 3 at each step, we might suspect that 3^n is involved in the answer.

After trial and error, we guess that $a_n = \frac{1}{2}(3^n - 1)$ and verify this by mathematical induction.

When $n = 1$, the formula gives $\frac{1}{2}(3^1 - 1) = 1$, which is indeed a_1, the first term in the sequence.

Now assume that $k \geq 1$ and that $a_k = \frac{1}{2}(3^k - 1)$.

We wish to prove that $a_{k+1} = \frac{1}{2}(3^{k+1} - 1)$

We have $a_{k+1} = 3a_k + 1 = 3\frac{1}{2}(3^k - 1) + 1$

Using the induction hypothesis.

Hence, $a_{k+1} = \frac{1}{2}3^{k+1} - \frac{3}{2} + 1 = \frac{1}{2}(3^{k+1} - 1)$ as required.

By the principle of mathematical induction, our guess is correct.

Problem 4.17. *A sequence is defined recursively by $a_0 = 1$, $a_1 = 4$ and $a_n = 4a_{n-1} - 4a_{n-2}$ for $n \geq 2$. Find the first six terms of this sequence. Guess a formula for a_n and establish the validity of your guess.*

Solution. Here there are two initial conditions $a_0 = 1$, $a_1 = 4$.

Also, the recurrence relation, $a_n = 4a_{n-1} - 4a_{n-2}$, defines the general term as a function of two previous terms.

The first six terms of the sequence are

$$a_0 = 1$$
$$a_1 = 4$$
$$a_2 = 4a_1 - 4a_0 = 4(4) - 4(1) = 12$$
$$a_3 = 4a_2 - 4a_1 = 4(12) - 4(4) = 32$$
$$a_4 = 4a_3 - 4a_2 = 4(32) - 4(12) = 80$$
$$a_5 = 4a_4 - 4a_3 = 4(80) - 4(32) = 192.$$

Finding a general formula for a_n requires some ingenuity.

Let us examine some of the first six terms.

We note that $a_3 = 32 = 24 + 8 = 3(8) + 4(8)$

and $a_4 = 80 = 64 + 16 = 4(16) + 16 = 5(16)$

and $a_5 = 192 = 6(32)$.

We guess that $a_n = (n + 1)\, 2^n$.

To prove this, we use the strong form of mathematical induction (with $n_0 = 0$).

When $n = 0$, we have $(0 + 1)\, 2^0 = 1(1) = 1$, in agreement with the given value for a_0.

When $n = 1$, $(1 + 1)\, 2^1 = 4 = a_1$. Now that the formula has been verified for $k = 0$ and $k = 1$.

We may assume that $k > 1$ and that $a_n = (n + 1)\, 2^n$ for all n in the interval $0 \le n < k$.

We wish to prove the formula is valid for $n = k$.

That is, we wish to prove that $a_k = (k + 1)\, 2^k$.

Since $k \ge 2$, we know that $a_k = 4a_{k-1} - 4a_{k-2}$.

Applying the induction hypothesis to $k - 1$ and to $k - 2$ (each of which is in the range $0 \le n < k$).

We have $a_{k-1} = k\, 2^{k-1}$ and $a_{k-2} = (k - 1)\, 2^{k-2}$.

Thus,
$$\begin{aligned} a_k &= 4(k\, 2^{k-1}) - 4(k - 1)\, 2^{k-2} \\ &= 2k\, 2^k - k\, 2^k + 2^k \\ &= k\, 2^k + 2^k = (k + 1)\, 2^k \quad \text{as required.} \end{aligned}$$

By principle of mathematical induction, the formula is valid for all $n \ge 0$.

Problem 4.18. *Solve the recurrence relation $a_n = a_{n-1} + 2$, $n \ge 2$ subject to initial condition $a_1 = 3$.*

Solution. We backtrack the value of a_n by substituting the expression of a_{n-1}, a_{n-2} and so on, until a pattern is clear.

Given
$$a_n = a_{n-1} + 2 \qquad \text{...(1)}$$

Replacing n by $n - 1$ in (1), we obtain
$$a_{n-1} = a_{n-2} + 2$$

From (1),
$$\begin{aligned} a_n &= a_{n-1} + 2 = (a_{n-2} + 2) + 2 \\ &= a_{n-2} + 2.2 \qquad \text{...(2)} \end{aligned}$$

Replacing n by $n - 2$ in (1), we obtain
$$a_{n-2} = a_{n-3} + 2$$

So, from (2), $a_n = (a_{n-3} + 2) + 2.2 = a_{n-3} + 3.2$

In general $a_n = a_{n-k} + k\,.\,2$

For $k = n - 1$,
$$\begin{aligned} a_n &= a_{n-(n-1)} + (n - 1)\,.\,2 \\ &= a_1 + (n - 1)\,.\,2 = 3 + (n - 1)\,.\,2 \end{aligned}$$

which is an explicit formula.

4.2. THE SECOND-ORDER LINEAR Homogeneous Recurrence Relation with constant coefficients

Let $k \in Z^+$ and $C_n\ (\ne 0), C_{n-1}, C_{n-2}, \ldots\ldots C_{n-k}\ (\ne 0)$ be real numbers. If a_n, for $n \ge 0$, is a discrete function, then

$$C_n a_n + C_{n-1}\, a_{n-1} + C_{n-2}\, a_{n-2} + \ldots\ldots + C_{n-k}\, a_{n-k} = f(x),\ n \ge k$$

is a linear recurrence relation (with constant coefficients) of order k. When $f(n) = 0$, for all $n \ge 0$, the relation is called homogeneous ; other wise, it is non-homogeneous.

The homogeneous relation of order two :

$$C_n a_n + C_{n-1} a_{n-1} + C_{n-2} a_{n-2} = 0,\ n \ge 2.$$

A solution of the form $a_n = Cr^n$, where $C \ne 0$ and $r \ne 0$ substituting $a_n = Cr^n$ into

$$C_n a_n + C_{n-1} a_{n-1} + C_{n-2} a_{n-2} = 0.$$

We obtain $C_n Cr^n + C_{n-1}Cr^{n-1} + C_{n-2}C_r^{n-2} = 0$,

with C, $r \neq 0$, this becomes

$$C_n r^2 + C_{n-1} r + C_{n-2} = 0,$$

a quadratic equation which is called the characteristic equation.

4.3. THE NON HOMOGENEOUS RECURRENCE RELATIONS

The recurrence relations

$$a_n + C_{n-1}a_{n-1} = f(n),\ n \geq 1 \quad ...(1)$$

$$a_n + C_{n-1}a_{n-1} + C_{n-2}a_{n-2} = f(n),\ n \geq 2 \quad ...(2)$$

where C_{n-1} and C_{n-2} are constants, $C_{n-1} \neq 0$, in (1), $C_{n-2} \neq 0$, and $f(n)$ is not identically 0.

Although there is no general method for solving all non homogeneous relations, for certain functions $f(n)$ we shall find a successful technique.

When $C_{n-1} = -1$, (1) gives, for the non homogeneous relation $a_n - a_{n-1} = f(n)$, we have

$$a_1 = a_0 + f(1)$$

$$a_2 = a_1 + f(2) = a_0 + f(1) + f(2)$$

$$a_3 = a_2 + f(3) = a_0 + f(1) + f(2) + f(3)$$

$$a_n = a_0 + f(1) + ... + f(n)$$

$$= a_0 + \sum_{i=1}^{n} f(i)$$

We can solve this type of relation in terms of n, if we find a suitable summation formula for $\sum_{i=1}^{n} f(i)$.

* The non homogeneous first-order relation

$$a_n + C_{n-1}a_{n-1} = kr^n$$

where k is a constant and $n \in Z^+$.

If r^n is not a solution of the associated homogeneous relation $a_n + C_{n-1}a_{n-1} = 0$, then $a_n(P) = Ar^n$, where A is a constant. When r^n is a solution of the associated homogeneous relation, then

$$a_n(P) = Bn\, r^n, \text{ for B a constant.}$$

* The non-homogeneous second order relation

$$a_n + C_{n-1}a_{n-1} + C_{n-2}a_{n-2} = k\, r^n,$$

where k is a constant. Here,

(*i*) $a_n^{(p)} = A\, r^n$, for A a constant, if r^n is not a solution of the associated homogeneous relation,

(*ii*) $a_n^{(p)} = Bn\, r^n$, where B is a constant, if $a_n^{(h)} = C_1 r^n + C_2 r_1^{\,n}$ where $r_1 \neq r$,

(*iii*) $a_n^{(p)} = Cn^2 r^n$, for C a constant, when $a_n^{(h)} = (C_1 + C_2 n)\, r^n$.

4.3.1. Characteristic Eqution Method :

This method can be used to solve any constant order linear recurrence equation with constant coefficient. This recurrence relation may be homogeneous or non-homogeneous. Before attempting to solve any such problem, let us first, understand what is characteristic equation for a given recurrence equation and how to find it.

A recurrence equation of the mentioned type can be arrranged in standard form as :

$$\mathbf{A_n + C_1A_{n-1} + C_2A_{n-2} + C_3\,A_{n-3} = R.H.S.} \quad ...(1)$$

Where C_1, C_2, C_3 are constant coefficients and R.H.S. has one of the following forms :

Form	**Examples**
Homogeneous	0
A constant to the nth power	$2^n, \pi^{-s}, 2^{-n}, \sqrt{2^n}$
A polynomial in n	$3, n^2, n^2 - n, n^3 + 2n - 1$
A product of a constant to the nth power and a polynomial in n	$2^n(n^2 + 2n - 1), (n - 1)\,n^6, n\,6^n$
A linear combination of any of the above	$(2^n + 3^{n/2})(n^2 + 2n - 1) + 5$

In the recurrence equation (1), assigning R.H.S. = 0, we get

$$A_n + C_1A_{n-1} + C_2A_{n-2} + C_3A_{n-3} = 0 \quad ...(2)$$

This equation (2) gives the homogeneous part of the given recurrence equation. Every recurrence equation has a homogeneous part. If the recurrence relation is homogeneous then it has only homogeneous part and solving such equation is one step process. On the other hand, if the given recurrence equation is non-homogeneous then its homogeneous part is obtained by assigning R.H.S. equal to zero.

A characteristic equation corresponds to homogeneous part of the given recurrence relation.

The characteristic equation of (2) is given as :

$$x^3 + C_1x^2 + C_2x + C_3 = 0 \quad ...(3)$$

This has been obtained by the following **procedure :**

(*i*) Find the order of the recurrence equation here it is 3.

(*ii*) Take any variable (say x) and substitute

A_n, A_{n-1}, A_{n-2}, by x^3, x^2, x respectively in the homogeneous part of the recurrence equation.

Equation so obtained is called **characteristic equation** of the given recurrence equation.

Example (1) : The characteristic equation of the recurrence equation $C_n = 3C_{n-1} - 2C_{n-2}$ is given by

$$x^2 - 3x + 2 = 0.$$

Example (2) : The characteristic equation of the recurrence equation $f_n = f_{n-1} + f_{n-2}$ is given by

$$x^2 - x - 1 = 0.$$

Example (3) : The characteristic equation of the recurrence equation

$$A_n - 5A_{n-1} + 6A_{n-2} = 2^n + n$$

is given by

$$x^2 - 5x + 6 = 0.$$

Theorem 4.1. **If the characteristic equation $x^2 - r_1x - r_2 = 0$ of the recurrence equation $a_n = r_1a_{n-1} + r_2a_{n-2}$ has two distinct roots S_1 and S_2 then $a_n = uS_1^n + vS_2^n$ is the closed form formula for the sequence where u and v depend on the initial condition.**

Proof. Since S_1 and S_2 are roots of

$$x^2 - r_1x - r_2 = 0 \quad ...(1)$$

We have
$$S_1^2 - r_1S_1 - r_2 = 0 \quad ...(2)$$
$$S_2^2 - r_1S_2 - r_2 = 0 \quad ...(3)$$

Since u and v are dependent on the initial conditions

We have $a_1 = uS_1 + vS_2$

and $a_2 = uS_1^2 + vS_2^2$

Now,
$$\begin{aligned} a_n &= uS_1^n + vS_2^n \\ &= uS_1^{n-2}S_1^2 + vS_2^{n-2}S_2^2 \\ &= uS_1^{n-2}[r_1S_1 + r_2] + vS_2^{n-2}[r_1S_2 + r_2] \quad \text{(from (2) and (3))} \\ &= r_1uS_1^{n-1} + r_2uS_1^{n-2} + r_1vS_2^{n-1} + r_2vS_2^{n-2} \\ &= r_1[uS_1^{n-1} + vS_2^{n-1}] + r_2[uS_1^{n-2} + vS_2^{n-2}] \\ &= r_1a_{n-1} + r_2a_{n-2} \end{aligned}$$

i.e., $a_n = uS_1^n + vS_2^n$ in an explicit formula for the given recurrence relation.

Theorem 4.2. **If the characteristic eqution $x^2 - r_1x - r_2 = 0$ of the recurrence equation $a_n = r_1a_{n-1} + r_2a_{n-2}$ has single roots then $a_n = uS^n + v_nS^n$ is the closed form formula for the sequence where u and v depend on the initial condition.**

Proof. Since S is roots of $x^2 - r_1x - r_2 = 0$ ...(1)

We have $S^2 - r_1S - r_2 = 0$ ...(2)

Now,
$$\begin{aligned} a_n &= uS^n + vnS^n \\ &= uS^n + v(n-1)S^n + vS^n \\ &= uS^{n-2}S^2 + v(n-1)S^{n-2}S^2 + vS^n \\ &= [uS^{n-2} + v(n-1)S^{n-2}]S^2 + vS^n \\ &= [uS^{n-2} + v(n-1)S^{n-2}](r_1S + r_2) + vS^n \\ &= r_1[uS^{n-1} + v(n-1)S^{n-1}] + r_2[uS^{n-2} + v(n-1)S^{n-2}] + vS^n \\ &= r_1[uS^{n-1} + v(n-1)S^{n-1}] + r_2[uS^{n-2} + v(n-2)S^{n-2}] + r_2vS^{n-2} + vS^n \\ &= r_1a_{n-1} + r_2a_{n-2} + vS^{n-2}[r_2 + S^2] \quad ...(3) \end{aligned}$$

From equation (2) we have

$$S = \frac{r_1 \pm \sqrt{r_1^2 + 4r_2}}{2} \qquad \because \quad x = \frac{-b \pm \sqrt{b^2 - 4ac}}{2a}$$

$\because$ S is an equal root, we have

$$\sqrt{r_1^2 + 4r_2} = 0 \qquad \text{or} \qquad S = \frac{r_1}{2} \qquad \text{or} \qquad r_1 = 2S.$$

Now substituting $r_1 = 2S$ in equation (2), we get

$$S^2 - 2S^2 - r_2 = 0$$
$$-S^2 - r_2 = 0$$
$$S^2 + r_2 = 0 \qquad ...(4)$$

Using the result from equation (4) in equation (3), we have

$$a_n = r_1 a_{n-1} + r_2 a_{n-2}$$

Therefore, the explicit formula for the recurrence equation $a_n = r_1 a_{n-1} + r_2 a_{n-2}$ is

$$a_n = uS^n + v_n S^n$$

with initial conditions

$$a_1 = uS + vS \quad \text{and} \quad a_2 = uS^2 + 2vS^2.$$

Note. (*i*) In general, if $S_1, S_2, S_3, ... S_r$ are r distinct roots of the characteristic equation of a rth order recurrence equation then its explicit formula is given by

$$a_n = u_1 S_1{}^n + u_2 S_2{}^n + u_3 S_3{}^n + ... + u_r S_r{}^n$$

where $\quad u_1, u_2, u_3, ... u_r$ depend on initial conditions.

(*ii*) If S is r times equal roots of the characteristic equation of a rth order recurrence equation then its explicit formula is given by

$$a_n = u_1 S^n + u_2 n S^n + u_3 n^2 S^n + ... + u_r n^{r-1} S^n$$

where $u_1, u_2, u_3, ... u_r$ depend on initial conditions.

(*iii*) A combination of 1 and 2 is also possible,

i.e., some roots of a characteristic equation are distinct and some are equal.

In that case, for distinct roots we use method outlined in 1 and for repeated roots we use method mentioned in 2.

** There are two parts to the total solution. The homogeneous part of the solution depends only on what is on the left of the total solution depends on what is on the R.H.S. and has the same form as the R.H.S. We will calculate the two parts separately and add them to form the total solution.

There are four steps in the process as listed below :

Step 1 : Find the homogeneous solution to the homogeneous equation. This results when you set the R.H.S. to zero. If it is already zero, skip the next two steps and go directly to the step 4. Your answer will contains one or more undetermined coefficients whose values cannot be determined until step 4.

Step 2 : Find the particular solution by guessing a form similar to the R.H.S. This step does not produce any additional undetermined coefficients, nor does it eliminate those from step 1.

Step 3 : Combine the homogeneous and particular solution.

Step 4 : Use boundary or initial conditions to eliminate the undetermined constants from the step 1.

We shall now discuss the way to find a particular solution : The form of a particular solution has nothing to do with the order of the recurrence relation. It only depends on the form of the R.H.S. of recurrence equation expressed in standard from. Guess a solution of the same form but with undetermined coefficients, which have to be calculated. We find their values by substituting the guessed particular solution into the recurrence equation. You may find the following table useful for guessing a particular solution :

R.H.S.	Guessed Particular Solution
17 (constant)	C (constant)
π^n (constant to the nth power)	$C\pi^n$ (constant to the nth power)
$2^n + 5^n + 3$ (linear combination)	$D2^n + E5^n + F$ (same linear combination)
$5n^3$ (polynomial in n)	$An^3 + Bn^2 + Cn + D$ (Decreating polynomial)
$5n^3 - 1$ (polynomial in n)	$An^3 + Bn^2 + Cn + D$ (Decreasing polynomial)
$3n^2 5^n$ (linear combination)	$5^n (Bn^2 + Cn + D)$ (linear combination)

Notice how a polynomial in n produces a decreasing polynomial with all orders, down to, and including the constant term. Without those extra terms, usually the coefficients cannot be determined successfully.

Theorem 4.3. If $\{a_n^{(P)}\}$ is a particular solution of the non-homogeneous linear recurrence relation with constant coefficients

$$a_n = C_1 a_{n-1} + C_2 a_{n-2} + \ldots\ldots + C_k a_{n-k} + F(n),$$ **then every solution is of the form $\{a_n^{(P)} + a_n^{(h)}\}$, where $\{a_n^{(h)}\}$ is a solution of the associated homogeneous recurrence relation**

$$a_n = C_1 a_{n-1} + C_2 a_{n-2} + \ldots\ldots + C_k a_{n-k}.$$

Proof. Since $\{a_n^{(P)}\}$ is a particular solution of the non-homogeneous recurrence relation.

We know that $a_n^{(P)} = C_1 a_{n-1}^{(P)} + C_2 a_{n-2}^{(P)} + \ldots\ldots + C_k a_{n-k}^{(P)} + F(n)$

Now suppose that $\{b_n\}$ is a second solution of the non-homogeneous recurrence relation, so that

$$b_n = C_1 b_{n-1} + C_2 b_{n-2} + \ldots\ldots + C_k b_{n-k} + F(n)$$

Substracting the first of these two equations from the second shows that

$$b_n - a_n^{(P)} = C_1(b_{n-1} - a_{n-1}^{(P)}) + C_2(b_{n-2} - a_{n-2}^{(P)}) + \ldots\ldots + C_k(b_{n-k} - a_{n-k}^{(P)})$$

It follows that $\{b_n - a_n^{(P)}\}$ is a solution of the associated homogeneous linear recurrence, say, $\{a_n^{(h)}\}$.

Consequently, $b_n = a_n^{(P)} + a_n^{(h)}$ for all n.

Theorem 4.4. Suppose that $\{a_n\}$ satisfies the linear non-homogeneous recurrence relation

$$a_n = C_1 a_{n-1} + C_2 a_{n-2} + \ldots\ldots + C_k a_{n-k} + F(n),$$

where $C_1, C_2, \ldots\ldots, C_k$ are real number and $F(n) = (b_t n^t + b_{t-1} n^{t-1} + \ldots\ldots + b_1 n + b_0)\, S^n$.

where $b_0, b_1, \ldots\ldots, b_t$ and S are real numbers. When S is a not root of the characteristic equation of the associated linear homogeneous recurrence relation, there is a particular solution of the form

$$(P_t n^t + P_{t-1} n^{t-1} + \ldots\ldots + P_1 n + P_0)\, S^n$$

When S is a root of this characteristic equation and its multiplicity is m, theorem is a particular solution of the form

$$n^m (P_t n^t + P_{t-1} n^{t-1} + \ldots\ldots + P_1 n + P_0)\, S^n.$$

Problem 4.19. *What is the solutions of recurrence relation*

$$a_n = a_{n-1} + 2a_{n-2} \quad \textit{with} \quad a_0 = 2 \textit{ and } a_1 = 7\ ?$$

Solution. The characteristic equation of the recurrence relation is $r^2 - r - 2 = 0$.

Its roots are $r = 2$ and $r = -1$.

Hence, the sequence $\{a_n\}$ is a solution to the recurrence relation if and only if

$$a_n = \alpha_1 2^n + \alpha_2(-1)^n \text{ for some constants } \alpha_1 \text{ and } \alpha_2.$$

From the initial conditions, it follows that

$$a_0 = 2 = \alpha_1 + \alpha_2$$

$$a_1 = 7 = \alpha_1 \cdot 2 + \alpha_2 \cdot (-1)$$

Solving these two equations shows that $\alpha_1 = 3$ and $\alpha_2 = -1$.

Hence, the solution to the recurrence relation and initial conditions is the sequence $\{a_n\}$ with

$$a_n = 3.2^n - (-1)^n.$$

Problem 4.20. *What is the solution of the recurrence relation*

$$a_n = 6a_{n-1} - 9a_{n-2} \textit{ with initial conditions } a_0 = 1 \textit{ and } a_1 = 6\ ?$$

Solution. The only root of $r^2 - 6r + 9 = 0$ is $r = 3$.

Hence, the solution to this recurrence relation is $a_n = \alpha_1 3^n + \alpha_2 n 3^n$ for some constants α_1 and α_2.

Using the initial conditions, it follows that

$$\alpha_0 = 1 = \alpha_1$$

$$a_1 = 6 = \alpha_1 \cdot 3 + \alpha_2 \cdot 3$$

Solving these two equations shows that $\alpha_1 = 1$ and $\alpha_2 = 1$.

Consequently, the solution to this recurrence relation and the initial conditions is

$$a_n = 3^n + n\, 3^n.$$

Problem 4.21. *Find the solution to the recurrence relation*

$$a_n = 6a_{n-1} - 11a_{n-2} + 6a_{n-3}$$

with initial conditions $a_0 = 2$, $a_1 = 5$ *and* $a_2 = 15$.

Solution. The characteristic polynomial of this recurrence relation is $r^3 - 6r^2 + 11r - 6 = 0$

The characteristic roots are $r = 1$, $r = 2$ and $r = 3$

Since $r^3 - 6r^2 + 11r - 6 = (r-1)(r-2)(r-3)$

Hence, the solutions to this recurrence relation are of the form $a_n = \alpha_1 \cdot 1^n + \alpha_2 \cdot 2^n + \alpha_3 \cdot 3^n$.

To find the constants α_1, α_2 and α_3, use the initial conditions.

This gives $a_0 = 2 = \alpha_1 + \alpha_2 + \alpha_3$

$$a_1 = 5 = \alpha_1 + \alpha_2 \cdot 2 + \alpha_3 \cdot 3$$

$$a_2 = 15 = \alpha_1 + \alpha_2 \cdot 4 + \alpha_3 \cdot 9$$

When these three simultaneous equations are solved for α_1, α_2 and α_3, we find that $\alpha_1 = 1$, $\alpha_2 = -1$ and $\alpha_3 = 2$.

Hence, the unique solution to this recurrence relation and the given initial conditions is the sequence $\{a_n\}$ with $a_n = 1 - 2^n + 2 \cdot 3^n$.

Problem 4.22. *Find the solution to the recurrence relation*

$$a_n = -3a_{n-1} - 3a_{n-2} - a_{n-3}$$

with initial conditions $a_0 = 1$, $a_1 = -2$ and $a_2 = -1$.

Solution. The characteristic equation of this recurrence relation is $r^3 + 3r^2 + 3r + 1 = 0$

Since $r^3 + 3r^2 + 3r + 1 = (r+1)^3$, there is a single root $r = -1$ of multiplicity three of the characteristic equation.

The solutions of this recurrence relation are of the form

$$a_n = \alpha_{1,0}(-1)^n + \alpha_{1,} 1^n (-1)^n + \alpha_{1,0} n^2 (-1)^n$$

To find the constant $\alpha_{1,0}$, $\alpha_{1,1}$ and $\alpha_{1,2}$, use the initial conditions.

This gives $a_0 = 1 = \alpha_{1,0}$

$$a_1 = -2 = -\alpha_{1,0} - \alpha_{1,1} - \alpha_{1,2}$$

$$a_2 = -1 = \alpha_{1,0} + 2\alpha_{1,1} + 4\alpha_{1,2}$$

The simultaneous solution of these three equations is

$$\alpha_{1,0} = 1,\ \alpha_{1,1} = 3, \text{ and } \alpha_{1,2} = -2.$$

Hence, the unique solution to this recurrence relation and the given initial conditions is the sequence $\{a_n\}$ with $a_n = (1 + 3n - 2n^2)(-1)^n$.

Problem 4.23. *Find the explicit formula for the Fibonacci numbers.*

Solution. Recall that the sequence of Fibonacci numbers satisfies the recurrence relation $f_n = f_{n-1} + f_{n-2}$ and also satisfies the initial conditions $f_0 = 0$ and $f_1 = 1$. The roots of the characteristic equation $r^2 - r - 1 = 0$ are $r_1 = \dfrac{(1+\sqrt{5})}{2}$ and $r_2 = \dfrac{(1-\sqrt{5})}{2}$.

Therefore, it follows that the Fibonacci numbers are given by

$$f_n = \alpha_1 \left(\frac{1+\sqrt{5}}{2}\right)^n + \alpha_2 \left(\frac{1-\sqrt{5}}{2}\right)^n$$

for some constant α_1 and α_2.

The initial conditions $f_0 = 0$ and $f_1 = 1$ can be used to find these constants.

We have $f_0 = \alpha_1 + \alpha_2 = 0$

$$f_1 = \alpha_1 \left(\frac{1+\sqrt{5}}{2}\right) + \alpha_2 \left(\frac{1-\sqrt{5}}{2}\right) = 1$$

The solution to these simultaneous equations for α_1 and α_2 is

$$\alpha_1 = \frac{1}{\sqrt{5}} \text{ and } \alpha_2 = -\frac{1}{\sqrt{5}}.$$

Conseqently, the Fibonacci numbers are given by

$$f_n = \frac{1}{\sqrt{5}} \left(\frac{1+\sqrt{5}}{2}\right)^n - \frac{1}{\sqrt{5}} \left(\frac{1-\sqrt{5}}{2}\right)^n.$$

Problem 4.24. *Find all solutions of the recurrence relation $a_n = 3a_{n-1} + 2n$. What is the solution with $a_1 = 3$?*

Solution. To solve this linear non-homogeneous recurrence relation with constant coefficiently we need to solve its associated linear homogeneous equation and to find a particular solution for the given non-homogeneous equation.

The associated linear homogeneous equation is $a_n = 3a_{n-1}$.

Its solutions are $a_n^{(h)} = \alpha 3^n$, where α is a constant.

We now find a particular solution. Since F(n) = $2n$ is a polynomial in n of degree one, a reasonable trial solution is a linear function in n, say, $P_n = cn + d$ where c and d are constants.

To determine whether there are any solutions of this form, suppose that $P_n = cn + d$ is such a solution.

Then the equation $a_n = 3a_{n-1} + 2n$ becomes $cn + d = 3\,(c(n-1) + d) + 2n$.

Simplifying and combining like terms gives $(2 + 2c)\,n + (2d - 3c) = 0$

It follows that $cn + d$ is a solution if and only if $2 + 2c = 0$ and $2d - 3c = 0$

This shows that $cn + d$ is a solution if and only if $c = -1$ and $d = -3/2$.

Consequently, $a_n^{(P)} = -n - \dfrac{3}{2}$ is a particular solution.

All solutions are of the form

$$a_n = a_n^{(P)} + a_n^{(h)} = -n - \frac{3}{2} + \alpha\,.\,3^n, \text{ where } \alpha_n \text{ is a cosntant.}$$

To find the solution with $a_1 = 3$, let $n = 1$ in the formula we obtained for the general solution. We find that $3 = -1 - \dfrac{3}{2} + 3\alpha$, which implies that $\alpha = \dfrac{11}{6}$.

The solution we seek is $a_n = -n - \dfrac{3}{2} + \left(\dfrac{11}{6}\right) 3^n$.

Problem 4.25. *Find all solutions of the recurrence relation*

$$a_n = 5a_{n-1} - 6a_{n-2} + 7^n.$$

Solution. This is a linear homogeneous recurrence relation. The solutions of its associated homogeneous recurrence relation $a_n = 5a_{n-1} - 6a_{n-2}$ are

$$a_n^{(h)} = \alpha_1\,.\,3^n + \alpha_2\,.\,2^n,$$

where α_1 and α_2 are constants.

Since F(n) = 7^n, a reasonable trial solution is $a_n^{(P)} = C\,.\,7^n$, where C is a constant.

Substituting the terms of this sequence into the recurrence relation implies that

$$C\,.\,7^n = 5C\,.\,7^{n-1} - 6C\,.\,7^{n-2} + 7^n$$

Factoring out 7^{n-2}, this equation becomes

$$49C = 35C - 6C + 49, \text{ which implies that } 20C = 49 \text{ or that } C = \frac{49}{20}$$

Hence, $a_n^{(P)} = \left(\frac{49}{20}\right) 7^n$ is a particular solution.

All solutions are of the form

$$a_n = \alpha_1 . 3^n + \alpha_2 . 2^n + \left(\frac{49}{20}\right) 7^n.$$

Problem 4.26. *What form does a particular solution of the linear non-homogeneous recurrence relation $a_n = 6a_{n-1} - 9a_{n-2} + F(n)$ have when $F(n) = 3^n$, $F(n) = n\, 3^n$, $F(n) = n^2 2^n$, and $F(n) = (n^2 + 1)\, 3^n$?*

Solution. The associated linear homogeneous recurrence relation is $a_n = 6a_{n-1} - 9a_{n-2}$.

Its characteristic equation $r^2 - 6r + 9 = (r - 3)^2 = 0$, has a single root 3 of multiplicity two with F(n) of the P(n) S^n, where P(n) is a polynomial and S is a constant, we need to ask whether S is a root of this characteristic equation.

Since S = 3 is a root with multiplicity $m = 2$ but S = 2 is not a root.

The particular solution has the form $P_0 n^2 3^n$ if $F(n) = 3^n$ the form $n^2 (P_1 n + P_0)\, 3^n$ if $F(n) = n\, 3^n$, the form $(P_2 n^2 + P_1 n + P_0)\, 2^n$ if $F(n) = n^2 2^n$ and the form $n^2 = (P_2 n^2 + P_1 n + P_0)\, 3^n$ if $F(n) = (n^1 + 1)\, 3^n$.

Problem 4.27. *Solve the recurrence equation*

$$a_r - 7a_{r-1} + 10a_{r-2} = 2^r \text{ with initial condition } a_0 = 0 \text{ and } a_1 = 6.$$

Solution. The general solution, also called **homogeneous** solution, to the problem is given by homogeneous part of the given recurrence equation.

The homogeneous part of the equation

$$a_r - 7a_{r-1} + 10a_{r-2} = 2^r \qquad ...(1)$$

is
$$a_r - 7a_{r-1} + 10a_{r-2} = 0 \qquad ...(2)$$

The characteristic equation of (2) is given by

$$x^2 - 7x + 10 = 0$$

$$\Rightarrow \quad (x - 2)(x - 5) = 0$$

$$\Rightarrow \quad x = 2 \text{ and } x = 5$$

Since the two roots 2 and 5 of characteristic equation are distinct, the homogeneous solution is given by

$$a_r = A2^r + B5^r \qquad ...(3)$$

Now, **particular solution** is given by

$$a_r = rC2^r$$

Substituting the value of a_r in equation (1), we get

$$C[r\, 2^r - 7(r - 1)\, 2^{r-1} + 10\, (r - 2)\, 2^{r-2}] = 2^r$$

or, $$C[4r - 7(r - 1)\, 2 + 10(r - 2)]\, 2^{r-2} = 2^r$$

or, $$C[4r - 14(r - 1) + 10r - 20] = 4$$

or, $$C[4r - 14r + 14 + 10r - 20] = 4$$

or, $$-6C = 4 \quad \Rightarrow \quad C = -\frac{2}{3}$$

Particular solution is $a_r = -\frac{2}{3} r\,2^r$.

The complete, also called total, solution is obtained by combining the homogeneous and particular solutions. This is given as

$$a_r = A2^r + B5^r - \frac{2}{3} r2^r \qquad ...(4)$$

The equation (4) contains two undetermined coefficients A and B, which are to be determined. To find this, we use the given initial conditions for $r = 0$ and $r = 1$.

Since values of a_0 and a_1 are given.

Putting $r = 0$ and $r = 1$ in equation (4), we get the following equations (5) and (6) respectively.

$$a_0 = A2^0 + B5^0 - \frac{2}{3} * 0 * 2^0$$

or $$0 = A + B \qquad ...(5) \quad (\because\ a_0 = 0 \text{ is given})$$

and $$a_1 = A2^1 + B5^1 - \frac{2}{3} * 1 * 2^1$$

or $$6 = 2A + 5B - \frac{4}{3} \qquad (\because\ a_1 = 6 \text{ is given})$$

or $$2A + 5B = \frac{22}{3} \qquad ...(6)$$

Solving equation (5) and (6), we get

$$A = -\frac{22}{9} \text{ and } B = \frac{22}{9}$$

Replacing A and B in equation (4) by its respective values, we get the closed form formula for the given recurrence equation. Thus,

$$a_r = -\frac{22}{9} 2^r + \frac{22}{9} 5^r - \frac{2}{3} r2^r$$

or $$a_r = \frac{22}{9} [5^r - 2^r] - \frac{2}{3} r2^r.$$

Problem 4.28. *Solve the recurrence equation*

$$A_n - A_{n-1} - A_{n-2} = 2n \text{ with } A_0 = 0 \text{ and } A_1 = 1.$$

Solution. The homogeneous part of the equation

$$A_n - A_{n-1} - A_{n-2} = 2n \qquad ...(1)$$

is $$A_n - A_{n-1} - A_{n-2} = 0 \qquad ...(2)$$

The characteristic equaion of (2) is given as

$$x^2 - x - 1 = 0$$

$$\Rightarrow \quad x = \frac{1 \pm \sqrt{1+4}}{2}$$

or $$x = \frac{1+\sqrt{5}}{2} \quad \text{and} \quad x = \frac{1-\sqrt{5}}{2}.$$

Thus, two roots of the characteristic equation are distinct. So, the homogeneous solution is given by

$$A_n = A\left[\frac{1+\sqrt{5}}{2}\right]^n + B\left[\frac{1-\sqrt{5}}{2}\right]^n \quad ...(3)$$

The particular solution is given by the guess

$$A_n = Cn + D$$

Substituting this value in equation (1), we get

$$Cn + D - C(n-1) - D - C(n-2) - D = 2n$$

or $$(Cn - Cn - Cn) + D + C - D + 2C - D = 2n$$

$$\Rightarrow \quad -Cn + 3C - D = 2n$$

$$\Rightarrow \quad -C = 2,\ 3C - D = 0$$

$$\Rightarrow \quad C = -2,\ D = -6$$

$\therefore$ $A_n = -2n - 6$ is the particular solution.

Combining homogeneous and particular solutions, we get total solution as

$$A_n = A\left[\frac{1+\sqrt{5}}{2}\right]^n + B\left[\frac{1-\sqrt{5}}{2}\right]^n - 2n - 6 \quad ...(4)$$

Equation (4) contains two undetermined coefficients A and B. Which are to be determined.

To find this we use the given initial conditions for $n = 0$ and $n = 1$, since values of A_0 and A_1 are given.

Putting $n = 0$ and $n = 1$ in equation (4), we get the following equations (5) and (6) respectively.

$$A_0 = A\left[\frac{1+\sqrt{5}}{2}\right]^0 + B\left[\frac{1-\sqrt{5}}{2}\right]^0 - 2 * 0 - 6$$

or $$0 = A + B = 0 \quad ...(5) \quad (\because \ A_0 = 0 \text{ is given})$$

or $$A_1 = A\left[\frac{1+\sqrt{5}}{2}\right]^1 + B\left[\frac{1-\sqrt{5}}{2}\right]^1 - 2 * 1 - 6$$

or $$1 = A\left[\frac{1+\sqrt{5}}{2}\right] + B\left[\frac{1-\sqrt{5}}{2}\right] - 8 \quad ...(6) \quad (\because \ A_1 = 1 \text{ is given})$$

Solving (5) and (6), we get

$$B - \frac{1-\sqrt{5}}{1+\sqrt{5}}\,B = 6 - \frac{18}{1+\sqrt{5}}$$

or $$B\,\frac{1+\sqrt{5}-1+\sqrt{5}}{1+\sqrt{5}} = \frac{6+6\sqrt{5}-18}{1+\sqrt{5}}$$

or $$B = \frac{3(\sqrt{5}-2)}{\sqrt{5}} \text{ and } A = \frac{3(\sqrt{5}+2)}{\sqrt{5}}.$$

Replacing A and B in equation (4) by its its respective values, we get the closed form formula for the given recurrence equation. Therefore, the final solution is

$$A_n = \frac{3(\sqrt{5}+2)}{\sqrt{5}}\left[\frac{1+\sqrt{5}}{2}\right]^n + \frac{3(\sqrt{5}-2)}{\sqrt{5}}\left[\frac{1-\sqrt{5}}{2}\right]^n - 2n - 6.$$

Problem 4.29. *Solve the recurrence equation*

$$a_n - 8a_{n-1} + 16a_{n-2} = 0 \text{ with } a_2 = 16 \text{ and } a_3 = 80.$$

Solution. The given equation is homogeneous, so only the homogeneous part of the solution is required.

In this case, we do not have to try for the particular solution and hence no combining of the particular part with the homogeneous part is needed.

Here, we will go to the step 4 from the step 1, skipping the steps 2 and 3.

The characteristic equation of the given recurrence equation is $x^2 - 8x + 16 = 0$

or $$(x-4)^2 = 0 \quad \Rightarrow \quad x = 4, 4.$$

Since the two roots of characteristic equation are equal, the homogeneous solution is given by

$$a_n = (u + vn)\, 4_n \qquad \text{...(1)}$$

The equation (1) contains two undetermined coefficients u and v, which are to be determined.

To find this, we use the given initial conditions for $n = 2$ and $n = 3$.

Since values of a_2 and a_3 are given.

Putting $n = 2$ and $n = 3$ in equation (1), we get the following equations (2) and (3) respectively. When $n = 2$,

$$a_2 = (u + 2v)\, 4^2 \quad \text{or} \quad 16 = 16\,(u + 2v)$$

or $$u + 2v = 1 \qquad \text{...(2)}$$

and when $n = 3$.

$$a_3 = (u + 3v)\, 4^3 \quad \text{or} \quad 80 = 64\,(u + 3v)$$

or $$u + 3v = \frac{5}{4} \qquad \text{...(3)}$$

Solving equation (2) and (3), we have

$$u = \frac{1}{2} \text{ and } v = \frac{1}{4}$$

Replacing u and v their respective values in (1), we have the solution in the form of closed formula.

Therefore, $$a_n = \left[\frac{1}{2} + \frac{n}{4}\right] 4^n.$$

Problem 4.30. *Solve the recurrence equation*

$$a_n = a_{n-1} + 2(n-1) \text{ with } a_0 = 1.$$

Solution. The given equation is non-homogeneous of order 1.

This can also be solved either by backtracking or by summation method.

Here, we shall solve it by characteristic equation method.

The characteristic equation of the given recurrence equation is $x - 1 = 0$ or $x = 1$.

The characteristic equation has only one root, so the homogeneous solution is given by

$$a_n = A * 1^n \quad ...(1)$$

Now to find particular solution, we guess

$$a_n = Bn + D$$

Now substituting this guess for a_n and a_{n-1} in the given recurrence equation, we get

$$Bn + D - B(n-1) - D = 2n - 2$$

or $$B = 2n - 2 \quad \Rightarrow \quad 0 = 2, \text{ this is impossible.}$$

Notice coefficient of n in L.H.S. is zero and in R.H.S., it is 2.

This has happened because of our assuption (guess).

Now we will modify our guess and take a polynomial in n of a degree one higher than what we took earlier as a guess to the particular solution.

Let $$a_n = Bn^2 + Cn + D$$

Now substituting this guess for a_n and a_{n-1} in the given recurrence equation, we get

$$Bn^2 + Cn + D - B(n-1)^2 - C(n-1) - D = 2n - 2$$

or $$2Bn + (C - B) = 2n - 2$$

$\Rightarrow$ $2B = 2$ and $C - B = -2$

$\Rightarrow$ $B = 1$ and $C = -1$ and $D = 0$

$\therefore$ $a_n = n^2 - n$, is the particular solution. Here, we have assigned $D = 0$.

The complete solution is then given by combining homogeneous and particular solution as

$$a_n = A + n^2 - n \quad ...(2)$$

To determine the value of A, we use the initial condition $a_0 = 1$.

We have from equation (2), $a_0 = A$ or $A = 1$ when $n = 0$.

Therefore $a_n = n^2 - n + 1$.

Problem 4.31. *Solve the recurrence equation*

$$a_n = 5a_{n-1} + 6a_{n-2} = 2^n + n \quad \textit{with}$$

initial condition $a_1 = 0$ *and* $a_2 = 10$.

Solution. The general to the problem is given by the homogeneous part of the given recurrence equation.

The homogeneous part of the equation

$$a_n - 5a_{n-1} + 6a_{n-2} = 2^n + n \quad ...(1)$$

is $$a_n - 5a_{n-1} + 6a_{n-2} = 0 \quad ...(2)$$

The characteristic equation of (2) is given as

$$x^2 - 5x + 6 = 0$$

$$\Rightarrow \quad (x-2)(x-3) = 0$$

$$\Rightarrow \quad x = 2 \text{ and } x = 3$$

The two roots 2 and 3 of characteristic equation are distinct. So the homogeneous solution is given by

$$a_n = \text{A}2^n + \text{B}3^n \qquad ...(3)$$

Let particular solution be

$$a_n = \text{C}n\,2^n + \text{D}n + \text{E}.$$

Substituting this particular solution in equation (1) for a_n, a_{n-1} and a_{n-2}, we get

$$\text{C}n\,2^n + \text{D}n + \text{E} - 5\,[\text{C}(n-1)\,2^{n-1} + \text{D}\,(n-1) + \text{E}]$$
$$+\,6[\text{C}(n-2)\,2^{n-2} + \text{D}(n-2) + \text{E}] = 2^n + n$$

or $$\text{C}[n\,2^n - 5(n-1)\,2^{n-1} + 6(n-2)\,2^{n-2}] + \text{D}[n - 5\,(n-1) + 6(n-2)]$$
$$+\,\text{E}[1 - 5 + 6] = 2^n + n$$

or $$\text{C}[4n - 10(n-1) + 6(n-2)]\,2^{n-2} + \text{D}\,[n - 5n + 5 + 6n - 12]$$
$$+\,2\text{E} = 2^n + n$$

or $$\text{C}[4n - 10n + 10 + 6n - 12]\,2^{n-2} + \text{D}[2n - 7] + 2\text{E} = 2^n + n$$

or $$\text{C}[-2]\,2^{n-2} + \text{D}[2n-7] + 2\text{E} = 2^n + n$$

or $$-2\text{C} = 4 \quad 2\text{D} = 1 \quad \text{and} \quad -7\text{D} + 2\text{E} = 0$$

or $$\text{C} = -2 \quad \text{D} = \frac{1}{2} \quad \text{and} \quad \text{E} = \frac{7}{4}.$$

Therefore, particular solution is

$$a_n = -2n\,2^n + \frac{n}{2} + \frac{7}{4}$$

Total solution is given by

$$a_n = \text{A}2^n + \text{B}3^n - 2n2^2 + \frac{n}{2} + \frac{7}{4} \qquad ...(4)$$

Using initial conditions for $n = 1$ and $n = 2$, we get the following two equations (5) and (6) respectively.

$$a_1 = \text{A}2^1 + \text{B}3^1 - 2 * 1 * 2^1 + \frac{1}{2} + \frac{7}{4}$$

or $$0 = 2\text{A} + 3\text{B} - 4 + \frac{1}{2} + \frac{7}{4} \qquad (\because \ a_1 = 0)$$

or $$2\text{A} + 3\text{B} = \frac{7}{4} \qquad ...(5)$$

and $$a_2 = \text{A}2^2 + \text{B}3^2 - 2 * 2 * 2^2 + \frac{2}{2} + \frac{7}{4}$$

or $$10 = 4A + 9B - 16 + 1 + \frac{7}{4} \qquad (\because \; a_2 = 10)$$

or $$4A + 9B = \frac{93}{4} \qquad ...(6)$$

$$A = -9 \qquad \text{and} \qquad B = \frac{79}{12}$$

Solving equation (5) and (6), we get

Replacing the values of A and B in equation (4), we get the closed form formula for the give recurrence equation

$$a_n = -9 * 2^n + \frac{79}{12} 3^n - 2n\, 2^n + \frac{n}{2} + \frac{7}{4}.$$

Problem 4.32. *Solve the recurrence equation*

$$C_n = 3C_{n-1} - 2C_{n-2} \textit{ for } n \geq 3 \textit{ with } C_1 = 5 \textit{ and } C_2 = 3.$$

Solution. The given equation is **homogeneous** so only homogeneous part of the solution is required.

The characteristic equation of the given recurrence equation is

$$x^2 - 3x + 2 = 0$$

or $$(x - 1)(x - 2) = 0$$

$\Rightarrow$ $$x = 1 \text{ and } x = 2.$$

The two roots of the characteristic equation are distinct. So the homogeneous solution is given by

$$C_n = u * 1^n + v * 2^n \qquad ...(1)$$

The equation (1) contains two undetermined coefficients, u and v, which are to be determined.

To find this, we use the given initial conditions for $n = 1$ and $n = 2$.

Since the values of C_1 and C_2 are given.

Putting $n = 1$ and $n = 2$ in equation (1), we get the following equation (2) and (3) respectively.

When $n = 1$

$$C_1 = u * 1 + v * 2 \qquad \text{or} \qquad 5 = u + 2v$$

or $$u + 2v = 5 \qquad ...(2)$$

and when $n = 2$,

$$C_2 = u * 1^2 + v * 2^2$$

or $$3 = u + 4v$$

or $$u + 4v = 3 \qquad ...(3)$$

By solving equation (2) and (3), we have

$$u = 7 \qquad \text{and} \qquad v = -1$$

Replacing u and v by their respective values in (1), we have the solution in the form of closed form formula.

Therefore $C_n = 7 - 2^n$.

Problem 4.33. *Solve the recurrence equation*

$$T_n = T_{\sqrt{n}} + C \log_2 n \quad \text{with } T = 1$$

by characteristic equation method.

Solution. The given equation can be written in standard form as

$$T_n - T_{\sqrt{n}} = C \log_2 n \qquad ...(1)$$

The homogeneous solution of (1) is given by the equation.

$$T_n - T_{\sqrt{n}} = 0 \qquad ...(2)$$

The characteristic equation of (2) is

$$x - \sqrt{x} = 0$$

$\Rightarrow \quad x^2 - x = 0 \quad$ or $\quad x(x-1) = 0$

$\Rightarrow \quad x = 0 \quad$ and $\quad x = 1$

The homogeneous solution is then given by

$$T_n = A * 0^n + B * 1^n \qquad ...(3)$$

Guess the particular solution as

$$T_n = ZC \log_2 n$$

Using this particular solution in equation (1), we get

$$ZC \log_2 n - ZC \log_2 n^{1/2} = C \log_2 n$$

or
$$\left(Z - \frac{Z}{2}\right) C \log_2 n = C \log_2 n$$

$$Z - \frac{Z}{2} = 1 \quad \text{or} \quad Z = 2$$

$\therefore$ Particular solution is $T_n = 2C \log_2 n$

Combining homogeneous and particular solution, we get the complete solution, which is :

$$T_n = A * 0^n + b * 1^n + 2C \log_2 n$$

$$\Rightarrow \qquad T_n = B + 2C \log_2 n$$

For $n = 1$, the initial condition is given.

Putting $n = 1$ in the above equation, we can find the value of undetermined coefficient B as below :

When $n = 1$, $T_1 = B + 2C \log_2 1$,

$\Rightarrow \quad 1 = B + 2C * 0 \qquad (\because \; T_1 = 1 \text{ and } \log_2 1 = 0)$

$\Rightarrow \quad B = 1$

Replacing B by 1, in the complete solution,

We have $\quad T_n = 1 + 2C \log_2 n.$

Problem 4.34. *Solve the recurrence equation*

$$a_r = \sqrt{a_{r-1} + \sqrt{a_{r-2} + \sqrt{a_{r-3} + \sqrt{\ldots\ldots}}}} \quad \textit{with } a_0 = 4.$$

Solution. By squaring the given equation, we have

$$a_r^2 = a_{r-1} + \sqrt{a_{r-2} + \sqrt{a_{r-3} + \sqrt{a_{r-4} + \sqrt{\ldots\ldots}}}}$$

$$\because \quad a_r^2 = a_{r-1} + a_{r-1} \qquad \left(\because \; a_{r-1} = \sqrt{a_{r-2} + \sqrt{a_{r-3} + \sqrt{a_{r-4} + \sqrt{\ldots\ldots}}}}\right)$$

$$a_r^2 = 2a_{r-1}$$

$$\Rightarrow \quad a_r^2 - 2a_{r-1} = 0 \qquad \ldots(1)$$

Let us repalce a_r with a new sequence term.

Let $\quad b_r = \log_2 a_r \qquad \Rightarrow \quad a_r = 2^{br}$

Substituting the value of a_r in terms of the new sequence variable b_r in equation (1), we get

$$(2^{br})^2 - 2 * 2^{br-1} = 0$$

$$2^{2br} - 2^{b_{r-1}+1} = 0$$

This is possible only if $2b_r - b_{r-1} + 1 = 0$ ...(2)

With initial condition $b_0 = \log_2 a_0 = \log_2 4 = 2$

$\Rightarrow \quad b_0 = 2$

Now, to solve the equation (2), which is of first order, linear and non-homogeneous.

The characteristic equation of (2) is

$$2x - 1 = 0 \quad \Rightarrow \quad x = \frac{1}{2}$$

$\therefore$ Homogeneous solution is

$$b_r = A\left(\frac{1}{2}\right)^r \qquad \ldots(3)$$

For particular solution, we guess (Assume) $b_r = B$. Substituting this value for sequence term b_r and b_{r-1} in equation (2), we have

$$2B - B = 1 \quad \Rightarrow \quad B = 1$$

$\therefore$ Particular solution is $b_r = 1$

Then, the total solution is

$$b_r = A\left(\frac{1}{2}\right)^r + 1$$

Using the initial condition $b_0 = 2$, we get $A = 1$

$\therefore$ The solution to the changed equation is

$$b_r = \left(\frac{1}{2}\right)^r + 1$$

Since $b_r = \log_2 a_r$, $a_r = 2^{br}$

$\therefore$ Final solution is given by

$$a_r = 2^{\left(\frac{1}{2}\right)^r + 1}.$$

Problem 4.35. *Solve the recurrence equation*

$$a_n + 9a_{n-2} = 0 \text{ with initial condition } a_0 = 0 \text{ and } a_1 = 1.$$

Solution. The characteristic equation for the given recurrence equation is given by

$$x^2 + 9 = 0 \quad \Rightarrow \quad x^2 = -9$$

$$\Rightarrow \quad x = \pm 3i$$

$\therefore$ The homogeneous solution is

$$a_n = A(-3i)^n + B(3i)^n$$

When the values of the coefficients are determined at the very end, the resulting sequence values a_n will be real.

Using the initial condition for $n = 0$ and $n = 1$, we get the following two equations.

$$0 = A(-3i)^n + B(3i)^0$$

$\Rightarrow$ $A + B = 0$ ...(1)

and $1 = A(-3i)^1 + B(3i)^1$

$\Rightarrow$ $-3iA + 3iB = 1$ ...(2)

Now solving equation (1) and (2), we get values of A and B which are

$$A = \frac{-1}{6i} = \frac{i}{6} \quad \text{and} \quad B = \frac{-i}{6}$$

Substituting the value of A and B in homogeneous solution we get

$$a_n = \frac{i}{6}(-3i)^n + \frac{-i}{6}(3i)^n.$$

Problem 4.36. *Find an explicit formula for the sequence defined by $C_n = 3C_{n-1} - 2C_{n-2}$ with initial conditions*

$$C_1 = 5 \text{ and } C_2 = 3.$$

Solution. The recurrence relation $C_n = 3C_{n-1} - 2C_{n-2}$ is a linear homogeneous relation of degree 2.

Its associated equation is $x^2 = 3x - 2$.

Rewriting this as $x^2 - 3x - 2 = 0$. We see there are two roots, 1 and 2.

We can find u and v so that $C_1 = u(1) + v(2)$ and $C_2 = u(1)^2 + v(2)^2$.

Solving this 2×2 system yields $u = 7$ and $v = -1$.

We have $C_n = 7.1^n + (-1) . 2^n$ or $C_n = 7 - 2^n$.

Note that using $C_n = 3C_{n-1} - 2C_{n-2}$ with initial conditions $C_1 = 5$ and $C_2 = 3$ gives 5, 3, – 1, – 9 as the first four terms of the sequence.

The formula $C_n = 7 - 2^n$ also produces 5, 3, – 1, – 9 as the first four terms.

Problem 4.37. *Solve the recurrence relation $d_n = 2d_{n-1} - d_{n-2}$ with initial conditions $d_1 = 1.5$ and $d_2 = 3$.*

Solution. The associated equation for this linear homogeneous relation is

$$x^2 - 2x + 1 = 0$$

This equation has one (multiple) root, 1.

Thus, $d_n = u(1)^n + vn(1)^n$.

Using this formula and the initial conditions,

$$d_1 = 1.5 = u + v(1) \text{ and}$$

$$d_2 = 3 = u + v(2)$$

We find that $u = 0$ and $v = 1.5$, then

$$d_n = 1.5\, n.$$

Problem 4.38. *The recurrence relation $f_n = f_{n-1} + f_{n-2}$, $f_1 = f_2 = 1$, defines the Fibonacci sequence 1, 1, 2, 3, 5, 8, 13, 21, ...*

The initial conditions are $f_1 = 1$ and $f_2 = 1$,

$f_n \le \left(\frac{5}{3}\right)^n$. *This gives a bound on how fast the Fibonacci numbers grow.*

Solution. (By strong induction)

Basic step : Here n_0 is 1, P(1) is $1 \le \frac{5}{3}$ and this is clearly true.

Induction step : We use P(j), $j \le k$ to show

$$P(k+1) : f_{k+1} \le \left(\frac{5}{3}\right)^{k+1}.$$

Consider the left-hand side of P(k + 1)

$$f_{k+1} = f_k + f_{k-1} \le \left(\frac{5}{3}\right)^k + \left(\frac{5}{3}\right)^{k-1}$$

$$= \left(\frac{5}{3}\right)^{k-1} \left(\frac{5}{3} + 1\right)$$

$$= \left(\frac{5}{3}\right)^{k-1} \left(\frac{8}{3}\right)$$

$$< \left(\frac{5}{3}\right)^{k-1} \left(\frac{5}{3}\right)^2$$

$$= \left(\frac{5}{3}\right)^{k+1}$$, the right-hand side of P(k + 1).

Problem 4.39. *Solve the recurrence relation $a_n = 5a_{n-1} - 6a_{n-2}$, $n \geq 2$ given $a_0 = 1$, $a_1 = 4$.*

Solution. The characteristic polynomial, $x^2 - 5x + 6$, has distinct roots $x_1 = 2$, $x_2 = 3$.

The solution is given by $a_n = C_1(2^n) + C_2(3^n)$.

Since $a_0 = 1$, we must have $C_1(2^0) + C_2(3^0) = 1$, and

since $a_1 = 4$. We have $C_1(2^1) + C_2(3^1) = 4$.

Therefore $C_1 + C_2 = 1$

$$2C_1 + 3C_2 = 4.$$

Solving, we have $C_1 = -1$, $C_2 = 2$.

The solution is $a_n = -2^n + 2(3^n)$.

Problem 4.40. *Solve the recurrence relation $a_n = 4a_{n-1} - 4a_{n-2}$, $n \geq 2$ with initial conditions $a_0 = 1$, $a_1 = 4$.*

Solution. The characteristic polynomial, $x^2 - 4x + 4$, has the repeated root $x = 2$.

Hence, the solution is $a_n = C_1(2^n) + C_2n(2^n)$.

The initial conditions yield $C_1 = 1$, $2C_1 + 2C_2 = 4$. So $C_2 = 1$.

Thus, $a_n = 2^n + n(2^n) = (n + 1)2^n$.

Problem 4.41. *Find a formula for the nth term of the Fibonacci sequence.*

Solution. To simplify the algebra which follows, we take for initial conditions

$$a_0 = a_1 = 1 \text{ rather than } a_1 = a_2 = 1.$$

Hence, the recurrence relation to be solved is

$$a_n = a_{n-1} + a_{n-2},\ n \geq 2.$$

We must remember that the nth term will be a_{n-1}.

The characteristic polynomial, $x^2 - x - 1$, has dinstinct root $\frac{1 \pm \sqrt{5}}{2}$.

Hence, the solution to our recurrence relation is

$$a_n = C_1 \left(\frac{1+\sqrt{5}}{2}\right)^n + C_2 \left(\frac{1-\sqrt{5}}{2}\right)^n.$$

The initial conditions give $C_1 + C_2 = 1$

$$C_1 \left(\frac{1+\sqrt{5}}{2}\right) + C_2 \left(\frac{1-\sqrt{5}}{2}\right) = 1$$

Yielding $C_1 = \frac{1}{\sqrt{5}}\left(\frac{1+\sqrt{5}}{2}\right)$ and $C_2 = -\frac{1}{\sqrt{5}}\left(\frac{1-\sqrt{5}}{2}\right)$.

Thus, the solution is

$$a_n = \frac{1}{\sqrt{5}}\left(\frac{1+\sqrt{5}}{2}\right)\left(\frac{1+\sqrt{5}}{2}\right)^n - \frac{1}{\sqrt{5}}\left(\frac{1-\sqrt{5}}{2}\right)\left(\frac{1-\sqrt{5}}{2}\right)^n$$

$$= \frac{1}{\sqrt{5}}\left(\frac{1+\sqrt{5}}{2}\right)^{n+1} - \frac{1}{\sqrt{5}}\left(\frac{1-\sqrt{5}}{2}\right)^{n+1}$$

The nth term of the Fibonacci sequence is

$$a_{n-1} = \frac{1}{\sqrt{5}}\left(\frac{1+\sqrt{5}}{2}\right)^n - \frac{1}{\sqrt{5}}\left(\frac{1-\sqrt{5}}{2}\right)^n.$$

Problem 4.42. *Solve $a_n = 2a_{n-1} + 3a_{n-2} + 5^n$, $n \geq 2$, given $a_0 = -2$, $a_1 = 1$.*

Solution. In looking for a particular solution, it seems reasonable to try $P_n = a(5^n)$. Substituting into the recurrence relation, we get

$$a(5^n) = 2a(5^{n-1}) + 3a(5^{n-2}) + 5^n$$

$$25a = 10a + 3a + 25$$

$$12a = 25, \quad \text{so } a = \frac{25}{12}$$

and $P_n = \frac{25}{12}(5^n)$ is a particular solution.

Next we solve the homogeneous recurrence relation

$$a_n = 2a_{n-1} + 3a_{n-2}$$

The characteristic polynomial, $x^2 - 2x - 3$, has disinct roots -1 and 3.

The solution $q_n = C_1(-1)^n + C_2(3^n)$.

The given recurrence relation has the solution

$$P_n + q_n = \frac{25}{12}(5^n) + C_1(-1)^n + C_2(3^n)$$

The initial conditions give

$$a_0 = -2 = \frac{25}{12} + C_1 + C_2$$

$$a_1 = 1 = \frac{25}{12}(5) - C_1 + 3C_2.$$

Hence $C_1 = -\frac{17}{24}, C_2 = -\frac{27}{8}$.

The solution is $a_n = \frac{25}{12}(5^n) - \frac{17}{24}(-1)^n - \frac{27}{8}(3^n)$.

Problem 4.43. *Solve the recurrence relation $a_n = -3a_{n-1} + n, n \geq 1$ where $a_0 = 1$.*

Solution. Since $f(n) = n$ is linear, we try a linear function for P_n.

That is, we set $P_n = a + b_n$ and attempt to determine a and b.

Putting this expression for P_n in the given recurrence relation, we obtain

$$\begin{aligned} a + b_n &= -3[a + b(n-1)] + n \\ &= -3a + 3b + (1 - 3b)\,n. \end{aligned}$$

This equation will hold if $a = -3a + 3b$ and $b = 1 - 3b$, which is the same as $a = \frac{3}{16}, b = \frac{1}{4}$

We conclude that $P_n = \frac{3}{16} + \frac{1}{4}n$ is a particular.

Solution to the recurrence, ignoring initial conditions.

The corresponding homogeneous recurrence relation in this case is $a_n = -3a_{n-1}$, whose characteristic polynomial is $x^2 + 3x$.

The characteristic roots are -3 and 0.

The solution to the homogeneous recurrence relation is $q_n = C_1(-3)^n + C_2(0^n) = C_1(-3)^n$.

Thus, $P_n + q_n = \frac{3}{16} + \frac{1}{4}n + C_1(-3)^n$.

Since $a_0 = 1$, $\frac{3}{16} + \frac{1}{4}(0) + C_1(-3)^0 = 1$

That is, $\frac{3}{16} + C_1 = 1$.

Thus, $C_1 = \frac{13}{16}$ and the solution is $a_n = \frac{3}{16} + \frac{1}{4}n + \frac{13}{16}(-3)^n$.

Problem 4.44. *Solve $a_{n+2} - 5a_{n+1} + 6a_n = 2$ with initial condition $a_0 = 1$ and $a_1 = -1$.*

Solution. The associated homogeneous recurrence relation is

$$a_{n+2} - 5a_{n+1} + 6a_n = 0 \qquad \ldots(1)$$

Let $a_n = r^n$ be a solution of (1)

The characteristic equation is $r^2 - 5r + 6 = 0$

$\Rightarrow \quad r = 3, 2$

So the solution of (1) is $a_n^{(h)} = C_1\, 3^n + C_2\, 2^n$.

To find the particular solution of the given equation, let $a_n^{(P)}$ = A. Substituting in the given equation

$$A - 5A + 6 = 2 \quad \Rightarrow \quad A = 1.$$

$a_n^{(P)} = 1$ which is a particular solution.

Hence the general solution is $a_n = a_n^{(h)} + a_n^{(P)} = C_1\, 3^n + C_2\, 2^n + 1$ ...(2)

To find C_1 and C_2, put $n = 0$ and $n = 1$ in (2)

or $\quad a_0 = C_1 + C_2 + 1$

$\quad 1 = C_1 + C_2 + 1$

$\Rightarrow \quad C_1 + C_2 = 0$ ...(3)

Again $\quad a_1 = 3C_1 + 2C_2 + 1$

$\quad -1 = 3C_1 + 2C_2 + 1$

or $\quad 3C_1 + 2C_2 = -2$ ...(4)

Solving (3) and (4), we get $C_1 = -2$ and $C_2 = 2$

Putting the values of C_1 and C_2 in (3), the required solution is

$$a_n = -2.3^n + 2.2^n + 1.$$

Problem 4.45. *What form does a particular solution of the linear homogeneous recurrence relation $a_{n+2} - 6a_{n+1} + 9a_n = f(n)$ have*

When $f(n) = 3^n$, $f(n) = n \,.\, 3^n$ and $f(n) = (n^2 + 1)\, 3^n$?

Solution. Let $a_n = r^n$ be a solution of the associated homogeneous recurrence relation.

$$a_{n+2} - 6a_{n+1} + 9a_n = 0$$

The characteristic equation $r^2 - 6r + 9 = 0 \;\Rightarrow\; (r-3)^2 = 0$,

has a single root 3 of multiplicity two.

Thus if $f(n) = 3^n$, the particular solution has the form $A_0 n^2 3^n$.

If $f(n) = n\, 3^n$, the particular solution has the form $n^2 (A_0 + A_1 n)\, 3^n$.

If $f(n) = a(n^2 + 1)\, 3^n$, the particular solution has the form $n^2 (A_0 + A_1 n + A_2 n^2)\, 3^n$.

Problem 4.46. *Solve $a_{n+2} - 4a_{n+1} + 4a_n = 2^n$.*

Solution. Let $a_n = r^n$ be a solution of the associated homogeneous recurrence relation

$$a_{n+2} - 4a_{n+1} + 4a_n = 0$$

The characteristic equation is

$$r^2 - 4r + 4 = 0 \quad \Rightarrow \quad (r-2)^2 = 0$$

has a single root 2 of multiplicity two. So the solution of associated homogeneous relation is

$a_n^h = (C_1 + C_2 n)\, 2^n$.

To find the particular solution of the given relation, we note $b = 2$ is a characteristic equation with multiplicity $s = 2$.

So, the particular solution has the form $a_n^{(P)} = A_0 n^2 \,.\, 2^n$.

Substituting in the given relation, we get

$$A_0(n+2)^2\,2^{n+2} - 4\,A_0(n+1)^2\,2^{n+1} + 4\,A_0n^2\,.\,2^n = 2^n$$

$$\Rightarrow \quad 4A_0(n+2)^2 - 8A_0(n+1)^2 + 4A_0n^2 = 1 \qquad \Rightarrow \quad A_0 = \frac{1}{8}$$

Hence the general solution is

$$a_n = a_n^{(h)} + a_n^{(P)} = (C_1 + C_2 n)\,2^n + \left(\frac{1}{8}\right) n^2\,.\,2^n.$$

Problem 4.47. *Solve the following* $y_{n+2} - y_{n+1} - 2y_n = n^2$.

Solution. Substituting $y_n = r^n$ in the associated homogeneous recurrence relation.

The characteristic equation is $r^2 - r - 2 = 0$

$\Rightarrow \quad (r+1)(r-2) = 0$

This gives $r = -1, r = 2$

The solution of the associated homogeneous recurrence relation is $y_b = C_1(-1)^n + C_2\,.\,2^n$.

Let the particular solution of the given equation is

$$A_n = A_0 + A_1 n + A_2 n^2 \qquad \text{(since } f(n) \text{ is a polynomial of degree 2)}$$

Substituting in the given equation, we have

$$A_0 + A_1(n+2) + A_2(n+2)^2 - [A_0 + A_1(n+1) + A_2(n+1)^2] - 2(A_0 + A_1 n + A_2 n^2) = n^2$$

$$(-2A_0 + A_1 + A_2) + (-2A_1 + 2A_2)\,n - 2A_2 n^2 = n^2$$

On comparing the coefficients of like powers of n,

We have
$$-2A_0 + A_1 + 3A_2 = 0 \qquad ...(1)$$
$$-2A_1 + 2A_2 = 0 \qquad ...(2)$$
$$-2A_2 = 1 \qquad ...(3)$$

From (3) $\quad A_2 = -\dfrac{1}{2}$

We have from (2) $\quad A_1 = A_2 = -\dfrac{1}{2}$

From (1) $\quad -2A_0 - \dfrac{1}{2} - \dfrac{3}{2} = 0 \qquad \Rightarrow \quad A_0 = -1$

Therefore, particular solution of given recurrence relation is

$$a_n^{(P)} = -1 - \left(\frac{1}{2}\right) n - \left(\frac{1}{2}\right) n^2$$

Hence the general solution of the given recurrence relation is

$$a_n = C_1(-1)^n + C_2\,.\,2^n - 1 - \frac{n}{2} - \left(\frac{1}{2}\right) n^2.$$

Problem 4.48. *Solve the recurrence relation*

$$a_n - 8a_{n-1} + 21a_{n-2} - 18a_{n-3} = 0.$$

Solution. Let $a_n = r^n$ be a solution of the given equation.

The characteristic equation is

$$r^3 - 8r^2 + 21r - 18 = 0$$

$$\Rightarrow \quad r^3 - 2r^2 - 6r^2 + 12r + 9r - 18 = 0$$

$$r^2(r-2) - 6r(r-2) + 9(r-2) = 0$$

$$(r-2)(r^2 - 6r + 9) = 0$$

$$(r-2)(r-3)^2 = 0$$

which gives $r = 2, 3, 3$.

So the general solution is

$$a_n = (b_1 + b_2 n)\, 3^n + b_2 2^n.$$

Problem 4.49. *Solve $a_n = a_{n-1} + 2a_{n-2}$, $n \geq 2$ with the initial conditions $a_0 = 0$, $a_1 = 1$.*

Solution. The given recurrence relation

$$a_n = a_{n-1} + 2a_{n-2}$$

i.e., $$a_n - a_{n-1} - 2a_{n-2} = 0 \qquad ...(1)$$

is a 2nd order linear homogeneous recurrence relation with constant coefficients.

Let, $a_n = r^n$ is a solution of (1).

The characteristic equation is

$$r^2 - r - 2 = 0$$

$$(r-2)(r+1) = 0$$

or $\quad r = 2, -1$ distinct real roots.

So the general solution is $a_n = b_1(2)^n + b_2(-1)^n$

Again $\quad a_0 = 0$ implies $b_1 + b_2 = 0$

Add $\quad a_1 = 1$ implies $2b_1 - b_2 = 1$

The solution of these two equations are $b_1 = \frac{1}{3}$ and $b_2 = -\frac{1}{3}$

Hence the explicit solution is given by $a_n = \left(\frac{1}{3}\right) 2^n - \left(\frac{1}{3}\right)(-1)^n$.

Problem 4.50. *Solve the recurrence relation of the Fibonacci sequence of numbers $f_n = f_{n-1} + f_{n-2}$, $n \geq 2$ with the initial condition $f_0 = 1$, $f_1 = 1$.*

Solution. Let $f_n = r^n$ is a solution of the given equation, the characteristic equation is $r^2 - r - 1 = 0$

$$r = \frac{(1 \pm \sqrt{5})}{2}$$

i.e., $\quad r_1 = \dfrac{1+\sqrt{5}}{2}, r_2 = \dfrac{1-\sqrt{5}}{2}$ are two distinct roots.

So the general solution is

$$a_n = b_1 \left(\frac{1+\sqrt{5}}{2}\right)^n + b_2 \left(\frac{1-\sqrt{5}}{2}\right)$$

Again $\quad a_0 = 1$ implies $b_1 + b_2 = 1$

$$a_1 = 1 \text{ implies } b_1 \left(\frac{1+\sqrt{5}}{2}\right) + b_2 \left(\frac{1-\sqrt{5}}{2}\right) = 1$$

Solving these two equations we get

$$b_1 = \frac{1}{\sqrt{5}},\ b_2 = -\frac{1}{\sqrt{5}}$$

Thus the nth Fibonacci number is given explicitly by

$$f_n = \frac{1}{\sqrt{5}} \left[\left(\frac{1+\sqrt{5}}{2}\right)^n - \left(\frac{1-\sqrt{5}}{2}\right)^n\right]$$

Problem 4.51. *Solve the recurrence relation* $a_n = 4(a_{n-1} - a_{n-2})$.

Solution. The given relation is $a_n - 4a_{n-1} + 4a_{n-2} = 0$ ...(1)

Let $a_n = r^n$ is a solution of (1)

Then the characteristic equation is $r^2 - 4r + 4 = 0$ which gives $r = 2, 2$

Thus the general solution is $a_n = (b_1 + nb_2)\, 2^n$

So, $a_0 = b_1$ and $a_1 = (b_1 + b_2)^2$

Now $a_1 = 1$ gives $b_1 = 1$ and $a_1 = 1$ gives $2(b_1 + b_2) = 1 \quad \Rightarrow \quad b_2 = -\frac{1}{2}$

So, the general solution is $a_n = \left(1 - \frac{1}{2}n\right) 2^n$.

4.4. THE METHOD OF GENERATING FUNCTIONS

One of the uses of generating function method is to find the closed form formula for a recurrence (equation) relation. Before using this method, ensure that the given recurrence equation is in linear form.

A non-linear recurrence equation cannot be solved by the Generating Function Method. Use substitution of **variable technique** to convert a **non-linear recurrence (equation) relation into linear.**

Solving a recurrence (equation) relation using generating function method involves two **steps process.**

Step 1 : Find generating function for the sequence for which the general term is given by recurrence (equation) relation.

Step 2 : Find coefficient of x^2 or $\frac{x^2}{n!}$ depending upon whether the generating function is binomial or exponential to get a_n, the general term of the sequence.

The value so obtained will be an algebraic formula for a_n, expressed in terms of n which is the position of a_n is sequence.

A generating function is a polynomial expression of the form

$$f(x) = a_0 + a_1x + a_2x^2 + a_3x^3 + ... + a_nx^n + ...$$

in which the coefficients a_i are all zero after a certain point, a generating function usually has infinitely many non-zero terms. There is an obvious correspondence between generating functions and sequences.

$$a_0, a_1, a_2, ...$$

$$a_0 + a_1x + a_2x^2 + a_3x^3 + ... \leftrightarrow a_0, a_1, a_2, a_3, ...$$

*If $f(x) = a_0 + a_1x + a_2x^2 + ...$

and $g(x) = b_0 + b_1x + b_2x^2 + ...$ then

$$f(x) + g(x) = (a_0 + b_0) + (a_1 + b_1)x + (a_2 + b_2)x^2 + ...$$

$$f(x)g(x) = (a_0b_0) + (a_1b_0 + a_0b_1)x + (a_0b_2 + a_1b_1 + a_2b_0)x^2 + ...$$

The coefficient of x^n in the product $f(x)g(x)$ is the infinite sum

$$a_0b_n + a_1b_{n-1} + a_2b_{n-2} + ... + a_nb_0.$$

Problem 4.52. *If* $f(x) = 1 + x + x^2 + ... + x^n + ...$ *and*

$$g(x) = 1 - x + x^2 - x^3 + ... + (-1)^n x^n + ...,$$

find $f(x) + g(x)$ *and* $f(x)g(x)$.

Solution. $f(x) + g(x)$

$$= (1 + x + x^2 + ... + x^n + ...) + (1 - x + x^2 - x^3 + ... + (-1)^n x^n + ...)$$

$$= (1 + 1) + (1 - 1)x + (1 + 1)x^2 + ... + (1 + (-1)^n)x^n + ...$$

$$= 2 + 2x^2 + 2x^4 + ...$$

$$f(x)g(x) = (1 + x + x^2 + ... + x^n + ...) . (1 - x + x^2 - x^3 + ... + (-1)^n x^n + ...)$$

$$= 1 + [1(-1) + 1(1)]x + [1(1) + 1(-1) + 1(1)]x^2 + ...$$

$$= 1 + x^2 + x^4 + x^6 + ...$$

Problem 4.53. *Solve the recurrence relation* $a_n = 3a_{n-1}, n \geq 1$, *given* $a_0 = 1$.

Solution. Consider the generating function

$$f(x) = a_0 + a_1x + a_2x^2 + ... + a_nx^n + ...$$

of the sequence $a_0, a_1, a_2, ...$ multiplying by $3x$ and writing the product $3xf(x)$ below $f(x)$ so that terms involving x^n match, we obtain

$$f(x) = a_0 + a_1x + a_2x^2 + ... + a_nx^n + ...$$

$$3xf(x) = 3a_0x + 3a_1x^2 + ... + 3a_{n-1}x^n + ...$$

Subtracting gives

$$f(x) - 3xf(x) = a_0 + (a_1 - 3a_0)x + (a_2 - 3a_1)x^2 + ... + (a_n - 3a_{n-1})x^n + ...$$

Since $a_0 = 1, a_1 = 3a_0$.

In general, $a_n = 3a_{n-1}$, this says that.

$$(1 - 3x) f(x) = 1.$$

Thus, $f(x) = \dfrac{1}{1 - 3x}$

We have $\dfrac{1}{1-x} = 1 + x + x^2 + x^3 + \ldots\ldots\ldots$ ...(∗)

Using (∗), $f(x) = 1 + 3x + (3x)^2 + \ldots\ldots + (3x)^n + \ldots\ldots.$

$= 1 + 3x + 9x^2 + \ldots\ldots + 3^n x^n + \ldots\ldots\ldots$

We conclude that a_n, which is the coefficient of x^n in $f(x)$, must equal 3^n.

We have $a_n = 3^n$ as the solution to our recurrence relation.

Problem 4.54. *Solve the recurrence relation $a_n = 2a_{n-1} - a_{n-2}$, $n \geq 2$, given $a_0 = 3$, $a_1 = -2$.*

Solution. Letting $f(x)$ be the generating function of the sequence in question.

We have $f(x) = a_0 + a_1 x + a_2 x^2 + \ldots\ldots + a_n x^n + \ldots\ldots$

$2xf(x) = 2a_0 x + 2a_1 x^2 + \ldots\ldots. + 2a_{n-1} x^n + \ldots\ldots$

$x^2 f(x) = a_0 x^2 + \ldots\ldots + a_{n-2} x^n + \ldots\ldots.$

Therefore, $f(x) - 2xf(x) + x^2 f(x)$

$= a_0 + (a_1 - 2a_0)x + (a_2 - 2a_1 + a_0) x^2 + \ldots\ldots. + (a_n - 2a_{n-1} + a_{n-2}) x^n + \ldots\ldots.$

$= 3 - 8x.$

Since $a_0 = 3$, $a_1 = -2$ and $a_n - 2a_{n-1} + a_{n-2} = 0$ for $n \geq 2$.

So, $(1 - 2x + x^2) f(x) = 3 - 8x$

$(1 - x)^2 f(x) = 3 - 8x$

$\Rightarrow \quad f(x) = \dfrac{1}{(1-x)^2} (3 - 8x)$

$= (1 + 2x + 3x^2 + \ldots\ldots. + (n + 1)x^n + \ldots\ldots) (3 - 8x)$

$= 3 - 2x - 7x^2 - 12x^3 + \ldots\ldots. + [3(n + 1) - 8n]x^n + \ldots\ldots$

$= 3 - 2x - 7x^2 - 12x^3 + \ldots\ldots. + (-5n + 3) x^n + \ldots\ldots$

Therefore $a_n = 3 - 5x$ is the desired solution.

Problem 4.55. *Solve the recurrence relation*

$$a_n = -3a_{n-1} + 10a_{n-2},\ n \geq 2, \text{ given } a_0 = 1,\ a_1 = 4.$$

Solution. Letting $f(x)$ be the generating function of the sequence in question.

We have $f(x) = a_0 + a_1 x + a_2 x^2 + \ldots\ldots + a_n x^n + \ldots\ldots.$

$3xf(x) = 3a_0 x + 3a_1 x^2 + \ldots\ldots + 3a_{n-1} x^2 + \ldots\ldots.$

$10x^2 f(x) = 10a_0 x^2 + \ldots\ldots + 10a_{n-2} x^n + \ldots\ldots$

Therefore, $f(x) = 3xf(x) - 10x^2 f(x)$

$= a_0 + (a_1 + 3a_0) x + (a_2 + 3a_1 - 10a_0) x^2 + \ldots\ldots.$

$+ (a_n + 3a_{n-1} - 10a_{n-2}) x^n + \ldots\ldots$

$= 1 + 7x$

Since $a_0 = 1, a_1 = 4$ and $a_n + 3a_{n-1} - 10a_{n-2} = 0$ for $n \geq 2$.

So $(1 + 3x - 10x^2) f(x) = 1 + 7x$

$$\Rightarrow \quad f(x) = \frac{1 + 7x}{1 + 3x - 10x^2} = \frac{1 + 7x}{(1 + 5x)(1 - 2x)}$$

At this point, it is useful to recall the method of partial fractions.

We get $$\frac{1}{(1 + 5x)(1 - 2x)} = \frac{A}{1 + 5x} + \frac{B}{1 - 2x} = \frac{A(1 - 2x) + B(1 + 5x)}{(1 + 5x)(1 - 2x)}$$

Equating numerators,

$$1 = (A + B) + (-2A + 5B)\,x$$

$$\Rightarrow \quad A + B = 1$$

$$-2A + 5B = 0$$

Solving for A and B, we get $A = \frac{5}{7}, B = \frac{2}{7}$

Therefore, $$\frac{1}{(1 + 5x)(1 - 2x)} = \frac{5}{7}\left(\frac{1}{1 + 5x}\right) + \frac{2}{7}\left(\frac{1}{1 - 2x}\right)$$

So $$f(x) = \frac{1 + 7x}{(1 + 5x)(1 - 2x)}$$

$$= \frac{5}{7}\left(\frac{1}{1 + 5x}\right)(1 + 7x) + \frac{2}{7}\left(\frac{1}{1 - 2x}\right)(1 + 7x)$$

$$= \frac{5}{7}\,[1 + (-5x) + (-5x)^2 + \ldots\ldots]\,(1 + 7x) + \frac{2}{7}\,[1 + (2x) + (2x)^2 + \ldots\ldots]\,(1 + 7x)$$

$$= \frac{5}{7}\,[1 - 5x + 25x^2 + \ldots + (-5)^n]\,x^n)\,(1 + 7x)] + \frac{2}{7}[1 + 2x + 4x^2 + \ldots\ldots 2^n x^n + \ldots\ldots]]\,(1 + 7x).$$

$$= \frac{5}{7}(1 + 2x - 10x^2 + \ldots\ldots + [(-5)^n + 7(-5)^{n-1}]\,x^n + \ldots\ldots)$$

$$+ \frac{2}{7}\,(1 + 9x + 18x^2 + \ldots\ldots + [2^n + 7(2^{n-1})\,x^n] + \ldots\ldots)$$

$$= \frac{5}{7}(1 + 2x - 10x^2 + \ldots\ldots + 2(-5)^{n-1}x^n + \ldots\ldots)$$

$$+ \frac{2}{7}\,(1 + 9x + 18x^2 + \ldots\ldots + 9(2^{n-1})\,x^n + \ldots\ldots)$$

$$= 1 + 4x - 2x^2 + \ldots\ldots + \left[-\frac{2}{7}(-5)^n + \frac{9}{7}(2^n)\right] x^n + \ldots\ldots$$

Hence $a_n = -\frac{2}{7}(-5)^n + \frac{9}{7}(2^n)$ is the desired solution.

Problem 4.56. *Solve the recurrence relation $a_n = -a_{n-1} + 2n - 3$, $n \geq 1$, given $a_0 = 1$.*

Solution. Let $f(x)$ be the generating function of the sequence $a_0, a_1, a_2, \ldots\ldots$

Then $f(x) = a_0 + a_1x + a_2x^2 + \ldots\ldots + a_nx^n + \ldots\ldots$

$xf(x) = a_0x + a_1x^2 + \ldots\ldots + a_{n-1}x^n + \ldots\ldots$

Therefore, $f(x) + xf(x)$

$$= a_0 + (a_1 + a_0)x + (a_2 + a_1)x^2 + \ldots\ldots + (a_n + a_{n-1})x^n + \ldots\ldots$$

given that, $a_0 = 1$ and $a_n + a_{n-1} = 2n - 3$.

Thus $a_1 + a_0 = 2(1) - 3 = -1$

$a_2 + a_1 = 2(2) - 3 = 1$, and so on.

We obtain $(1 + x) f(x) = 1 - x + x^2 + \ldots\ldots + (2n - 3)x^n + \ldots\ldots$

So $$f(x) = \frac{1}{1+x}\,(1 - x + x^2 + \ldots\ldots + (2x - 3)\,x^n + \ldots\ldots)$$

$$= (1 + (-x) + (-x)^2 + \ldots\ldots + (-x)^n + \ldots\ldots)\,.\,(1 - x + x^2 + \ldots\ldots + (2n-3)x^n + \ldots\ldots)$$
$$= (1 - x + x^2 - x^3 + \ldots\ldots + (-x)^n + \ldots\ldots)\,.\,(1 - x + x^2 + \ldots\ldots + (2n-3)\,x^n + \ldots\ldots)$$
$$= 1 - 2x + 3x^2 + [(2n-3) - (2n-5) + \ldots\ldots + (-1)^{n-1}(-1) + (-1)^n]\,x^n + \ldots\ldots$$

Now a_n is the coefficient of x^n, the term in the square brackets.

$$a_n = \left\{\sum_{k=0}^{n-1} (-1)^k\,[2n - (2k+3)]\right\} + (-1)^n$$

$$= (2n-3)\sum_{k=0}^{n-1}(-1)^k - 2\sum_{k=0}^{n-1}(-1)^k\,k + (-1)^n.$$

We have it to you to verify that

$$\sum_{k=0}^{n-1}(-1)^k\,k = \begin{cases} -\dfrac{n}{2} & \text{if } n \text{ is even} \\[2mm] \dfrac{n-1}{2} & \text{if } n \text{ is odd.} \end{cases}$$

So, if n is even, $a_n = 0 - 2\left(-\dfrac{n}{2}\right) + 1 = n + 1$,

if n is odd, $a_n = (2n-3) - 2\left(\dfrac{n-1}{2}\right) - 1$

$= 2n - 3 - n + 1 - 1 = n - 3$

The solution is $a_n = \begin{cases} n+1 & \text{if } n \text{ is even} \\ n-3 & \text{if } n \text{ is odd.} \end{cases}$

Note that this solution could also be written as

$$a_n = 2(-1)^n + n - 1.$$

Problem 4.57. *Suppose a is a real number. Show that* $\dfrac{1}{1-ax}$ *is the generating function for a certain geometric sequence.*

Solution. We have $\dfrac{1}{1-x} = 1 + x + x^2 + x^3 + \ldots\ldots$ (1)

Replacing x by ax in (1), we see that

$$\frac{1}{1-ax} = 1 + (ax) + (ax)^2 + (ax)^3 + \ldots\ldots$$

$$= 1 + ax + a^2x^2 + a^3x^3 + \ldots\ldots$$

From this, we see that $\dfrac{1}{1-ax}$ is the generating function for the sequence $1, a, a^2, a^3, \ldots\ldots$ which is the geometric sequence with first term 1 and common ratio a.

Problem 4.58. *Prove that* $\dfrac{1}{(1-x)^2} = 1 + 2x + 3x^2 + 4x^3 + \ldots\ldots + (n+1)x^n + \ldots\ldots$.

Solution. $(1-x)^2 (1 + 2x + 3x^2 + \ldots\ldots + (n+1)x^n + \ldots\ldots)$

$$= (1 - 2x + x^2)(1 + 2x + 3x^2 + \ldots\ldots + (n+1)x^n + \ldots\ldots)$$

$$= 1 + [1(2) - 2(1)]x + [1(3) - 2(2) + 1(1)]x^2 + \ldots\ldots$$

$$\ldots\ldots + [1(n+1) - 2(n) + 1(n-1)]x^n + \ldots\ldots$$

$$= 1, \quad \text{since } n + 1 - 2n + n = 0.$$

Problem 4.59. *Solve the following recurrence equation using generating function*

$$a_n = a_{n-1} + 2(n-1),\ a_0 = 1.$$

Solution. We will solve this using binomial generating function. Let $f(x)$ be the binomial generating function for the sequence a of which general term is given by the given recurrence equation.

Therefore, we have $f(x) = \sum_{n=0}^{\infty} a_n x^n = a_0 + \sum_{n=1}^{\infty} a_n x^n.$

or $$f(x) = a_0 + \sum_{n=1}^{\infty} [a_{n-1} + 2(n-1)]x^n .$$

$$= a_0 + \sum_{n=1}^{\infty} a_{n-1}x^n + 2\sum_{n=1}^{\infty} (n-1)x^n$$

$$= a_0 + x\sum_{n=1}^{\infty} a_{n-1}x^{n-1} + 2x\sum_{n=1}^{\infty} (n-1)x^{n-1}$$

$$= a_0 + x \sum_{m=0}^{\infty} a_m x^m + 2x\,[0 * x^0 + 1 * x^1 + 2 * x^2 + \ldots\ldots] \qquad [n-1=m]$$

$$= a_0 + x\,f(x) + 2x^2\,[1 + 2x + 3x^2 + \ldots\ldots]$$

$$\because \quad f(x) = \sum_{m=0}^{\infty} a_m x^m$$

or $$(1-x)\,f(x) = a_0 + \frac{2x^2}{(1-x)^2} = 1 + \frac{2x^2}{(1-x)^2} \qquad (\because \quad a_0 = 1)$$

$$\Rightarrow \quad f(x) = \frac{1 - 2x + 3x^2}{(1-x)^3}$$

$$f(x) = \frac{1}{(1-x)^3} - \frac{2x}{(1-x)^3} + \frac{3x^2}{(1-x)^3}$$

is the generating function of the equation.

The value of a_n is given by the coefficient of x^n is $f(x)$.

Therefore

Coeff. of x^n = coeff. of x^n in $\frac{1}{(1-x)^3}$ $- 2x$ * coeff. of x^{n-1} in $\frac{1}{(1-x)^3}$ $+ 3x^2$ * coeff of x^{n-2} in $\frac{1}{(1-x)^3}$

$$= {}^{3+n-1}C_n - 2 * {}^{3+(n-1)-1}C_{n-1} + 3 * {}^{3+(n-2)-1}C_{n-2}$$

or $$a_n = {}^{n+2}C_n - 2 * {}^{n+1}C_{n-1} + 3 * {}^{n}C_{n-2}$$

$$= \frac{n^2 + 3n + 2}{2} - (n^2 + n) + \frac{3(n^2 - n)}{2}$$

$$= \frac{n^2 + 3n + 2 - 2n^2 - 2n + 3n^2 - 3n}{2}$$

$$= \frac{2n^2 - 2n + 2}{2}$$

$$= n^2 - n + 1.$$

Therefore, $a_n = n^2 - n + 1$.

Problem 4.60. *Solve the following recurrence equation using generating function*

$$f_n = f_{n-1} + f_{n-2}, \text{ for } n \geq 2 \text{ and } f_0 = f_1 = 1.$$

Solution. We will solve this using binomial generating function.

Let $f(x)$ be the binomial generating function for the sequence f of which general term is given by the above recurrence equation.

Then, we have $f(x) = \sum_{n=0}^{\infty} f_n x^n$

$$= f_0 + f_1 x + \sum_{n=2}^{\infty} a_n x^n$$

$$= f_0 + f_1 x + \sum_{n=2}^{\infty} [f_{n-1} + f_{n-2}] x^n$$

$$= f_0 + f_1 x + \sum_{n=2}^{\infty} f_{n-1} x^n + \sum_{n=2}^{\infty} f_{n-2} x^n$$

$$= f_0 + f_0 x + \sum_{n=2}^{\infty} f_{n-1} x^n + \sum_{n=2}^{\infty} f_{n-2} x^n \qquad (\because\ f_0 = f_1)$$

$$= f_0 + x \sum_{n=1}^{\infty} f_{n-1} x^{n-1} + x^2 \sum_{n=2}^{\infty} f_{n-2} x^{n-2}$$

$$= f_0 + x \sum_{n-1=0}^{\infty} f_{n-1} x^{n-1} + x^2 \sum_{n-2=0}^{\infty} f_{n-2} x^{n-2}$$

$$= f_0 + x f(x) + x^2 f(x)$$

$$= 1 + x f(x) + x^2 f(x)$$

$$= \frac{1}{(1 - x - x^2)}$$ is the Generating Function of the equation.

Using partial fraction, we can write $f(x)$ as

$$f(x) = \frac{1}{\sqrt{5}} \left[\frac{A}{1 - x_1 x} - \frac{B}{1 - x_2 x} \right]$$

where $x_1 = \dfrac{1 + \sqrt{5}}{2}$ and $x_2 = \dfrac{1 - \sqrt{5}}{2}$

where $A = \dfrac{1 + \sqrt{5}}{2}$ and $B = \dfrac{1 - \sqrt{5}}{2}$

Therefore, coefficient of $x^n = \dfrac{1}{\sqrt{5}} \left[\left(\dfrac{1 + \sqrt{5}}{2} \right)^n - \left(\dfrac{1 - \sqrt{5}}{2} \right)^n \right]$

$$f_n = \frac{1}{\sqrt{5}} \left[\left(\frac{1 + \sqrt{5}}{2} \right)^n - \left(\frac{1 - \sqrt{5}}{2} \right)^n \right].$$

Problem 4.61. *Solve the following recurrence equation using generating function*

$$a_n - 8a_{n-1} + 16a_{n-2} = 0 \quad \text{for } n \geq 4$$

and $$a_2 = 16,\ a_3 = 80.$$

Solution. We will use binomial generating function to solve this problem.

Let $f(x)$ be the binomial generating function for the sequence a of which general term is given by the above recurrence equation. One thing to be noticed here is that the sequence term starts from $n = 2$, so summation will be used accordingly, $f(x)$ is given as,

$$f(x) = \sum_{n=n}^{\infty} a_n x^n$$

$$= a_2x^2 + a_3x^3 + \sum_{n=4}^{\infty} a_n x^n$$

$$= a_2x^2 + a_3x^3 + \sum_{n=4}^{\infty} [8a_{n-1} - 16a_{n-2}]\, x^n$$

$$= a_2x^2 + a_3x^3 + 8\sum_{n=4}^{\infty} a_{n-1}x^n - 16\sum_{n=4}^{\infty} a_{n-2}x^n$$

$$= a_2x^2 + a_3x^3 + 8\,[a_3x^4 + a_4x^5 + \ldots\ldots] - 16\,[a_2x^4 + a_3x^5 + \ldots\ldots]$$

$$= a_2x^2 + a_3x^3 + 8\,[a_2x^2 + a_3x^3 + \ldots\ldots]\,x - 8a_2x^3 - 16[a_2x^2 + a_3x^3 + \ldots\ldots]\,x^2$$

$$= a_2x^2 + a_3x^2 + 8x\,f(x) - 8a_2x^3 - 16x^2 f(x)$$

or $$(1 - 8x + 16x^2)\,f(x) = a_2x^2 + a_3x^3 - 8a_2x^3.$$

Now, substituting the value of a_2 and a_3 in the above equation, we get,

$$(1 - 8x + 16x^2)\,f(x) = 16x^2 + 80x^3 - 8 * 16x^3$$

$$\Rightarrow \qquad f(x) = \frac{16x^2 - 48x^3}{1 - 8x + 16x^2} = 16x^2\,\frac{1 - 3x}{(1 - 4x)^2}$$

Therefore, $f(x) = 16x^2\,\dfrac{1 - 3x}{(1 - 4x)^2}$

To find the general term a_n, we will have to find the coefficient of x^n from $f(x)$.

Using partial fraction method, we can write

$$\frac{1 - 3x}{(1 - 4x)^2} = \frac{A}{1 - 4x} + \frac{B}{(1 - 4x)^2}$$

Solving the above equation for A and B, we get A = 1, B = 1.

Therefore, the Generating Function can be written as

$$f(x) = 16x^2\left[\frac{1}{1 - 4x} + \frac{x}{(1 - 4x)^2}\right]$$

Hence, coefficient of x^n in $f(x)$ is given as

$$16 * (\text{coefficient of } x^{n-2} \text{ in first term} + \text{coefficient of } x^{n-3} \text{ in second term}).$$

This is equal to $\quad 16 * (4^{n-2} + (n-2)\, 4^{n-3})$

or
$$a_n = 16 * 4^{n-2}\left[1 + \frac{n-2}{4}\right]$$
$$= 4^n\left[\frac{1}{2} + \frac{n}{4}\right].$$

Problem 4.62. *Solve the following recurrence equation using generating function*

$$C_n = 3C_{n-1} - 2C_{n-2} = 0, \text{ for } n \geq 3,$$

and $\quad C_1 = 5,\ C_2 = 3.$

Solution. We will again use binomial generating function to solve this problem.

Let $f(x)$ be the binomial generating function for the sequence C of which general term is given by the above recurrence equation.

Since the sequence term starts from $n = 1$, so the summation will be used accordingly, $f(x)$ is given by,

$$f(x) = \sum_{n=1}^{\infty} C_n x^n = C_1 x + C_2 x^2 + \sum_{n=3}^{\infty} C_n x^n$$

$$= C_1 x + C_2 x^2 + \sum_{n=3}^{\infty} (3C_{n-1} - 2C_{n-2}) x^n, \qquad \text{(by definition of } C_n\text{)}$$

$$= C_1 x + C_2 x^2 + 3\sum_{n=3}^{\infty} C_{n-1} x^n - 2\sum_{n=3}^{\infty} C_{n-2} x^n.$$

$$= C_1 x + C_2 x^2 + 3[C_2 x^3 + C_3 x^4 + C_4 x^5 + \ldots\ldots] - 2[C_1 x^3 + C_2 x^4 + C_3 x^5 + \ldots\ldots]$$

$$= C_1 x + C_2 x^2 + 3[C_1 x + C_2 x^2 + C_3 x^3 + \ldots\ldots]x$$
$$- 3C_1 x^2 - 2[C_1 x + C_2 x^2 + C_3 x^3 + \ldots\ldots]x^2$$

$$= C_1 x + C_2 x^2 + 3x\, f(x) - 3C_1 x^2 - 2x^2 f(x).$$

or
$$(1 - 3x + 2x^2)\, f(x) = C_1 x + C_2 x^2 - 3C_1 x^2$$
$$= 5x + 3x^2 - 2 * 5x^2$$

Therefore,
$$f(x) = \frac{5x - 12x^2}{(1-x)(1-2x)} = x\,\frac{5 - 12x}{(1-x)(1-2x)}$$

To find the general term a_n, we shall have to find the coefficient of x^n from $f(x)$.

Using partial fraction method, we can write

$$\frac{5 - 12x}{(1-x)(1-2x)} = \frac{A}{1-x} + \frac{B}{1-2x}$$

Solving the above equation for A and B, we get A = 7, B = – 2

Therefore, the Generating Function can be written as

$$f(x) = x\left[\frac{7}{1-x} - \frac{2}{1-2x}\right]$$

Hence, coefficient of x^n in $f(x)$ is given as

(coefficient of x^{n-1} in first term + coefficient of x^{n-2} in second term)

This is equal to

$$7 - 2 * 2^{n-1}$$

or $$a_n = 7 - 2^n.$$

Problem 4.63. *Solve the following recurrence equation using generating function*

$$a_n - 7a_{n-1} + 10a_{n-2} = 2^n, \text{ for } n \geq 2, \quad \text{and} \quad a_0 = 0,\ a_1 = 6.$$

Solution. Let $f(x)$ be the binomial generating function for the sequence a of which general term is given by the above recurrence equation.

Then, $f(x)$ is given by

$$f(x) = \sum_{n=0}^{\infty} a_n x^n$$

$$= a_0x + a_1x + \sum_{n=2}^{\infty} a_n x^n .$$

$$= a_0x + a_1x + \sum_{n=2}^{\infty} (7a_{n-1} = 10a_{n-2} + 2^n)\ x^n \qquad \text{(by def. } n \text{ of } a_n)$$

$$= a_0x + a_1x + 7\sum_{n=2}^{\infty} a_{n-1}x^n - 10\sum_{n=2}^{\infty} a_{n-2}x^n + \sum_{n=2}^{\infty} 2^n x^n .$$

$$= a_0x + a_1x + 7x\sum_{n=1}^{\infty} a_{n-1}x^{n-1} - 10x^2\sum_{n=2}^{\infty} a_{n-2}x^{n-2} + \sum_{n=2}^{\infty} 2^n x^n .$$

$$= a_0x + a_1x + 7x f(x) - 10x^2 f(x) + \frac{4x^2}{1-2x} \qquad \left(\because \sum_{n=2}^{\infty} 2^n x^n = \frac{4x^2}{1-2x}\right)$$

or $$(1 - 7x + 10x^2) f(x) = a_0x + a_1x + \frac{4x^2}{1-2x}$$

$$= 6x + \frac{4x^2}{1-2x}$$

Therefore, $f(x) = \dfrac{6x - 8x^2}{(1-2x)^2\,(1-5x)} = x\,\dfrac{6 - 8x}{(1-2x)^2\,(1-5x)}$

We will have to find the coefficient of x^n from $f(x)$.

Using partial fraction method, we can write

$$\frac{6x - 8x^2}{(1-2x)^2\,(1-5x)} = \frac{A}{1-5x} + \frac{B}{1-2x} + \frac{Cx}{(1-2x)^2}$$

Solving the above equation for A, B and C, we get

$$A = \frac{22}{9},\quad B = -\frac{22}{9}\quad \text{and } C = -\frac{4}{3}$$

Therefore, the Generating Function can be written as,

$$f(x) = \frac{22}{9}\left[\frac{1}{1-5x}\right] - \frac{22}{9}\left[\frac{1}{1-2x}\right] - \frac{4}{3}\left[\frac{x}{(1-2x)^2}\right]$$

Hence, coefficient of x^n in $f(x)$ is given as

(coefficient of x^n in first term + coefficient of x^n in second term + coefficient of x^n in third term).

This is equal to $\dfrac{22}{9}5^n - \dfrac{22}{9}2^n - \dfrac{4}{3}n2^{n-1}$

Therefore, $a_n = \dfrac{22}{9}\,[5^n - 2^n] - \dfrac{2}{3}n2^n$.

Problem 4.64. *Solve the following recurrence equation using exponential generating function* $d_n = (n-1)(d_{n-1} + d_{n-2})$ *for* $n \geq 3$ *and* $d_1 = 0$, $d_2 = 1$.

Solution. Let $f(x)$ be the exponential generating function of the above recurrence relation, then $f(x)$ is given as

$$f(x) = \sum_{n=0}^{\infty} d_n \frac{x^n}{n!} \qquad ...(1)$$

Differentiating (1) with respect to x, we get

$$f'(x) = \sum_{n=0}^{\infty} d_n \frac{x^{n-1}}{(n-1)!} \qquad ...(2)$$

Replacing d_n by its definition, in the equation (2), we get

$$f'(x) = \sum_{n=0}^{\infty} (n-1)\,(d_{n-1} + d_{n-2})\,\frac{x^{n-1}}{(n-1)!}$$

$$= \sum_{n=0}^{\infty} (n-1)\,d_{n-1}\,\frac{x^{n-1}}{(n-1)!} + \sum_{n=0}^{\infty} (n-1)\,d_{n-2}\,\frac{x^{n-1}}{(n-1)!}$$

$$= \sum_{n=0}^{\infty} d_{n-1} \frac{x^{n-1}}{(n-2)!} + \sum_{n=0}^{\infty} d_{n-2} \frac{x^{n-1}}{(n-2)!}$$

$$= x \sum_{n=0}^{\infty} d_{n-1} \frac{x^{n-2}}{(n-2)!} + x \sum_{n=0}^{\infty} d_{n-2} \frac{x^{n-2}}{(n-2)!}$$

$$= x \left[d_0 \cdot \frac{x^{-1}}{(-1)!} + d_1 \cdot \frac{x^0}{(0)!} + d_2 \frac{x^1}{1!} + d_3 \cdot \frac{x^2}{2!} + \ldots\ldots \right]$$

$$+ x \left[d_0 + d_1 x + d_2 \cdot \frac{x^2}{2!} + d_3 \cdot \frac{x^3}{3!} + \ldots\ldots \right]$$

$$= xf'(x) + xf(x)$$

or $$(1 - x) f'(x) = xf(x)$$

$$\Rightarrow \quad \frac{f'(x)}{f(x)} = \frac{x}{1-x}$$

$$\Rightarrow \quad \frac{f'(x)}{f(x)} = \frac{1}{1-x} - 1 \qquad \ldots(3)$$

Integrating both sides of equation (3), we get

$$\log f(x) = -\log (1 - x) - x$$

$$\log f(x) (1 - x) = -x$$

$$f(x) (1 - x) = e^{-x}$$

$$\Rightarrow \quad f(x) = \frac{1}{1-x} \cdot e^{-x}$$

This is the generating function for the given recurrence equation.

To find the general term from this generating function we will have to find the coefficient of $\frac{x^n}{n!}$,

which is given as $d_n = n! \left(1 - \frac{1}{1!} + \frac{2}{2!} - \ldots\ldots (-1)^n \frac{1}{n!} \right)$.

Problem 4.65. *Use generating functions to solve the recurrence relation*

(i) $a_n = 3a_{n-1} + 2,\ a_0 = 1$

(ii) $a_n - 9a_{n-1} + 20a_{n-2} = 0,\ a_0 = -3,\ a_1 = -10$

(iii) $a_{n+2} - 2a_{n+1} + a_n = 2^{2n},\ a_0 = 2,\ a_1 = 1.$

Solution. Let $G(x) = \sum_{n=0}^{\infty} a_n x^n$, where G(x) is the generating function for the sequence $\{a_n\}$.

Multiplying each term in the given recurrence relation by x^n and summing from 1 to ∞, we get

$$\sum_{n=1}^{\infty} a_n x^n = 3 \sum_{n=1}^{\infty} a_{n-1} x^n + 2 \sum_{n=1}^{\infty} x^n$$

$$G(x) - a_0 = 3x\, G(x) + 2 \left[\frac{1}{1-x} - 1 \right]$$

Since $x\, G(x) = \sum_{n=0}^{\infty} a_n x^{n+1} = \sum_{n=1}^{\infty} a_{n-1} x^n$

$$\therefore \quad G(x) - 3xG(x) = 1 + \frac{2x}{1-x} \qquad (a_0 = 1)$$

or $$G(x) = \frac{1+x}{(1-x)(1-3x)}$$

$$= \frac{2}{1-3x} - \frac{1}{1-x} \qquad \text{(by partial fraction)}$$

$$\therefore \quad \sum_{n=0}^{\infty} a_n x^n = 2 \sum_{n=0}^{\infty} 3^n x^n - \sum_{n=0}^{\infty} x^n.$$

Hence $a_n = 2 \,.\, 3^n - 1$, which is the required solution.

(*ii*) Let $G(x) = \sum_{n=0}^{\infty} a_n x^n$, where G(*x*) is the generating function for the sequence $\{a_n\}$.

Multiplying each term in the given recurrence relation by x^n and summing from 2 to ∞, we get

$$\sum_{n=2}^{\infty} a_n x^n - 9 \sum_{n=2}^{\infty} a_{n-1} x^n + 20 \sum_{n=2}^{\infty} a_{n-2} x^n = 0$$

or $$[G(x) - a_0 - a_1 x] - 9x\, [G(x) - a_0] + 20x^2\, G(x) = 0$$

$$G(x)\, [1 - 9x + 20x^2] = a_0 + a_1 x + 9a_0 x$$

$$G(x) = \frac{a_0 + a_1 x - 9a_0 x}{1 - 9x + 20x^2}$$

$$= \frac{-3 - 10x + 27x}{1 - 9x + 20x^2}$$

$$= \frac{-3 + 17x}{(1-5x)(1-4x)} \qquad (\because \; a_1 = -3 \text{ and } a_1 = -10)$$

or $$G(x) = \frac{2}{1-5x} - \frac{5}{1-4x} \qquad \text{(by partial fraction)}$$

$$\therefore \quad \sum_{n=0}^{\infty} a_n x^n = 2\,\Sigma 5^n x^n - 5\,\Sigma 4^n x^n$$

Hence $a_n = 2.5^n - 5.4^n$, which is the required solution.

(*iii*) Let $G(x) = \sum_{n=0}^{\infty} a_n x^n$, where G($x$) is the generating function for the sequence $\{a_n\}$.

Multiplying each term in the given recurrence relation by x^n and summing from 1 to ∞, we get

$$\sum_{n=0}^{\infty} a_{n+2}x^n - 2\sum_{n=0}^{\infty} a_{n+1}x^n + \sum_{n=0}^{\infty} x^n = \sum_{n=0}^{\infty} 2^n x_n$$

or $$\frac{G(x) - a_0 - a_1 x}{x^2} - 2\left(\frac{G(x) - a_0}{x}\right) + G(x) = \frac{1}{1-2x}$$

$$\frac{G(x) - 2 - x}{x^2} - 2\left(\frac{G(x) - 2}{x}\right) + G(x) = \frac{1}{1-2x}$$

$$(x^2 - 2x + 1)\,G(x) = 2 + 3x + \frac{x^2}{1-2x}$$

$$G(x) = \frac{2}{(1-x)^2} + \frac{3x}{(1-x)^2} + \frac{x^2}{(1-2x)(1-x)^2}$$

By partial fraction, $$\frac{x^2}{(1-2x)(1-x^2)} = \frac{1}{1-2x} - \frac{1}{(1-x)^2}$$

$$G(x) = \frac{1}{(1-x)^2} + \frac{3x}{(1-x)^2} + \frac{1}{1-2x}$$

$$\Sigma a_n x^n = \Sigma(n+1)x^n + 3\Sigma n x^n + \Sigma 2^n x^n$$

Hence $$a_n = (n+1) + 3n + 2^n$$
$$= 1 + 4n + 2^n.$$

Problem 4.66. *Find the sequence $\{y_x\}$ having the generating function G given by*

$$G(x) = \frac{3}{1-x} + \frac{1}{1-2x}.$$

Solution. If the generating function

$$G(x) = \frac{1}{1-x}, \text{ the sequence is } \{y_x\} = \{1\}$$

Therefore, if the generating function is

$$G(x) = \frac{3}{1-x}, \text{ the sequence is } \{y_n\} = \{3\}$$

Also, if the generating function is

$$G(x) = \frac{1}{1-ax}, \text{ the sequence is } \{y_n\} = \{a^n\}$$

Therefore, if the generating function is

$$G(x) = \frac{1}{1-2x}, \text{ the sequence is } \{y_n\} = \{2^n\}$$

Hence the required sequence is $\{y_n\} = \{3 + 2^n\}$

Second Method :

We have $$G(x) = \frac{3}{1-x} + \frac{1}{1-2x}$$

$$= 3(1-x)^{-1} + (1-2x)^{-1}$$

$$= 3(1 + x + x^2 + \ldots\ldots + x^n + \ldots\ldots) + (1 + 2x + 2^2x^2 + \ldots\ldots + 2^nx^n + \ldots\ldots)$$

$$= (3 + 1) + (3 + 2)x + (3 + 2^2) + \ldots\ldots + (3 + 2^n)\,x^n + \ldots\ldots$$

$$= \sum_{n=0}^{\infty} (3 + 2^n)\,x^n$$

$$= \sum_{n=0}^{\infty} y_n x^n, \text{ where } y_n = 3 + 2^n$$

Therefore, the required sequence is $\{y_n\} = \{3 + 2^n\}$.

Problem 4.67. *Find the generating function which will give the number of solutions in integers of $x + y + z = 5$ if*

(i) each variable is non negative.

(ii) each variable is non negative and atmost 3.

(iii) each variable is at least 2 and at most 4.

(iv) $0 \le x \le 5$, $2 \le y \le 6$, $5 \le z \le 8$, x is even and y is odd.

Solution. (*i*) The number of solutions is the coefficient of x^5 in the generating function

$$(1 + x + x^2 + x^3 + x^4 + x^5)^3$$

(*ii*) The number of solution is the coefficient of x^5 in the generating function $(1 + x + x^2 + x^3)$

(*iii*) The number of solution is the coefficient of x^5 in the generating function $(x^2 + x^3 + x^4)^3$

(*iv*) The number of solution is the coefficient of x^5 in the generating function

$$(1 + x^2 + x^4)(x^3 + x^5)(x^5 + x^6 + x^7 + x^8).$$

Problem 4.68. *Use generating functions to find the number of ways to select r objects of n different kinds if at least one object of each kind is selected.*

Solution. Let us consider the sequence $\{a_n\}$ where a_n denotes the number of ways to select r objects of n different kinds which includes at least one object of each kind.

Let $G(x) = \Sigma a_r x^r$ be the generating function. Each of the n kind objects constitutes the factor

$$(x + x^2 + \ldots\ldots)$$

Now, $(x + x^2 + x^3 + \ldots\ldots)^n = x^n (1 + x + x^2 + \ldots\ldots)^n$

$$= x^n (1 - x)^{-n} = x^n \Sigma C(n + r - 1, r)x^r$$

$$= \Sigma C(n + r - 1, r)x^{n+r}$$

$$= \Sigma C(z - 1, z - n)x^z$$

$$= \Sigma C(r - 1, r - n)x^r$$

Hence the number of ways to select r objects of n different kinds which includes atleast one object of each kind is $C(r - 1, r - n)$.

Problem 4.69. *Solve the recurrence relation $a_k = 3a_{k-1}$ for $k = 1, 2, 3, \ldots\ldots$ and initial condition $a_n = 2$.*

Solution. Let G(x) be the Generating Function for the sequence $\{a_k\}$, that is,

$$G(x) = \sum_{k=0}^{\infty} a_k x^k.$$

First note that $\quad xG(x) = \sum_{k=0}^{\infty} a_k x^{k+1} = \sum_{k=1}^{\infty} a_{k-1} x^k$

Using the recurrence relation, we see that

$$G(x) - 3x\,G(x) = \sum_{k=0}^{\infty} a_k x^k - 3 \sum_{k=1}^{\infty} a_{k-1} x^k$$

$$= a_0 + \sum_{k=1}^{\infty} (a_k - 3a_{k-1})x^k$$

$$= 2,$$

Since $a_0 = 2$ and $a_k = 3a_{k-1}$.

Thus $\quad G(x) - 3xG(x) = (1 - 3x)\,G(x) = 2.$

Solving for G(x) shows that $G(x) = \dfrac{2}{(1-3x)}$.

Using the identity $\dfrac{1}{(1-ax)} = \sum\limits_{k=0}^{\infty} a^k x^k$

We have $\quad G(x) = 2 \sum\limits_{k=0}^{\infty} 3^k a^k = \sum\limits_{k=0}^{\infty} 2 . 3^k x^k$.

Consequently, $a_k = 2.3^k$.

Problem 4.70. *Suppose that a valid codeword is an n-digit number in decimal notation containing an even number of 0s. Let a_n denote the number of valid codewords to length n. The sequence $\{a_n\}$ satisfies the recurrence relation.*

$a_n = 8a_{n-1} + 10^{n-1}$ and the initial condition $a_1 = 9$. Use the generating functions do find an explicit formula for a_n.

Solution. To make our work with generating functions simpler, we extend this sequence by setting $a_0 = 1$; when we assign this value to a_0 and use the recurrence relation.

We have $a_1 = 8a_0 + 10^0 = 8 + 1 = 9$, which is consistent with our original initial condition.

(It also makes sense because there is one codeword of length 0-the emply string.)

We multiply both sides of the recurrence relation by x^n to btain $\quad a_n x^n = 8a_{n-1}x^n + 10^{n-1}x^n$.

Let $G(x) = \sum\limits_{n=0}^{\infty} a_n x^n$ be the generating function of the sequence $a_0, a_1, a_2, \ldots\ldots$

We sum both sides of the last equation starting with $n = 1$, to find that

$$G(x) - 1 = \sum_{n=1}^{\infty} a_n x^n = \sum_{n=1}^{\infty} (8a_{n-1}x^n + 10^{n-1}x^n)$$

$$= 8 \sum_{n=1}^{\infty} a_{n-1}x^n + \sum_{n=1}^{\infty} 10^{n-1}x^n$$

$$= 8x \sum_{n=1}^{\infty} a_{n-1}x^{n-1} + x \sum_{n=1}^{\infty} 10^{n-1}x^{n-1}$$

$$= 8x \sum_{n=0}^{\infty} a_n x^n + x \sum_{n=0}^{\infty} 10^n x^n$$

$$= 8x\, G(x) + \frac{x}{(1-10x)}$$

Therefore, we have

$$G(x) - 1 = 8x\, G(x) + \frac{x}{(1-10x)}$$

Solving for G(x) shows that

$$G(x) = \frac{1-9x}{(1-8x)(1-10x)}$$

Expanding the right-hand side of this equation into partial fractions (as is done in the integration of rational functions studied in calculus) gives

$$G(x) = \frac{1}{2}\left(\frac{1}{1-8x} + \frac{1}{1-10x}\right).$$

Using a, with $a = 8$ and $a = 10$, once, which gives

$$G(x) = \frac{1}{2}\left(\sum_{n=0}^{\infty} 8^n x^n + \sum_{n=0}^{\infty} 10^n x^n\right)$$

$$= \sum_{n=0}^{\infty} \frac{1}{2}(8^n + 10^n)x^n .$$

Consequently, we have shown that

$$a_n = \frac{1}{2}\ (8^n + 10^n).$$

Problem 4.71. *Use generating functions to show that*

$$\sum_{k=0}^{n} C(n, k)^2 = C(2n, n)$$

Whenever n is a positive integer.

Solution. First note that by the Binomial Theorem C($2n$, n) is the coefficient of x^n in $(1 + x)^{2n}$. However, we also have

$$(1 + x)^{2n} = [(1 + x)^n]^2$$

$$= [C(n, 0) + C(n, 1)x + C(n, 2)x^2 + \ldots\ldots + C(n, n)x^n]^2$$

The coefficient of x^n in this expression is

$$C(n, 0)C(n, n) + C(n, 1)\ C(n, n-1) + C(n, 2)\ C(n, n-2) + \ldots\ldots + C(n, n)\ C(n, 0).$$

This equals $\sum_{k=0}^{n} C(n, k)^2$, since $C(n, n-k) = C(n, k)$

Since both C($2n$, n) and $\sum_{k=0}^{n} C(n, k)^2$ represent the coefficient of x^n in $(1 + x)^{2n}$, they must be equal.

(Table of equivalent)

Expression for Generating Functions

If $\quad A(x) = \sum_{n=0}^{\infty} a_n x^n$, then

$$\sum_{n=k}^{\infty} a_n x^n = A(x) - a_0 - a_1 x - \ldots\ldots - a_{k-1}x^{k-1}$$

$$\sum_{n=k}^{\infty} a_{n-1} x^n = x(A(x) - a_0 - a_1 x - \ldots\ldots\ldots - a_{k-2}x^{k-2})$$

$$\sum_{n=k}^{\infty} a_{n-2} x^n = x^2(A(x) - a_0 - a_1 x - \ldots\ldots - a_{k-3}x^{k-3})$$

$$\sum_{n=k}^{\infty} a_{n-3} x^n = x^3(A(x) - a_0 - a_1 x - \ldots\ldots - a_{k-4}x^{k-4})$$

– – – – – – – –

$$\sum_{n=k}^{\infty} a_{n-k} x^n = x^k(A(x)).$$

Theorem 4.5. *If $\{a_n\}_{n=0}^{\infty}$ is a sequence of numbers which satisfy the linear recurrence relation with constant coefficients.*

$a_n + C_1 a_{n-1} + \ldots\ldots + C_k a_{n-k} = 0$ *where $C_k \neq 0$, and $n \geq k$, then the generating function $A(x)$ $= \sum_{n=0}^{\infty} a_n x^n$ equals $\dfrac{P(x)}{Q(x)}$, where $P(x) = a_0 + (a_1 + C_1 a_0)x + \ldots\ldots + (a_{k-1} + C_1 a_{k-2} + \ldots\ldots$ $+ C_{k-1} a_0)x^{k-1}$ and $Q(x) = 1 + C_1 x + \ldots\ldots + C_k x^k$.*

Conversely, given such polynomials P(x) and Q(x), where P(x) has degree less than k, and Q(x) has degree k, there is a sequence $\{a_n\}_{n=0}^{\infty}$ whose generating function is $A(x) = \dfrac{P(x)}{Q(x)}$.

Table of Generating Functions

Sequence a_n	*Generating Function A(x)*
1. $C(k, n)$	$(1 + x)^k$
2. 1	$\dfrac{1}{1 - x}$
3. a^n	$\dfrac{1}{1 - ax}$

Contd.

4. $(-1)^n$	$\frac{1}{1+x}$
5. $(-1)^n a^n = (-a)^n$	$\frac{1}{1+ax}$
6. $C(k-1+n, n)$	$\frac{1}{(1-x)^k}$
7. $C(k-1+n, n)a^n$	$\frac{1}{(1-ax)^k}$
8. $C(k-1+n, n)(-a)^n$	$\frac{1}{(1+ax)^k}$
9. $n+1$	$\frac{1}{(1-x)^2}$
10. n	$\frac{1}{(1-x)^2}$
11. $(n+2)(n+1)$	$\frac{2}{(1-x)^3}$
12. $(n+1)(n)$	$\frac{2x}{(1-x)^3}$
13. n^2	$\frac{x(1+x)}{(1-x)^3}$
14. $(n+3)(n+2)(n+1)$	$\frac{6}{(1-x)^4}$
15. $(n+2)(n+1)(n)$	$\frac{6x}{(1-x)^4}$
16. n^3	$\frac{x(1+4x+x^2)}{(1-x)^2}$
17. $(n+1)a^n$	$\frac{1}{(1-ax)^2}$
18. na^n	$\frac{ax}{(1-ax)^2}$
19. n^2a^n	$\frac{(ax)(1+ax)}{(1-ax)^3}$
20. n^3a^n	$\frac{(ax)(1+4ax+a^2x^2)}{(1-ax)^4}$

Problem 4.72. *Solve the recurrence relation $a_n - 7a_{n-1} + 10a_{n-2} = 0$ for $n \geq 2$.*

Solution. Let $A(x) = \sum_{n=0}^{\infty} a_n x^n$

Next multiply each term in the recurrence relation by x^n and sum from 2 to ∞.

$$\sum_{n=2}^{\infty} a_n x^n - 7 \sum_{n=2}^{\infty} a_{n-1} x^n + 10 \sum_{n=2}^{\infty} a_{n-2} x^n = 0$$

Replace each infinite sum by an expression from the table of equivalent expressions :

$$[A(x) - a_0 - a_1 x] - 7x [A(x) - a_0] + 10x^2[A(x)] = 0$$

Then simplify

$$A(x)(1 - 7x + 10x^2) = a_0 + a_1 x - 7a_0 x$$

or

$$A(x) = \frac{a_0 + (a_1 - 7a_0)x}{1 - 7x + 10x^2} = \frac{a_0 + (a_1 - 7a_0)x}{(1 - 2x)(1 - 5x)}$$

Decompose A(x) as a sum of partial fractions :

$$A(x) = \frac{C_1}{1 - 2x} + \frac{C_2}{1 - 5x}$$

where C_1 and C_2 are constants, as yet undetermined Express A(x) as a sum of familiar series :

$$A(x) = \frac{C_1}{1 - 2x} + \frac{C_2}{1 - 5x} = C_1 \sum_{n=0}^{\infty} 2^n x^n + C_2 \sum_{n=0}^{\infty} 5^n x^n .$$

Express a_n as coefficient of x^n in A(x) and in the sum of the other series :

$$a_n = C_1 2^n + C_2 5^n$$

Now the constants C_1 and C_2 are uniquely determined once values for a_0 and a_1 are given.

For example, if $a_0 = 10$, and $a_1 = 41$, we may use the form of the general solution $a_n = C_1 2^n + C_2 5^n$, and left $n = 0$ and $n = 1$ to obtain the two equations.

$C_1 + C_2 = 10$ and $2C_1 + 5C_2 = 41$ which determine the values $C_1 = 3$ and $C_2 = 7$. Thus, in this case the unique solution of the recurrence relation is

$$a_n = (3)2^n + (7)5^n.$$

Problem 4.73. *Solve the recurrence relation*

$$a_n - 9a_{n-1} + 26a_{n-2} - 24a_{n-3} = 0 \text{ for } n \geq 3.$$

Solution. Let $A(x) = \sum_{n=0}^{\infty} a_n x^n$. Then multiply by x^n and sum from 3 to ∞ since $n \geq 3$.

Thus $$\sum_{n=3}^{\infty} a_n x^n - 9 \sum_{n=3}^{\infty} a_{n-1} x^n + 26 \sum_{n=3}^{\infty} a_{n-2} x^n - 24 \sum_{n=3}^{\infty} a_{n-3} x^n = 0$$

Replace the infinite sums by equivalent expressions

$$(A(x) - a_0 - a_1x - a_2x^2) - 9x(A(x) - a_0 - a_1x) + 26x^2(A(x) - a_0) - 24x^3A(x) = 0$$

Simplify, $A(x)(1 - 9x + 26x^2 - 24x^3) = a_0 + a_1x + a_2x^2 - 9a_0x - 9a_1x^2 + 26a_0x^2$

or

$$A(x) = \frac{a_0 + (a_1 - 9a_0)x + (a_2 - 9a_1 + 26a_0)x^2}{1 - 9x + 26x^2 - 24x^3}$$

Now $1 - 9x + 26x^2 - 24x^3 = (1 - 2x)(1 - 3x)(1 - 4x)$ so that there are constants C_1, C_2, C_3 such that

$$A(x) = \frac{C_1}{1-2x} + \frac{C_2}{1-3x} + \frac{C_3}{1-4x}$$

$$= C_1 \sum_{n=0}^{\infty} 2^n x^n + C_2 \sum_{n=0}^{\infty} 3^n x^n + C_3 \sum_{n=0}^{\infty} 4^n x^n$$

$$= \sum_{n=0}^{\infty} (C_1 2^n + C_2 3^n + C_3 4^n) x^n$$

Thus, $a_n = C_12^n + C_23^n + C_34^n$ and C_1, C_2 and C_3 can be determined once the initial conditions for a_0, a_1 and a_2 are specified.

Assume that the initial conditions are $a_0 = 0$, $a_1 = 1$ and $a_2 = 10$

Then $A(x) = \dfrac{x + x^2}{(1-2x)(1-3x)(1-4x)} = \dfrac{C_1}{(1-2x)} + \dfrac{C_2}{(1-3x)} + \dfrac{C_3}{(1-4x)}$

and $C_1(1 - 3x)(1 - 4x) + C_2(1 - 2x)(1 - 4x) + C_3(1 - 2x)(1 - 3x) = x + x^2$

Let $x = \frac{1}{2}, \frac{1}{3}, \frac{1}{4}$ and find that $C_1 = \frac{3}{2}$, $C_2 = -4$ and $C_3 = \frac{5}{2}$

Thus, in this case, $a_n = \frac{3}{2}(2^n) - 4(3^n) + \frac{5}{2}(4^n)$.

Problem 4.74. *Solve* $a_n - 6a_{n-1} + 12a_{n-2} - 8_{n-3} = 0$ *by generating functions.*

Solution. $A(x) = \sum_{n=0}^{\infty} a_nx^n = \dfrac{P(x)}{Q(x)}$

$$= \frac{a_0 + (a_1 - 6a_0)x + (a_2 - 6a_1 12a_0)x^2}{1 - 6x + 12x^2 - 8x^3}$$

But since $1 - 6x + 12x^2 - 8x^3 = (1 - 2x)^3$

We use partial fractions to calculate that there are constants C_1, C_2, C_3 such that

$$A(x) = \frac{C_1}{(1-2x)} + \frac{C_2}{(1-2x)^2} + \frac{C_3}{(1-2x)^3}$$

$$= C_1 \sum_{n=0}^{\infty} 2^n x^n + C_2 \sum_{n=0}^{\infty} \binom{n+1}{n} 2^n x^n + C_3 \sum_{n=0}^{\infty} \binom{n+2}{n} 2^n x^n$$

$$= \sum_{n=0}^{\infty} \left[C_1 2^n + C_2 (n+1) 2^n + C_3 \frac{(n+2)(n+1)}{2} 2^n \right] x^n$$

so that $$a_n = C_1 2^n + C_2 (n+1) 2^n + C_3 \frac{(n+2)(n+1)}{2} 2^n.$$

Problem 4.75. *Solve* $a_n - 8a_{n-1} + 21a_{n-2} - 18a_{n-3} = 0$ *for* $n \geq 3$.

Solution. Here, if $A(x) = \sum_{n=0}^{\infty} a_n x^n$, then

$$\sum_{n=3}^{\infty} a_n x^n - 8 \sum_{n=3}^{\infty} a_{n-1} x^n + 21 \sum_{n=3}^{\infty} a_{n-2} x^n - 18 \sum_{n=3}^{\infty} a_{n-3} x^n = 0,$$

$$(A(x) - a_0 - a_1 x - a_2 x^2) - 8x(A(x) - a_0 - a_1 x) + 21x^2(A(x) - a_0) - 18x^2 A(x) = 0$$

or $$A(x) = \frac{a_0 + (a_1 - 8a_0)x + (a_2 - 8a_1 + 21a_0)x^n}{1 - 8x + 21x^2 - 18x^3}$$

But since $1 - 8x + 21x^2 - 18x^3 = (1 - 2x)(1 - 3x)^2$

We see that there are constants C_1, C_2, C_3 such that

$$A(x) = \frac{C_1}{(1-2x)} + \frac{C_2}{(1-3x)} + \frac{C_3}{(1-3x)^2}$$

Then, $$A(x) = \sum_{n=0}^{\infty} [C_1 2^n + C_2 3^n + C_3 n^3 \, C(n+1, n)] \, x^n$$

or $$a_n = C_1 2^n + C_2 3^n + C_3 (n+1) 3^n.$$

Problem Set 4.1

1. Find the first four terms of the sequence defined by each of the recurrence relations and initial conditions

(*a*) $a_n = 2a_{n-1} + n, \quad n \geq 2, a_1 = 1$

(*b*) $a_n = a_{n-1}^2, \quad n \geq 2, a_1 = 12$

(*c*) $a_n = n(a_{n-1})^2, \quad n \geq 1, a_0 = 1$

(*d*) $b_k = b_{k-1} + 2b_{k-2}, \quad k \geq 2, b_0 = 1, b_1 = 1$

(*e*) $b_k = kb_{k-1} + k^2 b_{k-2}, \quad k \geq 2, b_0 = 1, b_1 = 1.$

2. Show that the sequence 2, 3, 4, 5, ..., $2n + 1$, ... for $n \geq 0$, satisfies the recurrence relation $a_k = 2a_{k-1} - a_{k-2}$, $k \geq 2$.

3. Show that the sequence 0, 1, 3, 7, ..., $2^m - 1$, ... for $n \geq 0$, satisfies the recurrence relation $a_k = 3a_{k-1} - 2a_{k-2}$, $k \geq 2$.

4. Show that $a_n = C_1 + C_2 2^n - n$ is a solution of the recurrence relation $a_n - 3a_{n-1} + 2a_{n-2} = 1$.

5. Show that $a_n = 0$, $a_n = 4^n$, $a_n = n4^n$ and $a_n = 2.4^n + 3n4^n$ are all solutions of the same recurrence relation

$$a_n = 8a_{n-1} - 16a_{n-2}.$$

6. Show that $a_n = C_1 2^n + C_2 4^n$ is a solution of the recurrence relation $a_n - 6a_{n-1} + 8a_{n-2} = 0$.

7. Show that $a_n = (C_1 + C_2 n) . 2^n$ is a solution of the recurrence relation $a_n - 4a_{n-1} + 4a_{n-2} = 0$.

8. Prove that if $S_n = aS_{n-1}$ for $n \geq 1$ and $a \neq 0$ then $S_n = a^n S_0$.

9. Use the iterative method to find the solution to each of the following recurrence relations and initial condtions.

(*a*) $a_n = a_{n-1} + 2$, $a_0 = 1$ (*b*) $a_n = 2a_{n-1} + 1$, $a_1 = 7$

(*c*) $a_n = a_{n-1} + n$, $a_1 = 4$ (*d*) $a_n = na_{n-1}$, $a_0 = 5$.

10. Solve the recurrence relations together with initial conditions given

(*a*) $a_n = a_{n-1} + 6a_{n-2} : n \geq 2$, $a_0 = 1$, $a_1 = 1$

(*b*) $a_n = 3a_{n-1} - 3a_{n-2} + a_{n-3}$, $a_0 = 0$, $a_3 = 3$, $a_5 = 10$

(*c*) $a_n = 7a_{n-1} - 10a_{n-2}$, $a_0 = 4$, $a_1 = 17$

(*d*) $a_n - 8a_{n-1} + 16a_{n-2} = 0$, $a_2 = 16$, $a_3 = 80$

(*e*) $a_n = a_{n-1} + 2a_{n-2}$, $a_0 = 2$, $a_1 = 7$.

11. Solve each of the following recurrence relations that satisfies the initial conditions.

(*a*) $a_n - a_{n-1} = n$, $a_0 = 2$

(*b*) $a_n - 2a_{n-1} = 6n$, $a_1 = 2$

(*c*) $a_n - a_{n-1} = 6_{n+1}$, $a_0 = 1$

(*d*) $a_n - a_{n-1} = 3^n$, $a_0 = 1$

(*e*) $a_{n+1} - 2a_n = 3 + 4^n$, $a_0 = 2$

(*f*) $a_{n+2} + 2a_{n+1} - 15a_n = 6n + 10$, $a_r = 1$, $a_1 = -\dfrac{1}{2}$

(*g*) $a_n - 4a_{n-1} + 3a_{n-2} = 2^n$, $a_1 = 1$, $a_2 = 11$.

12. Given that $a_0 = 0$, $a_1 = 1$, $a_2 = 4$ and $a_3 = 12$ satisfy the recurrence relation $a_n + C_1 a_{n-1} + C_2 a_{n-2} = 0$, determine a_n.

13. Solve the following recurrence equations (relations) :

(*a*) $a_r - 7a_{r-1} + 10a_{r-2} = 0$ with $a_0 = 0$ and $a_1 = 3$

(*b*) $a_r - 4a_{r-1} + 4a_{r-2} = 0$ with $a_0 = 1$ and $a_1 = 6$

(*c*) $a_r - 7a_{r-1} + 10a_{r-2} = 3^r$ with $a_0 = 0$ and $a_1 = 1$

(*d*) $a_r + 6a_{r-1} + 9a_{r-2} = 3$ with $a_0 = 0$ and $a_1 = 1$

(*e*) $a_r = a_{r-1} + a_{r-2} = 0$ with $a_0 = 0$, $a_1 = 2$

(*f*) $a_r - a_{r-1} - a_{r-2} = 0$ with $a_0 = 1, a_1 = 1$

(*g*) $a_r - 2a_{r-1} + 2a_{r-2} - a_{r-3} = 0$ with $a_0 = 2, a_1 = 1$ and $a_2 = 1$.

14. Use the backtracking method to find an explicit formula for the sequence defined by the recurrence equation and initial conditions.

(*a*) $a_n = 2.5 * a_{n-1}$ with $a_1 = 4$ (*b*) $b_n = b_{n-1} - 2$ with $b_1 = 0$

(*c*) $C_n = C_{n-1} + n$ with $C_1 = 4$ (*d*) $d_n = -1.1 * d_{n-1}$ with $d_1 = 5$

(*e*) $e_n = 5e_{n-1} + 3$ with $e_1 = 2$ (*f*) $g_n = ng_{n-1}$ with $g_1 = 6$.

15. Solve each of the recurrence equations :

(*a*) $a_n = 4a_{n-1} + 5a_{n-2}$ with $a_1 = 2$ and $a_2 = 6$

(*b*) $b_n = -3b_{n-1} - 2b_{n-2}$ with $b_1 = -2$ and $b_2 = 4$

(*c*) $C_n = -6C_{n-1} - 9C_{n-2}$ with $C_1 = 2.5$ and $C_2 = 4.7$

(*d*) $d_n = 4d_{n-1} - 4d_{n-2}$ with $d_1 = 1$ and $d_2 = 7$

(*e*) $e_n = 2e_{n-2}$ with $e_1 = \sqrt{2}$ and $e_2 = 6$

(*f*) $g_n = 2g_{n-1} - 2g_{n-2}$ with $g_1 = 1$ and $g_2 = 4$.

16. Solve the difference (recurrence) equation

$$a_r - ra_{r-1} = r! \quad \text{for } r \geq 1 \text{ and given that } a_0 = 2.$$

(**Hint :** Use change of variable method. Let $b_r = a_r/r!$)

17. Solve the difference equation

$$a_r^2 - 2a_{r-1}^2 = 1 \quad \text{for } r \geq 1 \text{ and given that } a_0 = 2.$$

(**Hint :** Use change of variable method. Let $b_r = a_r^2$)

18. Solve the difference equation

$$ra_r + ra_{r-1} - a_{r-1} = 2^r \quad \text{for } r \geq 1 \text{ and given that } a_0 = 273$$

(**Hint :** Use change of variable method. Let $b_r = ra_r$)

19. Find the closed formula for the Fibonacci sequence

$$f_n = f_{n-1} + f_{n-2}$$

20. Given that $a_0 = 0, a_1 = 1, a_2 = 4$ and $a_3 = 12$ satisfy the recurrene equation $a_r + C_1a_{r-1} + C_2a_{r-2} = 0$ and determine a_r.

21. Determine the particular solution for the difference equations

(*a*) $a_n - 3a_{n-1} + 2a_{n-2} = 2^n$ (*b*) $a_n - 4a_{n-1} + 4a_{n-2} = 2^n$

(*c*) $a_n - 2a_{n-1} = 7n^2$ (*d*) $a_n - 2a_{n-1} = 7^n$.

22. Solve the recurrence relation $a_n = a_{n-1} + 6a_{n-2}, n \geq 2$, given $a_0 = 1, a_1 = 3$.

23. Solve the recurrence relation $a_n = 6a_{n-1} - 9a_{n-2}, n \geq 2$ given $a_0 = -5, a_1 = 3$.

24. Solve the recurrence relation $a_n = -8a_{n-1} - a_{n-2}, n \geq 2$ given $a_0 = 0, a_1 = 1$.

25. Solve the recurrence relation $a_{n+1} = 2a_n + 3a_{n-1}, n \geq 1$ given $a_0 = 0, a_1 = 8$.

26. (*a*) Solve the recurrence relation $a_n = -2a_{n-1} + 15a_{n-2}, n \geq 2$ given $a_0 = 1, a_1 = -1$.

(*b*) Solve the recurrence relation $a_n = -2a_{n-1} + 15a_{n-2} + 24, n \geq 2$ given $a_0 = 1, a_1 = -1$.

27. (*a*) Solve the recurrence relation $a_n = 4a_{n-1}, n \geq 1$ given $a_0 = 1$.

(*b*) Solve the recurrence relation $a_n = 4a_{n-1} + 8^n, n \geq 1$, given $a_0 = 1$.

28. (*a*) Solve the recurrence relation $a_n = 4a_{n-1} - 9, n \geq 1$ given $a_0 = 4$.

(*b*) Solve the recurrence relation $a_n = 4a_{n-1} + 3n2^n, n \geq 1$, given $a_0 = 4$.

29. Solve the recurrence relation $a_n = 5a_{n-1} - 2a_{n-2} + 3n^2, n \geq 2$ given $a_0 = 0, a_1 = 3$.

30. Solve the recurrence relation $a_n = -6a_{n-1} - 9a_{n-2}, n \geq 2$ given $a_0 = 1, a_1 = -4$.

31. (*a*) Solve the recurrence relation $a_n = 5a_{n-1} - 6a_{n-2}, n \geq 2$ given $a_0 = 2, a_1 = 11$.

(*b*) Solve the recurrence relation $a_n = 5a_{n-1} - 6a_{n-2} + 3n, n \geq 2$, given $a_0 = 2, a_1 = 14$.

32. Solve the recurrence relation $a_n = 4a_{n-1} - 4a_{n-2} + n, n \geq 2$, given $a_0 = 5, a_1 = 9$.

33. Solve the following recurrence relation

(*a*) $a_{n+1} = -8a_n - 16a_{n-1}, n \geq 1$, given $a_0 = 5, a_1 = 17$

(*b*) $a_{n+1} = -8a_n - 16a_{n-1} + 5, n \geq 1$, given $a_0 = 2, a_1 = -1$

(*c*) $9a_n = 6a_{n-1} - a_{n-2}, n \geq 2$, given $a_0 = 3, a_1 = -1$

(*d*) $a_n = 2a_{n-1} - a_{n-2}, n \geq 2$, given $a_0 = 40, a_1 = 37$

(*e*) $a_n = -5a_{n-1} + 6a_{n-2}, n \geq 2$, given $a_0 = 5, a_1 = 19$

(*f*) $a_{n+1} = 7a_n - 10a_{n-1}, n \geq 2$, given $a_1 = 10, a_2 = 29$

(*g*) $a_n = -6a_{n-1} + 7a_{n-2}, n \geq 2$, given $a_0 = 32, a_1 = -17$.

34. Find a recurrence relation for the balance B(*k*) owed at the end of *k* months on a loan of \$5000 at a rate of 7% if a payment of \$100 is made each month.

[**Hint :** Express B(*k*) in terms of B(*k* – 1), the monthly interest is (0.07/12 B(*k* – 1)]

35. (*a*) Find a recurrence relation for the balance B(*k*) owed at the end of *k* months on a loan at a rate of *r* if a payment P is made on the loan each month.

[**Hint.** Express B(*k*) in terms of B(*k* – 1) and note that the monthly interest rate is *r*/12.]

(*b*) Determine what the monthly payment P should be so that the loan is paid off after T months.

36. (*a*) Find a recurrence relation for the number of bit strings of length *n* that contain three of consecutive 0's.

(*b*) What are the initial conditions ?

(*c*) How many bit strings of length seven contain three consecutive 0'*s* ?

37. (*a*) Find a recurrence relation for the number of bit strings of length *n* that contain a pair ofi consecutive 0's.

(*b*) What are the initial conditions ?

(*c*) How many bit strings of length seven conatin two consecutive 0's ?

38. (*a*) Find a recurrence relation for the number of bit strings of length *n* that do not contain three consecutive 0's.

(*b*) What are the initial conditions ?

(*c*) How many bit strings of length seven do not contain three consecutive 0's ?

39. (*a*) Find a recurrence relation for the number of ways to climb n stairs if the person climbing the stairs can take one stair or two stairs at a time.

(*b*) What are the initial conditions ?

(*c*) How many ways can this person climb a flight of eight stairs ?

Problem Set 4.2

1. Apply the generating function technique to solve the following recurrence relation with given initial conditions

(*a*) $a_{n+1} - (n+1)a_n = 0$, $a_0 = 1$ (*b*) $a_n - 4a_{n-2} = 0$, $a_0 = 0$, $a_1 = 1$

(*c*) $a_{n+2} - 3a_{n+1} + 2a_n = 0$, $a_0 = 2$, $a_1 = 3$ (*d*) $a_{n+2} - 5a_{n+1} + 6a_n = 2$, $a_0 = 1$, $a_1 = 2$

(*e*) $a_{n+2} + 4a_{n+1} + 4a_n = 0$, $a_0 = 1$, $a_1 = 0$ (*f*) $a_n - 7a_{n-1} + 10a_{n-2} = 0$, $a_0 = 3$, $a_1 = 0$.

2. Apply the generating function technique to solve the following recurrence relations :

(*a*) $a_{n+2} + 2a_{n+1} + a_n = 1 + n$

(*b*) $a_{n+2} - 2a_{n+1} + a_n = 2^n$

(*c*) $a_n = 4(a_{n-1} - a_{n-2}) + 2^n$

(*d*) $a_n = 3a_{n-1} - 4n + 3(2^n)$

(*e*) $a_n = 6a_{n-1} - 12a_{n-2} + 8a_{n-2} + 3^n$.

3. Solve the following recurrence equations using generating function method :

(*a*) $A_n = A_{n-1} + A_{n-2} + 2n$ given that $A_0 = 0$ and $A_1 = 1$

(*b*) $a_n + a_{n-1} = 3n2^n$ given that $a_0 = 0$

(*c*) $a_n = 5a_{n-1} - 3$ given that $a_0 = 1$

(*d*) $h_n = 4h_{n-2}$ given that $h_0 = 0$ and $h_1 = 1$

(*e*) $a_n = \begin{cases} 0 & \text{for } n = 0 \\ 2a_{n-1} + 3 & \text{for } n > 0 \end{cases}$

(*f*) $a_n = \begin{cases} 16 & \text{for } n = 2 \\ 60 & \text{for } n = 0 \\ 8a_{n-1} - 16a_{n-2} & \text{for } n > 0 \end{cases}$

(*g*) $a_n - 2a_{n-1} - 4a_{n-2} = 4^n$ given that $a_0 = 1$ and $a_1 = 7$

(*h*) $a_n - 5a_{n-1} + 6a_{n-2} = 2^n + n$ given that $a_1 = 0$ and $a_2 = 10$

(*i*) $3a_{n+2} - 8a_{n+1} - 3a_n = 3^n - 2n + 1$

(*j*) $a_{n+2} - 6a_{n+1} + 8a_n = 3n^2 + 2 - 5 * 3^n$.

4. Using the method of generating functions, sovle the recurrence relation $a_n = 2a_{n-1}$, $n \geq 1$, given $a_0 = 1$.

5. Use the method of generating function to sovle the recurrence relation $a_n = 5a_{n-1} - 6a_{n-2}$, given $a_0 = 1$, $a_1 = 4$.

6. Use the generating function to find a formula for a_n given $a_0 = 5$ and $a_n = a_{n-1} + 2^n$ for $n \geq 1$.
7. Use the method of generating function to sovle the recurrence relation $a_n = 3a_{n-1} + 1, n \geq 1$, given $a_0 = 1$.
8. Use the method of generating functions to solve the recurrence relation $a_n = 4a_{n-1} + 4a_{n-2}, n \geq 2$, given $a_0 = 1, a_1 = 4$.
9. Solve the recurrence relation, using generating function method, $a_n = 4a_{n-1} - 4a_{n-2} + 4^n, n \geq 2$, with the initial conditions $a_0 = 2, a_1 = 8$.
10. Solve each of the following using generating functions
    (a) $a_n = -5a_{n-1}, n \geq 1$, given $a_0 = 2$
    (b) $a_n = -5a_{n-1} + 3, n \geq 1$, given $a_0 = 2$.
    (c) $a_n = 4a_{n-1} - 3a_{n-2}, n \geq 2$, given $a_0 = 2, a_1 = 5$.
    (d) $a_n = -10a_{n-1} - 25a_{n-2}, n \geq 2$, given $a_0 = 1, a_1 = 25$
    (e) $a_n = 2a_{n-1} + a_{n-2} - 2a_{n-3}, n \geq 3$, given $a_0 = 1, a_1, = 3, a_2 = 6$.
    (f) $a_n = a_{n-1} + a_{n-2} - a_{n-3}, n \geq 3$, given $a_0 = 2, a_1 = -1, a_2 = 3$.
11. let a_n denote the number of n-digit numbers, each of whose digits is 1, 2, 3, or 4 and in which the number of 1's even
    (a) Find a recurrence relation for a_n (b) Find a_n explicit formula for a_n.
12. Let $a_n = r\ a_{n-1} + S\ a_{n-2} + f(n),\ n \geq 2$, be a second order recurrence relation with constant coefficients show that if P_n satisfies this recurrence relation for $n \geq 2$ and q_n satisfies the associated homogeneous recurrence relation $a_n = ra_{n-1} + Sa_{n-2}$ for $n \geq 2$, then $P_n + q_n$ satisfies the given relation for $n \geq 2$.
13. Find the general solutions to each of the following non-linear recurrence relation :
    (a) $a_n - 5a_{n-1} + 6a_{n-2} = 1$ (b) $a_n - 6a_{n-1} + 8a_{n-2} = 3$
    (c) $a_{n+2} - 2a_{n+1} + a_n = 3n + 5$ (d) $a_{n+2} - 5a_{n+1} + 6a_n = n^2$
    (e) $a_n + 4a_{n-1} + 4a_{n-2} = n^2 - 3n + 5$ (f) $2a_n - 7a_{n-1} + 3a_{n-2} = 2^n$
    (g) $a_n - 4a_{n-1} + 4a_{n-2} = (n + 1)\ 2^n$.
14. Show that $a_n = \left(\frac{1}{6}\right) 5^n$ is a solution of the homogeneous linear recurrence relation

    $a_{n+2} - 5a_{n+1} + 6a_n = 2^n$.
15. Show that $a_n = -2^{n+1}$ is a solution of the non homogeneous linear recurrence relation $a_n = 3a_{n-1} + 2^n$.
16. Solve the recurrence relation $a_n = 7a_{n-1}, n \geq 1$ and $a_2 = 98$.
17. Find a_{12} is $a_{n+1}^2 = 5a_n^2$, $a_n > 0$ for $n \geq 0$ and $a_0 = 2$.
18. A bank pays 6% (annual) interest on savings, compounding the interest monthly. In Bonnie deposits \$1000 on the first day to May, how much will this deposit be worth a year later ?
19. Solve the relation $a_n = n\ .\ a_{n-1}$, where $n \geq 1$ and $a_0 = 1$.
20. Solve the recurrence relation $a_n + a_{n-1} - 6a_{n-2} = 0$ where $n \geq 2$ and $a_0 = -1, a_1 = 8$.
21. Solve the recurrence relation $F_{n+2} = F_{n+1} + F_n$ where $n \geq 0$ and $F_0 = 0, F_1 = 1$.

22. For $n \geq 0$, let S = {1, 2, 3, ...} (when $n = 0$, S ≠ 0) and let a_n denote the number of subsets of S that contain no consecutive integers. Find and solve a recurrence relation for a_n.

23. Find a recurrence relation for the number of binary sequences of length n that have no consecutive 0's.

24. Solve the recurrence relation $2a_{n+3} = a_{n+2} + 2a_{n+1} - a_n$, $n \geq 0$, $a_0 = 0$, $a_1 = 1$, $a_2 = 2$.

25. Determine $(1 + \sqrt{3i})^{10}$.

26. Solve the recurrence relation $a_n = 2(a_{n-1} - a_{n-2})$, $n \geq 2$ and $a_0 = 1$, $a_1 = 2$.

27. Solve the recurrence relation $a_{n+2} = 4a_{n+1} - 4a_n$, $n \geq 0$ and $a_0 = 1$, $a_1 = 3$.

28. Solve the recurrence relation $a_n - a_{n-1} = 3n^2$, where $n \geq 1$ and $a_0 = 7$.

29. Solve and recurrence relation $a_n - 3a_{n-1} = 5(7^n)$ where $n \geq 1$ and $a_0 = 2$.

30. Solve the recurrence relation $a_n - 3a_{n-1} = 5(3^n)$ where $n \geq 1$, and $a_0 = 2$.

31. Pauline takes out a loan of S dollars that is to be paid back in T periods of time. If r is the interest rate per period for the loan, what (constant) payment P must she make at the end of each period ?

32. Solve the recurrence relation $a_{n+2} - 4a_{n+1} + 3a_n = -200$, $n \geq 0$, $a_0 = 3000$, $a_1 = 3300$.

33. Solve the recurrence relation $a_n - 3a_{n-1} = n$, $n \geq 1$, $a_0 = 1$.

34. Solve the recurrence relation $a_n = a_{n-1} + f(n)$ for $n \geq 1$ for substitution.

Answers 4.1

1. (*a*) 1, 4, 11, 26 (*b*) 2, 4, 16, 256 (*c*) 1, 1, 2, 12
(*d*) 1, 1, 3, 5 (*e*) 1, 1, 6, 27

9. (*a*) $1 + 2n$ (*b*) $2^{n+2} - 1$ (*c*) $3 + n(n+1)/2$
(*d*) $5n\,!$

10. (*a*) $a_n = 3^n$ (*b*) $n(n-1)/2$ (*c*) $a_n = 2^n + 3.5^n$
(*d*) $a_n = (2 + n)4^{n-1}$ (*d*) $a_n = 3.2^n - (-1)^n$

11. (*a*) $a_n = n(n+1)/2 + 2$ (*b*) $a_n = 10.2^n - 6n - 12$ (*c*) $a_n = (n+1)^2$
(*d*) $a_n = 3^{n+1} - 1/2$ (*e*) $a_n = (5/6)(-2)^n + 1 + 1/6 \,.\, 4^n$
(*f*) $a_n = 5/8\,(-5)^n + (11/8)3^n - 1/2\,n - 1$ (*g*) $a_n = 3^{n+1} - 2^{n+2}$.

12. $a_n = n \,.\, 2^{n-1}$ **22.** $a_n = 3^n$ **23.** $a_n = -5(3^n) + 6n(3^n)$

24. $a_n = \dfrac{1}{2\sqrt{15}}(-4 + \sqrt{5}\,)^n - \dfrac{1}{2\sqrt{15}}(4 - \sqrt{15}\,)^n$

25. $a_n = -2(-1)^n + 2(3^n) = 2[(-1)^{n+1} + 3^n]$

26. (*a*) $a_n = \dfrac{1}{2}(3^n) + \dfrac{1}{2}(-5)^n$ (*b*) $a_n = 2(3^n) + (-5)^n - 2$

27. (*a*) $a_n = 4^n$, (*b*) $a_n = 2(8^n) - 4^n$

28. (*a*) $a_n = 4^n + 3$ (*b*) $a_n = 10(4^n) - (3n + 6)2^n$

34. $B(k) = (1 + (0.07/12))\, B(k-1) - 100$, with $B(0) = 5000$

37. (a) $a_n = a_{n-1} + a_{n-2} + 2^{n-2}$ for $n \geq 2$. (b) $a_0 = 0, a_1 = 0$ (c) 94.

38. (a) $a_n = a_{n-1} + a_{n-2} + a_{n-3}$ for $n \geq 3$ (b) $a_0 = 1, a_1 = 2, a_2 = 4$ (c) 81

39. (a) $a_n = a_{n-1} + a_{n-2}$ for $n \geq 2$ (b) $a_0 = 1, a_1 = 1$ (c) 34.

Answers 4.2

1. (a) $a_n = n\,!$ (b) $a_n = 0$ if n is even $= 2^{n-1}$ if n is odd

(c) $a_n = 1 + 2^n$ (d) $a_n = 1 - 2^n + 3^n$

(e) $a_n = 2^n(1 - n)$ (f) $a_n = 13(5)^n - 10(2)^n$.

2. (a) $a_n = (C_0 + C_7 n)(-1)^n + \frac{1}{6}(2n - 1)$ (b) $a_n = C_0 + C_7 n + 2^n$

(c) $a_n = \left(C_0 + C_1 n + \frac{1}{2}n^2\right) 2^n$ (d) $a_n = C3^n + 2n + 3(1 - 2^{n+1})$

(e) $a_n = (C_0 + C_1 n + C_2 n^2)\, 2^n + 27.3^n$

4. $a_n = 2^n$ **5.** $a_n = -2^n + 2.3^n$

6. $a_n = 2^{n+1} + 3$

10. (a) $a_n = 2(-5)^n$ (e) $a_n = -\frac{1}{2} + \left(-\frac{1}{6}\right)(-1)^n + \frac{5}{3}(2^n)$

(f) $a_n = \frac{1}{4}(2n + 1) + \frac{7}{4}(-1)^n$.

13. (a) $a_n = C_1 3^n + C_2 n^2 + \frac{1}{2}$ (b) $a_n = C_1 2^n + C_2 4^n + 1$

(c) $a_n = C_1 + C_2 n + \frac{1}{2}n^3 + n^2$ (d) $a_n = C_1 2^n + C_2 3^n + \frac{1}{2}n^2 + \frac{3n}{2} + \frac{5}{2}$

(e) $a_n = C_1 + C_2 n + \frac{1}{2}n^3 + n^2$ (f) $a_n = \frac{C_1}{2^n} + C_2 3^n - \left(\frac{4}{3}\right) 2^n$

(g) $a_n = (C_1 + C_2 n)\, 2^n + n^2 \,.\, 2\left(\frac{n}{6} + 1\right)$.

CHAPTER

5

Group Theory

5.1 GROUP

A non empty set G with a binary operation • defined on it constitutes a group (G, o) if the following four properties hold.

(*i*) For all x and y in G, $x \, o \, y$ is in G. (In multiplicative notation one writes xy instead of $x \bullet y$)

(*ii*) There exists an identity element e in G such that $x \, o \, e = e \, o \, x = x$ for all x in G.

(*iii*) Corresponding to each element x in G, there exists an inverse element x^{-1} in G such that $x \, o \, x^{-1} = x^{-1} \, o \, x = e$.

(*iv*) For every x, y and z in G the elements

$x \bullet (y \bullet z)$ and $(x \bullet y) \bullet z$ are identical.

The associativity property (*iv*) allows us to write $x \bullet y \bullet z$ for the triple product. We usually write $a \bullet b$ as ab and (G, o) as G if there is no risk of ambiguity.

5.2 SUBGROUP

A subset H of G is called a subgroup of (G, o), if (H, o) is a group.

5.3 FINITE GROUP

If G is a finite set with $| G | = n$ then (G, o) is a finite group of order n.

For example, the symmetric difference of sets A and B is defined by $A * B = (A \cup B) - (A \cap B)$ that is, $A * B$ is the set of elements that belong to A or to B but not to both.

5.4 PERMUTATION

Suppose that G is a fixed subgroup of the symmetric group of a finite set X and x is a given element of X.

Let $Gx \equiv \{g(x) : g \in G\}$

$G_x \equiv \{g \in G : g(x) = x\}$

$F(g) \equiv \{z \in X : g(z) = z\}.$

In words, Gx (the orbit of x with respect to G) is the set of all images of the given element x under the permutations in G ; G_x (the stabilizer of x in G) is the set of all permutations in G that have x as a fixed point ; F(g) (the permutation character of g in X) is the set of all fixed points of a given permutation $g \in G$.

5.5 PERMUTATION GROUPS AND THEIR CYCLE INDICES

A permutation of a finite set X is a bijective (one-to-one and onto) mapping from X to X. Suppose f is a permutation of X and x is any element of X. Define recursively, $f'(x) \equiv f(x), f^2(x) \equiv f(f'(x)), \ldots f^i(x) \equiv f(f^{i-1}(x))$, ... since X is finite, there exists a positive integer r such that $f^r(x) = x$.

The sequence $\langle x, f^1(x), f^2(x), \ldots\ldots, f^{r-1}(x) \rangle$ is called a cycle of order (or length) r of the permutation f.

Obviously, every permutation of X can be represented as a composition of k disjoint cycles, where k is atleast 1 and atmost the cardinality of X.

The concept of the cycle representation of a permutation f of X = {1, 2, ... n}. The following algorithm produces this representation :

(*i*) Choose an element i of X (usually $i = 1$). Find the image of i under the mapping f, then the image of the image, then ......, until the image j appears such that $f(j) = i$. Thus the cycle (i ...... j) has been generated.

(*ii*) Choose an element of X not found in any one of the cycles already generated, and use this element as element i in step (*i*), thereby generating a new cycle.

(*iii*) Repeat step (*ii*) until X has been exhausted.

The cycle representation of a permutation is unique upto the order of the cycles in the composition and upto the choice, within each cycle, of the leading element.

For example, Given X = {1, 2, ..., 8} and 1 2 3 4 5 6 7 8 $\xrightarrow{f}$ 3 2 5 1 4 8 6 7

Starting with 1 : $f(1) = 3, f(3) = 5, f(5) = 4, f(4) = 1$.

Thus we have a cycle of length 4 which may be denoted by (1 3 5 4) [or (3 5 4 1) or (5 4 1 3) or (4 1 3 5)]

Starting with 2 : $f(2) = 2$. We have the cycle (2) of length 1.

Starting with 6 : $f(6) = 8, f(8) = 7, f(7) = 6$. Now we have a cycle of length 3 which may be denoted by (6 8 7) [or (8 7 6) or (7 6 8)].

The sum of the lengths has reached 4 + 1 + 3 = 8 = | X | which means we are finished : the cycle representation of f is (1 3 5 4) (2) (6 8 7) (or ......).

5.6 WEIGHT

Let the cycle representation of f, a permutation of an n set, consist of a_1 cycles of length 1, a_2 cycles of length 2, ... a_i cycles of length i, ... Then the type of f is the vector $[a_1\ a_2 \ldots\ldots a_n]$, and the weight of the type is the positive integer $W = 1^{a_1} 2^{a_2} \ldots\ldots n^{a_n}$.

For example, the permutation of above example, has 1 cycle of length 1, 1 cycle of length 3, and 1 cycle of length 4. The type of this permutation is [1 0 1 1 0 0 0 0]. The weight of this type is $1' 3' 4' = 12$.

5.7 CYCLE INDEX

Let G denote a group, of order m, of permutations of an n-set and length $g \in$ G be of type $[a_1, a_2, \ldots\ldots, a_n]$. The cycle index of g is the monic multinomial

$$Z(g\,; x_1, x_2, \ldots\ldots, x_n) \equiv x_1^{a_1}\, x_2^{a_2} \ldots\ldots x_x^{a_n}$$ and the cycle index of G is the multinomial.

$$Z(G\,; x_1, x_2, \ldots\ldots, x_n) \equiv \frac{1}{m} \sum_{g \in G} Z\,(g\,; x_1, x_2, \ldots\ldots, x_n)$$

For example, suppose the 4 vertices of a square are labeled 1, 2, 3 and 4, clockwise. A clockwise rotation through an angle of 0°, 90°, 180° or 270° takes the square into itself. Thus there are 4 circular or cyclic symmetries. In addition, there are 4 dihedral symmetries that are obtained by reflection of the square in the 2 diagonals and in the 2 lines bisecting opposite sides.

Conversely, the symmetries of the square compose a subgroup G of order 8 of S_4; the elements of G are as follows :

(*i*) The permutation induced by rotating the square clockwise through 0° is $g_1 = e = (1)\,(2)\,(3)\,(4)$, with cycle index $x_1^{\,4}$.

(*ii*) The permutation induced by rotating the square clockwise through 90° is $g_2 = (1\ 2\ 3\ 4)$ with cycle index $x_4^{\,1}$.

(*iii*) The permutation induced by rotating the square clockwise through 180° is $g_3 = (1\ 3)\,(2\ 4)$, with cycle index $x_2^{\,2}$.

(*iv*) The permutation induced by rotating the square clockwise through 270° is $g_4 = (1\ 4\ 3\ 2)$, with cycle index $x_4^{\,1}$.

(*v*) The permutation induced by reflection in the line joining the midpoints of $\overline{1\,2}$ and $\overline{3\,4}$ is $g_5 = (1\ 2)\,(3\ 4)$, with cycle index $x_2^{\,2}$.

(*vi*) The permutation induced by reflection in the line joining the midpoints of $\overline{1\,4}$ and $\overline{2\,3}$ is $g_6 = (1\ 4)\,(2\ 3)$ with cycle index $x_2^{\,2}$.

(*vii*) The permutation induced by reflection in the diagonal joining corners 2 and 4 is $g_7 = (2)\,(4)\,(1\ 3)$, with cycle index $x_1^{\,2}\,x_2^{\,1}$.

(*viii*) The permutation induced by reflection in the diagonal joining corners 1 and 3 is $g_8 = (1)\,(3)\,(2\ 4)$, with cycle index $x_1^{\,2}\,x_2^{\,1}$.

The cycle index of G is therefore

$$Z\,(G\,;x_1, x_2, x_3, x_4) = \frac{1}{8}\,(x_1^{\,4} + 2x_1^{\,2}\,x_2 + 3x_2^{\,2} + 2x_4).$$

5.8 COLORING AND EQUIVALENT w.r.t. GROUP OF PERMUTATION

A function *f* from a finite set X to a finite set of colors Y is called a coloring of X. Two colorings *f* and ϕ in the set C of all colorings of X are said to be equivalent (indistinguishable) with respect to a group G of permutations of X if there exists a permuation π in G such taht $f(x) = \phi(\pi(x))$ for all x in X.

In other words, if we attach names to the elements of X, so that G may be considered a group of 'renamings', then we do not distinguish between 2 colorings of X that become identical under some renaming in G.

Clearly, the relation of indistinguishability is reflexive, symmetric, and transitive, *i.e.,* an equivalence relation.

5.9 PATTERN

The equivalence classes into which C partitioned by the indistinguishability relation are called the patterns in C (with respect to the group G).

For example, If G = $\{e\}$ then any 2 colorings are distinguishable, so that the number of patterns is the number of colorings. Because

$$Z(G\ ;x_1, x_2, \ldots\ x_n) = Z(e\ ; x_1, x_2, \ldots\ x_n) = x_1^{\ n}.$$

5.10 PATTERN INVENTORY

Let the weight function w map Y into a set of r colors, $\{w(y_1), w(y_2), \ldots\ w(y_r)\}$. The pattern inventory (of C) with respect of G is the multinomial.

PI (G ; $w(y_1), w(y_2), \ldots\ w(y_r)$)

$$\equiv \sum_{\substack{n_1+n_2+\ldots+n_r=n \\ n_i \geq 0}} \tau(n_1, n_2, \ldots, n_r)\,[w(y_1)]^{n_1}\,[w(y_2)]^{n_2} \ldots [w(y_r)]^{n_r}$$

The coefficient $\tau(n_1, n_2, \ldots, n_r)$ gives the number of distinguishable (with respect to G) colorings (= numebr of patterns) that assign color $w(y_1)$ to n_1 elements of X ; color $w(y_2)$ to n_2 elements ; ......, color $w(y_r)$ to n_r elements. The summation is over the sizes of the color classes into which X is divided, the sum consists of $C(n + r - 1, r - 1)$ terms.

5.11 ISOMORPHIC GROUP

Groups (G, o) and (G′, o') are isomorphic (identical in structure) if there exists a one-to-one correspondence f between G and G′ such that $f(x\ o\ y) = f(x)\ o'\ f(y)$, for all x and y in G.

5.12 CYCLIC GROUP

If x is an element in a group (G, o), we write $x\ o\ x$ as x^2, $x\ o\ x^2 = x^2\ o\ x$ as x^3, and so on.

Similary, $x^{-1}\ o\ x^{-1}$ is written as x^{-2}, $x^{-1}\ o\ x^{-2}$ as x^{-3} etc.

Thus the kth power, x^k, of the element x is well defined when k is any non-zero integer, we make the natural definition $x^{\circ} \equiv e$. The group G is said to be a cyclic group if it contains an element x such that every element of G is a power of x. In this case we say that G is generated by x, and we write $G = \langle x \rangle$. If x generates G and if the powers of x are all distinct, G is an infinite cyclic group.

5.13 ABELIAN GROUP

A group (G, o) is abelian if $x\ o\ y = y\ o\ x$ for every x and y in G.

5.14 ORDER OF AN ELEMENT

If x is an element of (G, o) and if there exists a positive integer m such that x^m is the identity element e in G, then x is said to be of finite order. If x is of finite order, the smallest positive integer k such that $x^k = e$ is the order of x in G.

5.15 DIRECT PRODUCT

If G and G′ are two groups, the direct product of G and G′ is the set of all ordered pairs

$$G \times G' = \{g, g'\} : g \in G, g' \in G'\}$$

endowed with the binary operation defined by

$$(g_1, g_1')\,(g_2, g_2') = (g_1g_2, g_1'g_2').$$

5.16 LEFT AND RIGHT COSET

If H is a subgroup of G and x is an element of G, the set $xH \equiv \{xh : h \in H\}$ is called the left coset of H with respect to x, the right coset of H with respect to x is $Hx \equiv \{hx : h \in H\}$.

5.17 CONJUGATE (PERMUTATION)

Two permutations f and g of X are said to be conjugate if there exists a permutation h of X such that $hf = gh$.

5.18 REGULAR ICOSAHEDRON

A regular polytope (a solid in which all faces are congruent polygons and each vertex is incident with the same number of faces) with 12 vertices, 20 faces (congruent equilateral triangles) and 30 edges is called a regular icosahedron.

Theorem 5.1. *Suppose that a finite set X possesses exactly k distinct orbits with respect to a group G of permutations of X. Then*

(*i*) *For every* $x \in X$, $|Gx||G_x| = |G|$

(*ii*) $\Sigma_{x \in X} |G_x| = k|G|$

(*iii*) $\Sigma_{x \in X} |G_x| = \Sigma_{g \in G} |F(g)|$

Proof. (*i*) Let g and h be in G_x. Then $g(h(x)) = g(x) = x$ which implies gh is in G_x.
Therefore G_x is a subgroup of G.

To produce a bijection between Gx and the set L of distinct left cosets of G_x.

Let $u \in Gx$, *i.e.*, $u = g(x)$ for some $g \in G$.

Consider the mapping $u \to g\,G_x$ from Gx to L

(1) The mapping is onto. In fact, if $l\,G_x \in L$, we have l being a permutation of X, $l(x) = y (y \in X)$. This means that $y \in Gx$ and $y \to l\,G_x$.

(2) The mapping is one-one. Let u and v belong to $Gx : u = g(x)$ and $v = h(x)$, for $g, h \in G$.
Suppose that in L, $g\,G_x = h\,G_x$. Then $h^{-1}g \in G_x$ which implies $h^{-1}(g(x)) = x$ or $g(x) = h(x)$ or $u = v$.
Thus our mapping is the desired bijection.

(*ii*) There exists elements $x_1, x_2, ..., x_k$ such that $\{Gx_1, Gx_2, ..., Gx_k\}$ is a partition of X. This let us write $\sum_{x \in X} |G_x| = \sum_{i=1}^{k} \sum_{x \in Gx_i} |G_x|$

But, $|G_x|$ has the constant value $|G_{x_i}|$ over G_{x_i} (since $x_i \in G_{x_i}$)

Hence, $\sum_{x \in X} |G_x| = \sum_{i=1}^{k} |G_{x_i}||G_{x_i}| = \sum_{i=1}^{k} |G| = k|G|$

(*iii*) In the sum $\sum_{g \in G} |F(g)|$ the count of any $x \in X$ is $|G_x|$

Therefore $\sum_{g \in G} |F(g)| = \sum_{x \in X} |G_x|$.

5.19 Theorem 5.2. *Burnside-Frobenius Theorem*

$$\sum_{g \in G} |F(g)| = k|G|$$

5.20 Theorem 5.3. *Pólya's First Enumeration Theorem*

Let C be the set of all functions (colorings) from an n-set X to an r set Y ($n \geq 2$). Let G be a group of permutations of X, with cycle index Z (G ; x_1, x_2, ..., x_n). Then the number of patterns in C with respect to G is Z (G ; r, r, ..., r).

Proof. The patterns in C with respect to G (a permutation group on X) are the distinct orbits in C with respect to G, and these are the distinct orbits in C′ with respect to G′ (a permutation group on C). Their number is given by the Burnside-Frobenius theorem as

$$k = \frac{1}{|G'|} \sum_{\pi' \in G'} |F(\pi')| \qquad ...(1)$$

where $F(\pi') = \{f \in C : \pi'(f) = f\}$

Now, because $\pi'(f) = f$ if and only $f(\pi(x)) = f(x)$ for all $x \in X$ and because $|G'| = |G|$, one can convert (1) back to X and G :

$$k = \frac{1}{|G|} \sum_{\pi \in G} |\{f \in C : f(\pi(x)) = f(x) \text{ for all } x \in X\}| \qquad ...(2)$$

Now, if $f(\pi(x)) = f(x)$ and if $(x_1\ x_2 \ldots\ldots x_j)$ is a cycle of π, $f(x_1) = f(x_2) = \ldots\ldots f(x_j)$ that is, f is constant over each cycle of π.

Conversely, if f is constant over each cycle of π and if $(xx_t \ldots\ldots x_u)$ is the cycle involving the arbitrary element $x \in X$, $f(\pi(x)) = f(x_t) = f(x)$.

It follows that the summa in the right hand side of (2) is just the number of ways of coloring X with $r \geq 2$ colors so that elements in the same cycle of the permutation π are given the same color.

If Π is of type $[a_1\ a_2 \ldots\ldots a_n]$, this number of ways is $r^{a_1 + a_2 + \ldots\ldots + a_n}$;

Equation (2) becomes

$$k = \frac{1}{|G|} \sum_{\pi \in G} r^{a_1 + a_2 + \ldots\ldots + a_n} \equiv \frac{1}{|G|} \sum_{\pi \in G} Z(\pi ; r, r, \ldots\ldots, r) \equiv Z(G ; r, r, \ldots\ldots, r).$$

5.21 Theorem 5.4. *Pólya's second Enumeration Theorem*

The pattern inventory, PI (G ; $w(y_1)$, $w(y_2)$, ..., $w(y_r)$), is the value of the cycle index, Z(G ; x_1, x_2, ..., x_n), at $x_i = [w(y_1)]^i + [w(y_2)]^i + \ldots\ldots + w[w(y_r)]^i$ (i = 1, 2, ... n).

Proof. We note that, the weights function $w(f)$ has the required constancy property needed for an application of the weighted Burnside-Frobenius theorem to the orbits in C with respect to the permutation group G′.

Now, we have $k = \frac{1}{|G'|} \sum_{\pi' \in G'} |F(\pi')|$

$$PI(G\,;\, w(y_1), w(y_2), \ldots\ldots, w(y_r)) = \Sigma w\,(C_i) = \frac{1}{|G'|} \sum_{\pi' \in G'} w(\pi') \qquad \ldots(1)$$

Where $w(\pi') = \sum_{f \in F(\pi')} w(f)$

Convert back to X and G,

$$PI = \frac{1}{|G|} \sum_{\pi \in G} \left\{ \sum_{\substack{f \in C : f(\pi(x)) = f(x) \\ (all\ x)}} [w(f(x_1))][w(f(x_2))] \ldots\ldots [wf(x_n))] \right\} \qquad \ldots(2)$$

The inner summation in (2) may be taken over all functions $f(x)$ that are constant over each cycle of π.

Let π be of type $[a_1\ a_2\ \ldots\ldots\ a_n]$ and define a horrendacs multinomial in the $w(y_i)$ as

$$\Omega \equiv \overbrace{[w(y_1) + w(y_2) + \ldots\ldots + w(y_r)] \ldots\ldots [w(y_1) + w(y_2) + \ldots\ldots + w(y_r)]}^{a_1 \text{ factors}}$$

$$\text{X}\, \overbrace{[w(y_1)^2 + w(y_2)^2 + \ldots\ldots + w(y_r)^2] \ldots\ldots [w(y_1)^2 + w(y_2)^2 + \ldots\ldots + w(y_r)^2]}^{a_2 \text{ factors}}$$

$$\text{X}\, \overbrace{[w(y_1)^3 + w(y_2)^3 + \ldots\ldots + w(y_r)^3] \ldots\ldots [w(y_1)^3 + w(y_2)^3 + \ldots\ldots + w(y_r)^3]}^{a_3 \text{ factors}}$$

$$\text{X} \ldots\ldots\ldots\ldots\ldots\ldots\ldots\ldots\ldots\ldots$$

$$\text{X}\, \overbrace{[w(y_1)^n + w(y_2)^n + \ldots\ldots + w(y_r)^n] \ldots\ldots [w(y_1)^n + w(y_2)^n + \ldots\ldots + w(y_r)^n]}^{a_n \text{ factors}}$$

The expansion of Ω consists of $r^{a_1 + a_2 + \ldots\ldots a_n}$ terms, which number is also the number of functions $f(x)$ that are constant over each cycle of π. The equality is no accident, we now demonstrate that the individual terms in the expansion are precisely the weights w of the individual functions $f(x)$.

Suppose that the cycles in the representation of π are put into one-to-one correspondence with the factors of Ω in the natural way : the one-cycles correspond one-to-one with the first a_1 factors, the 2 cycles, with the next a_2 factors, and so on.

If $f(x)$ maps a given j cycle T into y_v, draw a circle around the quantity $w(y_v)^j = \pi_{x \in T}\ w(f(x))$.

The expansion term given by the product of all circled quantities (one in each factors of Ω) will equal $\pi_U \left[\pi_{x \in U}\ w(f(x)) \right]$ in which U runs through all cycles of π.

But these cycles effect a partition of X, so our expansion term is just $\underset{x \in X}{\pi} w(f(x)) = W(f)$.

We have just proved that the inner sum in (2) has the value Ω. But, by construction.

$\Omega = Z(\pi\,;x_1, x_2, ..., x_n)|_{x_j} = w(y_1)^j + w(y_2)^j + ... + w(y_r)^j \; (j = 1, 2, ... n).$

5.22 Theorem 5.5. *Lagrange's theorem*

The order of a finite group is divisible by the order of any subgroup.

Proof. Let the group be of order n and let a given subgroup, of order S, have r distinct left cosets.

Let x be any element of G. Then x is an element of xH, since $x = xe$ and e is in H.

Thus every element of G is in at least 1 left coset of H. Two distinct left cosets have no elements in common.

Thus the left cosets of H make up a partition of G.

Let $H = \{h_1, h_2, h_3, ... h_k\}$ and let x be any element of G. Then $xH = \{xh_1, xh_2, ... xh_k\}$.

The elements of xH must be distinct, for $xh_i = xh_j$ would imply $h_i = h_j$. Hence $|xH| = k$.

Therefore $rS = n$.

5.23 Theorem 5.6. *Characterization theorem for cyclic groups.*

If G is a group of order $n \geq 2$, the following are equivalent :

(i) G is a cyclic group

(ii) For each divisor d of n, the cardinality of $\{x \in G : x^d = e\}$ is d.

(iii) For each divisor d of n, the cardinality of $\{x \in G :$ the order of x is $d\}$ is $\phi(d)$.

Proof. $(i) \Rightarrow (ii)$

Suppose G is generated by x. Let $n = dk$ and consider the collection $Y \equiv \{x^{\circ}, x^k, x^{2k}, x^{3k}, ..., x^{(d-1)k}\}$.

The d elements in this collection are distinct (because x is of order n). The typical element x^{ik} of Y satisfies $(x^{ik})^d = (x^{dk})^i = (x^n)^i = e^i = e$.

Thus Y is a subset of $\{x \in G : x^d = e\}$.

Conversely, let y be any element of G such that $y^d = e$.

Since x is a generator of G, there exists an integer $0 \leq m \leq n - 1$ such that $y = x^m$.

Therefore $x^{md} = e$.

But x is of order n ; so that, for some integer r,

$$md = rn = rdk \text{ or } m = rk.$$

Thus $y = x^{rk}$, with $0 \leq r \leq d - 1$ (because $r/d = m/n$) which means that $y \in Y$.

Consequently, $\{x \in G : x^d = e\}$ and Y are identical sets. So that $|\{x \in G : x^d = e\}| = |Y| = d$.

$(ii) \Rightarrow (iii)$

Let y be an element of G, of order C. Then $y^d = e$ if and only if $\frac{c}{d}$ (c divides d).

Consequently, $\{x \in G : x^d = e\}$ may be partitioned in such manner that the ith cell consists of all elements of G whose order equals the ith divisor of d. Define $f(c)$ to be the number of elements of order c, and specialize d to a divisor of n. Then by (*ii*),

$$\sum_{\frac{c}{d}} f(c) = d$$

The möbius formula yields $f(d) = \sum_{\frac{c}{d}} \mu(c)\left(\frac{d}{c}\right) = \phi(d)$

(*iii*) $\Rightarrow$ (*i*)

By (*iii*), with $d = n$, there exist $\phi(n) \geq 1$ elements of order n in G. Hence $G = C_n$.

Problem 5.1. *A group (G, o) is abelian if x o y = y o x for every x and y in G. Show that every cyclic group is abelian.*

Solution. Let x and y be 2 elements in a cyclic group G generated by g. Because $x = g^m$ and $y = g^n$ for some integers m and n.

$$x \text{ o } y = g^m \text{ o } g^n = g^{m+n} = g^{n+m} = g^n \text{ o } g^m = y \text{ o } x.$$

Problem 5.2. *Prove that, in any group (a) the identity element is unique ; and (b) the inverse of any element is unique.*

Solution. (*a*) Suppose there existed two identities, e and f.

Then, since e is a right-identity and f is a left identity, $f = f \text{ o } e = e$.

(*b*) If element x had 2 inverses, y and z, the associative law would give

$$(y \text{ o } x) \text{ o } z = y \text{ o } (x \text{ o } z) \text{ or } e \text{ o } z = y \text{ o } e \quad \text{or} \quad z = y.$$

Problem 5.3. *Show that the set of all integers under the binary operation of addition is an infinite cyclic group.*

Solution. If z is the set of all integers, $(z, +)$ is a group because all 4 group axioms are satisfied by the structure.

Let $G = < a >$ be an infinite cyclic group.

The mapping $f : Z \rightarrow G$ defined by $f(z) = a^z$ is obviously a bijection, and we have

$$f(z + w) = a^{z+w} = a^z \text{ o } a^w = f(z) \text{ o } f(w)$$

Therefore the 2 groups are isomorphic, which means that $(z, +)$ is the infinite cyclic group $< 1 >$.

Problem 5.4. *Show that, under an isomorphism,*

(a) the identity elements of G and G′ correspond

(b) if u and v are inverses in G, then f (u) and f (v) are inverses in G′ and

(c) give an example of 2 isomorphic groups of order n.

Solution. (*a*) In G, $x \text{ o } e = e \text{ o } x = x$;

Whence $f(x \text{ o } e) = f(e \text{ o } x) = f(x)$ or $f(x)\ 0'\ f(e) = f(e)\ 0'\ f(x) = f(x)$

which shows that the idenity in G′ is $e' = f(e)$.

(*b*) From $u \circ v = v \circ u = e$ and part (*a*).

$$f(u)\, o'\, f(v) = f(v)\, o'\, f(v) = e'.$$

(*c*) One such pair is composed of the group of rotational symmetries of a regular *n*-gon and the group ({0, 1, 2, ..., *n* – 1}, +), where the operation + is addition modulo *n*.

Problem 5.5. *Show that if H (| x | = k) is a finite subgroup of G, then every left (right) coset of H has cardinality k.*

Solution. Let H = $\{h_1, h_2, h_3, \ldots\ldots, h_k\}$ and let x be any element of G. Then xH = $\{xh_1, xh_2, \ldots\ldots, xh_k\}$.

The elements of xH must be distinct, for $xh_i = xh_j$ would imply $h_i = h_j$.

Hence | xH | = k.

Problem 5.6. *Show that if H is a subgraph of G and if x and y are in G, then either xH ∩ yH is empty or xH = yH.*

Solution. If xH ∩ yH is not empty, there exists an element z which is in xH and also in yH.

Hence there exist h and h' in H such that

$z = xh = yh'$, which in turn implies $y^{-1}x = h'h^1$ is in H, which gives xH = yH.

Problem 5.7. *Show that a subset H of a finite group is a subgroup if and only if H is closed with respect to multiplication.*

Solution. Let x and y be 2 elements in H, and let the order of y be m. Then $y^m = e$ implies $y^{m-1} = y^{-1}$ and by hypothesis, y^{m-1} is in H.

Thus x and y^{-1} are in H.

Problem 5.8. *Show that the class of distinct left cosets of a subgroup H of a group G constitutes a partition of the group.*

Solution. Let x be any element of G. Then x is an element of xH, since $x = xe$ and e is in H.

Thus every element of G is in at least 1 left coset of H.

Two distinct left cosets have no elements in common.

Thus the left cosets of H make up a partition of G.

Problem 5.9. *Prove that a subgroup of a cyclic group is cyclic.*

Solution. Let G′ be a subgroup of G = <*x*>.

Every element in G′ is of the form x^k, let m be the smallest positive k for which x^k is in G′.

Now, for any integer k, the division algorithm gives $k = qm + r$, where $0 \le r < m$.

Hence, $x^k = x^{qm+r} = (x^m)^q\, x^r = ux^x$ in which $u \in$ G′ because $x^m \in$ G′ and G′ is closed under multiplication. It follows that u^{-1} also belongs to G′, so that, if $x^k \in$ G, $x^r = u^{-1}\, x^k \in$ G′.

It r were positive, this would violate the minimality of m.

Therefore, $r = 0$ and each x^k in G′ may be written as $(x^m)^q$; *i.e.*, G′ = < x^m >.

Problem 5.10. *If G is a cyclic group of order* n, *find the number of distinct generators of G.*

Solution. Suppose G = <*x*> = $\{e, x^1, x^2, \ldots\ldots, x^{n-1}\}$.

Let m be any positive integer less than, and relatively prime to, n, consider the cyclic group

$$G' = \langle x^m \rangle = \{e, x^{1m}, x^{2m}, \ldots\ldots\, x^{(n-1)m}\}.$$

To establish that G′ = G it suffices to show that the elements of G′ are distinct.

Suppose, on the contrary, that for some $0 \le b < a \le n - 1$ we had $x^{am} = x^{bm}$. Then necessarily $(a - b)m = cn$ for some integer c.

But, m and n being relatively prime, this equation would require that n divide $a - b$, an impossibility since $0 < a - b < n$.

If, on the other hand, $m = rp$ and $n = sp$ $(r < s < n)$ then $(s^m)^s = (x^n)^r = e$ so that the order of x^m, and consequently the order of $<x^m>$ is smaller than n.

The conclusion is that G has just as many generators as there are positive integers less than, and relatively prime to n, *i.e.*, $\phi(n)$ generators.

Problem 5.11. *If G and G′ are two groups, the direct product of G and G′ is the set of all ordered pairs,* $G \times G' = \{(g, g') : g \in G, g' \in G'\}$ *endowed with the binary operation defined by* $(g_1, g_1')(g_2 \cdot g_2') = (g_1g_2, g_1' g_2')$ *show that* $G \times G'$ *is a group.*

Solution. By definition, the product of 2 elements in $G \times G'$ is in $G \times G'$.

Further, $(g, g')(e, e') = (ge, g'e') = (g, g') = (eg, e'g') = (e, e')(g, g')$.

Thus (e, e') is the identity in the direct product.

Also, $(g, g')(g^{-1}, g'^{-1}) = (g^{-1}, g'^{-1})(g, g') = (e, e')$.

So each element in the direct product has an inverse element. The associativity rule is obviously satisfied in the product structure. Thus $G \times G'$ is a group.

5.24. Theorem 5.7. *The Burnside-Frobenius Theorem with weights*

Suppose that $X_1, X_2, \ldots\ldots, X_k$ *are the distinct orbits in the set* $X = \{x_1, x_2, \ldots\ldots, x_n\}$ *with respect to the permutation group* $G = \{g_1, g_2, \ldots\ldots g_m\}$. *On X define a weight function w(x)-weights may be numbers or algebraic symbols, with the property that whenever* x_r *and* x_s *are in the same orbit,* $w(x_r) = w(x_s)$. *Use the following reçipe to induce a weight function on G :*

$$W(g_i) = \sum_{x \in G(g_i)} w(x) \qquad (i = 1, 2, \ldots\ldots, m)$$

that is, the weight of permutation in G is the total weight of its fixed points in X, then $\sum_{i=1}^{m} w(g_i) = \left(\sum_{P=1}^{k} w_P\right)$ *m in which* w_P $(P = 1, 2, \ldots\ldots, k)$ *is the unique value assumed by w(x) over* X_P.

Proof. Let t be an element of X_P, so that $X_P = Gt$ and $w(t) = w_P$.

By definition of the stabilizer, t contributes its weight to exactly $|G_t|$ summands on the left side of (1)

$$k = \frac{1}{|G|} \sum_{\pi \in G} [\{f \in C : f(\pi(x)) = f(x) \text{ for all } x \in X\}] \qquad \ldots(1)$$

its contribution is thus $|G_t| w_P$. Because $|G_x|$ is constant over X_P, any other element of X_P makes the same cobstribution.

Consequently, the net contribution of X_P to the left side is

$$|G_t| \times |G_t| w_P = mw_P$$

As this weight is precisely reflected in the right side (1) the equation is proved.

5.25 Theorem 5.8. *Cauchy's Formula*

The number of permutations of $X = \{1, 2, ..., n\}$ *that are of type* $[a_1 a_2 ...... a_n]$ *is* $\dfrac{n!}{Wa_1!\, a_2! ...... a_n!}$

where $W = 1^{a_1} 2^{a_2} ..... a^{a_n}$ *is the weight of the type.*

Proof. The number of ways of partitioning X into a_1 cells of cardinality 1, a_2 cells of cardinality 2, ... a_n of cardinality n, is given by

$$N = \frac{n!}{[a_1!(1!)^{a_1}][a_2!(2!)^{a_2}]......[a_n!(n!)^{a_n}]}$$

But a cell is not the same as a cycle.

Infact, a cell with q elements gives rise to $(q - 1)!$ distinct cycles of length $q - 1$ for each circular permutation of the elements. Hence the desired number is

$$N[(1-1)!]^{a_1} [(2-1)!]^{a_2} ...... [(n-1)!]^{a_n} = \frac{n!}{[a_1!1^{a_1}][a_2!2^{a_2}]......[a_n!n^{a_n}]}$$

$$= \frac{n!}{Wa_1!\,a_2!......a_n!}.$$

5.26 Theorem 5.9. *Cayley's Theorem*

Every finite group is isomorphic to a group of permutations.

Proof. The idea behind the proof is very simple, because each element of a group G has its inverse, the rows of the multiplication table for G must be distinct permutations of G. Thus, given the finite group (G, o), where $G = (g_1, g_2, ..., g_m\}$, define m distinct permutations of G by

$$\pi_1(g) = g_1 \, o \, g \quad \pi_2(g) = g_2 \, o \, g \; ...... \; \pi_m(g) = g_m \, o \, g$$

Consider the group (G′, o'), where $G' = \{\pi_1, \pi_2, ..., \pi_m\}$ and where o' denotes multiplication of permutations as defined. The mapping $f : G \to G'$ defined by

$$f(g_i) = \pi_i \; (i = 1, 2, ..., m)$$

is obviously a bijection. Indeed, it is an isomorphism, for, if $g_i \, o \, g_j = g_k$, then for each $g \in G$,

$$\pi_k(g) = g_k \, o \, g = (g_i \, o \, g_j) \, o \, g = g_i \, o \, (g_j \, o \, g)$$

$$= g_i \, o \, \pi_j(g) = \pi_i(\pi_j(g)) = (\pi_i \, o' \, \pi_j)(g)$$

i.e., $\pi_i \, o' \, \pi_j = \pi_k$ (f preserves group multiplication)

Problem 5.12. *Show that there are precisely 17, 824 distinguishable (under rotations) vertex colorings of the regular dodecahedron using 1 or 2 colors.*

Solution. The regular icosahedron will have as its geometric dual a solid with 20 vertices and 12 faces, each of which is a regular pentagon ; this is the regular dodecahedron.

Therefore, the cycle index as

$$Z(G\, ; x_1, x_2, ..., x_{20}) = \frac{1}{60}(x_1^{20} + 15x_2^{10} + 20x_1^2x_3^6 + 24x_5^4)$$

The number of vertex colorings is then

$$Z(G\ ;\ 2, 2, \ldots\ldots, 2) = \frac{1}{60}(1, 069, 440) = 17{,}824.$$

Problem 5.13. *Find the number of ways, under the rotational group, of coloring the vertices and faces of a regular octahedron so that 4 vertices are red, 2 vertices are blue, 4 faces are green, and 4 faces are yellow.*

Solution. Because the vertex coloring and the face coloring are independent, we may treat them separately and then use the product rule.

The cycle index of the group of vertex permutation is

$$\frac{1}{24}\ (x_1^{\,6} + 3x_1^{\,2}x_2^{\,2} + 6x_2^{\,3} + 6x_1^{\,2}x_4 + 8x_3^{\,2})$$

Therefore, the pattern inventory for red (R) and blue (B) is

$$\frac{1}{24}\ [(R + B)^6 + 3(R + B)^2\ (R^2 + B^2)^2 + 6(R^2 + B^2)^3 + 6(R + B)^2\ (R^4 + B^4) + 8(R^3 + B^3)^2]$$

The coefficient R^4B^2 in the pattern inventory is

$$\frac{1}{24}\left[\frac{6!}{4!2!} + 3(3) + 6\left(\frac{3!}{2!1!}\right) + 0\right] = 2$$

The cycle index of the group of face permutation is

$$\frac{1}{24}\ (x_1^{\,8} + 9x_2^{\,4} + 8x_1^{\,2}x_3^{\,2} + 6x_4^{\,2})$$

The pattern inventory for green (G) and yellow (Y) is

$$\frac{1}{24}\ [(G + Y)^8 + 9(G^2 + Y^2)^4 + 8(G + Y)^2\ (G^3 + Y^3)^2 + 6(G^4 + Y^4)]$$

The coefficient of G^4Y^4 in the pattern inventory is

$$\frac{1}{24}\left[\frac{8!}{4!4!} + 9\left(\frac{4!}{2!2!}\right) + 8(4) + 6(2)\right] = 7$$

Thus there are (2)(7) = 14 ways of coloring.

Problem 5.14. *Find the number of distinguishable necklaces consisting of 7 stones, of which 2 stones are red, 3 stones are blue, 2 stones are green, when (a) only rotational symmetrices (of a regular polygon with 7 vertices) are considered ; and (b) both rotational and reflectional symmetrices are considered.*

Solution. (*a*) The group here is cyclic and of prime order

$$Z\ (C_7\ ;\ x_1, x_2, \ldots\ldots, x_7) = \frac{1}{7}\ (x_1^{\,7} + 6x_7)$$

$$\text{PI}(C_7; R, B, G) = \frac{1}{7}[(R+B+G)^7 + 6(R^7B^7 + G^7)]$$

The number we seek $\tau(2, 3, 2)$ will be $\frac{1}{7}$ times the coefficient of $R^2B^3G^2$ in $(R + B + G)^7$.

The multinomial theorem gives

$$\tau(2, 3, 2) = \frac{1}{7}\frac{7!}{2!3!2!} = 30.$$

(*b*) The group is the dihedral group H_{14}.

$$\text{PI}(H_{14}; R, B, G) = \frac{1}{2}\text{ PI}(C_7; R, B, G) + (R+B+G)(R^2+B^2+G^2)^3$$

using the result of (*a*), we have

$$\tau(2, 3, 2) = \frac{1}{2}(30) + \frac{1}{2}\left(0 + \frac{3!}{1!1!1!} + 0\right) = 18.$$

Problem 5.15. *Let $X = \{1, 2, 3, 4\}$, $Y = \{y_1, y_2\}$; $w(y_1) = R$, $w(y_2) = B$; and $G = \{(1)(2)(3)(4), (1\ 2)(3\ 4), (1\ 3)(2\ 4), (1\ 4)(2\ 3)\}$.*

Find the pattern inventory for the set C of all functions from X to Y.

Solution. The cycle index is $\frac{1}{4}(x_1^4 + 3x_2^2)$.

By pólya's first theorem, with $r = |Y| = 2$, the number of pattern in C is

$$k = \frac{1}{4}(2^4 + 3 \,.\, 2^2) = 7.$$

To visualize these 7 patterns, it is helpful to have a concrete model of X and G.

Fortunately, we have several available. If X is identified with the vertex set of the square, then G will be the subgroup $\{g_1, g_3, g_5, g_6\}$ of the full symmetry group D_8.

Now, there are 5 possible values of the assignment vector (n_1, n_2) : (0, 4) (1, 3) (2, 2) (3, 1) (4, 0)

Obviously, (0, 4) and (4, 0) each determine a single pattern (there's only one way to paint all vertices the same color) ; the respective weights of these patterns are $w(C_1) = B^4$ and $w(C_2) = R^4$.

Similarly, (1, 3) and (3, 1) generate 1 pattern apiece (there is a reflection or rotation in G that will give the odd-colored vertex any desired location) ; $w(C_3) = RB^3$, $w(C_4) = R^3B$.

By elimination (2, 2) must give rise to 7 – 4 = 3 patterns, with

$$w(C_5) = w(C_6) = w(C_7) = R^2B^2$$ (as shown figure below)

Thus, $\text{PI}(G; R, B) = \sum_{i=1}^{7} w(C_i)\ R^4 + R^3B + 3R^2B^2 + RB^3 + B^4.$

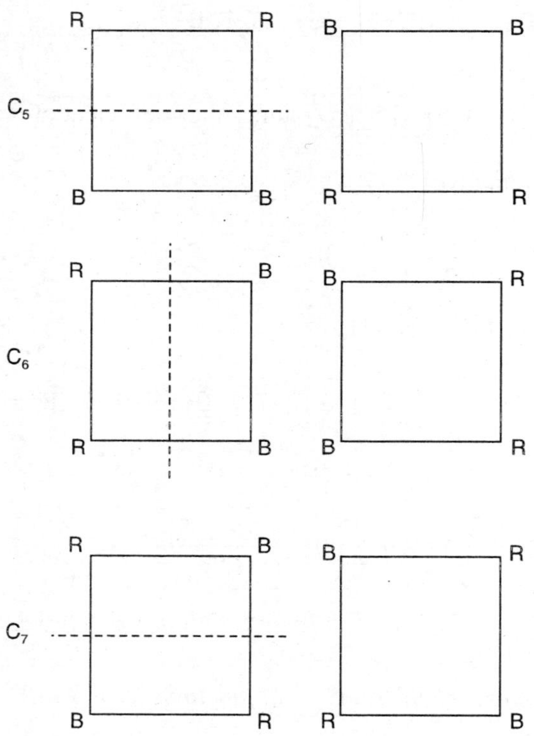

Fig. 1.

Problem 5.16. *Let C be the (finite) set of all functions f from a finite set X to a finite set Y, and let G be a group of permutation of X. For each π in G, define a mapping π′ from C to C by* $\pi'(f(x)) = f(\pi(x))$ *(for each* $x \in X$ *and each* $f \in C$*)*

Prove that (a) *π′ is a permutation of C, and*

(b) $G' = \{\pi' : \pi \in G\}$ *is a group.*

Solution. (*a*) If $\pi'(f_1) = \pi'(f_2)$, then $f_1(\pi(x)) = f_2(\pi(x))$ for every $x \in$ X, which implies that $f_1(t) = f_2(t)$ for every $t \in$ X.

So $f_1 = f_2$ (π′ is injective). Also, π′ is surjective, infact, for any $f \in$ C,

$$f(x) = f(\pi(\pi^{-1}(x))) = \pi'(f(\pi^{-1}(x))) \equiv \pi'((f\pi^{-1})(x))$$

Hence, as a bijection, π′ is a permutation of C.

(*b*) To show that G′ is closed with respect to multiplication (composition).

Let π_1 and π_2 in G respectively determine π_1' and π_2' in G′.

Our assertion is that $\pi_1\pi_2$ in G determines $\pi_1'\,\pi_2'$ in G.

i.e.,
$$(\pi_1\pi_2)' = \pi_1'\,\pi_2'.$$

$$(\pi_1\pi_2)'(f(x)) = f(((\pi_1\pi_2)(x)) = f((\pi_1(\pi_2(x)))$$
$$= \pi_1'(f(\pi_2(x))) = \pi_1'(\pi_2'f(x))) = (\pi_1'\,\pi_2')(f(x)).$$

Problem 5.17. *Let G be a group of permutations of $X = \{x_1, x_2, ... x_n\}$ and let C be the set of all functions from X to $Y = \{y_1, y_2, ... y_n\}$. If w(y) is a given weight function on Y, we induce a weight function w(f) on C by the formula*

$$w(f) = [w(f(x_1))]\,[w(f(x_2))] \;......\; [w(f(x_n))].$$

(a) If f and ϕ in C are equivalent with respect to G, prove that $w(f) = w(\phi)$.

(b) Denote by $C_1, C_2 ..... C_k$ the distinct patterns in C ; let $w(C_i)$ (i = 1, 2, ..., k) stand for the constant value of w over C_i. Show that the pattern inventory of C can be expressed as

$$PI\,(G\,;\, w(y_1), w(y_2), ..., w(y_r)) = \sum_{i=1}^{k} w(C_i).$$

Solution. (*a*) Since *f* and ϕ are equivalent, there exists a permutation π of *x* such that $f(x) = \phi(\pi(x))$ for all *x* in X.

Therefore $$\begin{aligned} w(f) &= [w(f(x_1))]\,[w(f(x_2))] \;.....\; [w(f(x_n))] \\ &= [w(\phi(\pi(x_1)))]\,[w(\phi(\pi(x_2)))] \;......\; [w(\phi(\pi(x_n)))] \\ &= [w(\phi(x_1'))]\,[w(\phi(x_2'))] \;......\; [w\phi(x_n'))] = w(\phi). \end{aligned}$$

(*b*) In a permutation of colored objects, the numbers of red objects of green objects, etc. clearly do not change. It follows that all colorings *f* making up a given pattern C_i in C are characterized by the same "assignment vector" $(n_1, n_1, ..., n_r)$.

This means that any $f \in C_i$ maps n_1 elements of X into y_1, n_2 elements into y_2, ..., n_r elements into y_r, so that the weight of C_i is

$$w(C_i) = w(f) = [w(y_1)]^{n_1}\,[w(y_2)]^{n_2} \;......\; [w(y_r)]^{n_r}$$

Now, in the definition of the pattern inventory, the coefficient $\tau\,(n_1, n_2, ..., n_r)$ is defined to be the number of pattern answering to the vector $(n_1, n_2, ..., n_r)$.

Hence we can write.

$$\sum_{i=1}^{k} w(C_i) \equiv \text{total weight of the } k \text{ patterns}$$

$$= \sum_{(n_1, n_2, ...\, n_r)}^{k} [\text{total weight of patterns answering to } (n_1, n_2, ..., n_r)]$$

$$= \sum_{(n_1, n_2, ...\, n_r)} \tau\,(n_1, n_2, ...\, n_r)\,[w(y_1)]^{n_1}\,[w(y_2)]^{n_2} \;......[w(y_r)]^{n_r}$$

$$\equiv \text{PI}(G\,;\, w(y_1), w(y_2), ...\, w(y_r)).$$

Problem 5.18. *(a) How may one define a cycle index for an orbitrary finite group ?*

(b) Illustrate the procedure for the group of subsets of $X = \{a, b\}$ under the symmetric difference.

Solution. (*a*) Take the cycle index of the permutation group G′ to be the cycle index of the abstract group G.

(*b*) The multiplication table for G = {ϕ, {*a*}, {*b*}, X} is

*	ϕ	{*a*}	{*b*}	X	
ϕ	ϕ	{*a*}	{*b*}	X	$\to \pi_1$: (*) ({*a*})({*b*})(X)
{*a*}	{*a*}	ϕ	X	{*b*}	$\to \pi_2$: (ϕ{*a*}) ({*b*}X)
{*b*}	{*b*}	X	ϕ	{*a*}	$\to \pi_3$: (ϕ{*b*})({*a*}X)
X	X	{*b*}	{*a*}	ϕ	$\to \pi_4$: (ϕX)({*a*}{*b*})

To the right of each row is shown the cycle representation of the permutation is G′ generated by that row.

These 4 permutations have the respective cycle indices

$x_1^4, x_2^2, x_2^2, x_2^2$ hence

$$Z = (G ; x_1, x_2, x_3, x_4) = Z(G' : x_1, x_2, x_3, x_4) = \frac{1}{4}(x_1^4 + 3x_2^2).$$

Problem 5.19. *Obtain the cycle index of the group of permutations of the 6 edges induced by the rotational symmetrices of the regular tetrahedron.*

Solution. The tetrahedron is ABCD (vertices), with edges AB, AC, AD,BC, BD, and CD marked as 1, 2, 3, 4, 5 and 6 respectively. The 12 rotational symmetries listed the following edges permutations.

(*i*) *e* = (1)(2)(3)(4)(5)(6)

(*ii*) (1 2 3)(4 6 5) and (1 3 2)(4 5 6) about vertex A, and a similar pair for each of the other 3 vertices.

(*iii*) (1)(6)(2 5)(3 4) and 2 similar permutations.

Hence $Z(G ; x_1, x_2, ... x_6) = \frac{1}{12}(x_1^6 + 3x_1^2x_2^2 + 8x_3^2).$

Problem 5.20. *In a military mess the food trays are rectangular and divided into 4 equal rectangular compartments. Find the number of distinguishable ways of filling a tray with 4 foods if the long dimension must be parallel to the table edge.*

Solution. Label the corners of the tray 1, 2, 3 and 4 (clockwise), where $\overline{1\,2}$ is a longer side. The symmetry group G of the rectangle is composed of the following permutations :

(1) (2) (3) (4)	[the zero rotation]
(1 3) (2 4)	[180° rotation]
(1 2) (3 4)	[reflection in perpendicular axis]
(1 4) (2 3)	[reflection in parallel axis]

Hence, $Z(G ; x_1, x_2, x_3, x_4) = \frac{1}{4}(x_1^4 + 3x_2^2)$ and

$$Z(G ; 4, 4, 4, 4) = \frac{1}{4}(4^4 + 3.4^2) = 76 \text{ ways.}$$

Problem 5.21. *Two permutations f and g of X are said to be conjugate if there exists a permutation h of X such that hf = gh. Show that 2 permutations are conjugate if and only if they are of the same type.*

Solution. If f and g are conjugate permutations, there exists a permutation h such that $g = hfh^{-1}$.

Suppose $C = (x_1 x_2 \ldots\ldots x_r)$ is a cycle of f, of length r.

Then $\quad f(x) = x_2 \qquad f(x_2) = x_3 \ldots\ldots f(x_r) = x_1$

Let $h(x_i) = y_i$ for each i. Then

$$g(y_i) = h(f(h^{-1}(y_i)) = h(f(x_i)) = h(x_{i+1}) = y_{i+1}$$

in which the subscript is to be evaluated modulo r.

Thus every cycle of length r corresponds to a cycle of g of length r, and vice versa. So f and g are of the same type.

On the other hand, assume that f and g are of the same type, and let $C = (x_1 x_2 \ldots\ldots x_r)$ be a cycle of f.

Then g has a cycle of the form $C' = (y_1 y_2 \ldots\ldots y_r)$.

Define $h(x_i) = y_i$ over C and similarly over every other cycle of f, this makes h a bijection from X to X, or a permutation of X. We have

$$h(f(x_i)) = h(x_{i+1}) = y_{i+1} = g(y_i) = g(h(x_i))$$

so f and g are conjugate.

Problem 5.22. *Find the number of ways of coloring the corners of a regular pentagon using 3 colors if indistinguishability is with respect to the subgroup of rotational symmetries.*

Solution. There are $3^5 = 243$ ways of coloring the 5 corners if rotational symmetries are ignored, thus we have a set C of 243 elements.

The group G′ of rotational symmetries has 5 elements.

Let g_1' be the identity, then $F(g_1')$ has 243 elements.

The other rotations will preserve a color configuration if and only if it involves a single color.

This means that $F(g_2')$, $F(g_3')$, $F(g_4')$, $F(g_5')$ have 3 elements each.

Thus, the number of colorings is $\left(\frac{1}{5}\right)(243 + 12) = 51$.

Problem 5.23. *Show that if G is a finite group of order n, then all elements of G are of finite order and no order exceeds n.*

Solution. If $x \in G$, the elements in $\{x^k : k = 0, 1, \ldots\ldots\}$ cannot be all distinct, since G is finite.

Hence there must exist integers p and q, where $p > q \geq 0$, such that $x^p = x^q$, which implies $x^{p-q} = e$. So, x is of finite order, say, k.

Because x generates a subgroup of order k, $k \leq n$.

Problem 5.24. *A finite cyclic group of order m is denoted by C_m. Show that if m and n are relatively prime, the direct product $C_m \times C_n$ is a cyclic group.*

Solution. As the direct product is a finite group of order mn, it suffices to prove that it contains an element of order mn.

Let C_m be generated by x and C_n be generated by y, and let k be the order of the element $Z = (x, y)$ of the direct product.

Then $Z^k = (e, e') = (x^k, y^k)$, which implies that k is a multiple of m and a multiple of n.

Since k is the smallest positive integer P such that $Z^P = (e, e')$, it follows that k is the least common multiple of m and n.

But, if m and n are relatively prime, their least common multiple is their product.

Thus the order of Z is mn.

Problem 5.25. *If $f = (1\ 2\ \ldots\ldots\ n)$ is a permutation of $X = \{1, 2, \ldots\ldots\ n\}$ then the cyclic group (of permutations) $G = <f>$ is of order n. Prove that in the cycle representation of any element of G all cycles are of same length.*

Solution. The type of $f^\circ = e$ is $[n\ 0\ 0\ 0\ \ldots\ldots\ 0]$.

Let $m = m(i)$ be the length of the shortest cycle in the cycle representation of f^i $(1 \le i \le n-1)$ and let x be an element in some cycle of f^i of length m. Then $f^{im}(x) = (f^i)^m(x) = x$.

Now, if $y \in X$, both x and y belong to the same cycle of the permutation $f = (1\ 2\ \ldots\ldots\ n)$. This implies that there exists r such that $f^r(x) = y$.

Consequently, $(f^i)^m(y) = f^{im} f^r(x) = f^r f^{im}(x) = f^r(x) = y$. So the element y belongs to a cycle in f^i whose length divides m. But m is the length of the shortest cycle in f^i. Thus, every cycle in f^i is of length $m(i)$, which common length must be a divisor of n.

Problem 5.26. *If $X = \{1, 2, 3, 4\}$ and $G = \{g_1, g_2, g_3, g_4\}$ is a group of permutations of X, where*

$$g_1 = (1)(2)(3)(4) \qquad g_3 = (1)(2)(3\ 4)$$
$$g_2 = (1\ 2)(3)(4) \qquad g_4 = (1\ 2)(3\ 4)$$

find the cycle index of G.

Solution. Since X has 4 elements, the index has 4 variable x_i, where $i = 1, 2, 3, 4$. The element g_1 has 4 cycles of length 1 ; so its contribution is $x_1^{\,4}$.

Both g_2 and g_3 have 2 cycles of length 1 and 1 cycle of length 2, their contribution is $2x_1^{\,2}x_2$.

The contribution of g_4 is $x_2^{\,2}$. Thus the cycle index of the group is $(|\,G\,| = 4)$

$$Z = (G\ ; x_1, x_2, x_3, x_4) = \frac{1}{4}(x_1^{\,4} + 2x_1^{\,2}x_2 + x_2^{\,2}).$$

Problem 5.27. *The length of a stick is n feet. The individual feet are marked consecutively 1, 2, 3, ... n. The only symmetries are rotations about the center through 0° and 180°. Obtain the cycle index of this permutation group.*

Solution. Here $G = \{e, g\}$

(*i*) If $n = 2k$, $g = (\overline{1\ 2k})\ (\overline{2\ 2k-1})\ (\overline{3\ 2k-2})\ \ldots\ldots\ (\overline{k\ k+1})$

and so the cycle index of $\frac{1}{2}\,(x_1^{\,2k} + x_2^{\,k})$.

(*ii*) If $n = 2k + 1$, $g = (\overline{1\ 2k+1})\ (\overline{2\ 2k})\ \ldots\ldots\ (\overline{k\ k+2})\ (k+1)$

and so the cycle index is $\frac{1}{2}\,(x_1^{\,2k+1} + x_1x_2^{\,k})$.

Problem Set 5.1

1. Let G denote the set of all 2×2 matrices such that the first row is [1 m], where m is an integer, and the second row is [0 1]. Show that G is an infinite cyclic group under matrix multiplication. Find the generator of this group.
2. Prove that a subgroup of a cyclic group is cyclic.
3. Given a finite set X and a group G of permutation of X, prove that the distinct orbits with respect to G constitute a partition of X.
4. Show that the Burnside-Frobenius theorem holds for $X = \{a, b, c, d\}$ and $G = \{g_1, g_2, g_3, g_4\}$, where g_1 maps each element into itself ; g_2 maps a and b into each other and c and d into each other, g_3 maps a and c into each other and b and d into each other, g_4 maps a and d into each other and b and c into each other.
5. From the Burnside-Frobenius theorem, obtain the number of ways of seating n people around a circular table.
6. Use the Burnside-Frobenius theorem to find the number of distinguishable colorings, with respect to the symmetry group of the square, of a 3×3 chessboard if 2 cells must be colored black and the others white.
7. Show that if f and g are permutations, fg and gf are of the same type.
8. Prove that conjugate permutations have the same number of fixed points.
9. If $A = \{\alpha, \beta, \gamma, \delta\}$ and $H = \{h_1, h_2, h_3, h_4\}$ is a group of permutations of A, where

$$h_1 = (\alpha)(\beta)(\gamma)(\delta) \qquad h_3 = (\alpha\gamma)(\beta\delta)$$
$$h_2 = (\alpha\beta)(\gamma\delta) \qquad h_4 = (\alpha\delta)(\beta\gamma)$$

find the cycle index of H.
10. Find the cycle index of the group of face permutations induced by the rotational symmetries of the cube.
11. Find the cycle index of the group of vertex permutations induced by the rotational symmetries of the cube.
12. Display the complete group of symmetries of a regular $(2m + 1)$-gon.
13. Display the complete group of symmetries of a regular $2m$-gon.
14. Obtain the cycle index of the dihedral group H_{2n}.
15. A regular tetrahedron has 4 vertices, 4 faces (congruent equilateral triangles), and 6 edges. Find the cycle index of the group of permutations of the 4 vertices (or 4 faces) induced by the rotational symmetries of the regular tetrahedron.
16. Let G be the set of all 3×3 matrices A that have [1 a b] as the first row, [0 1 c] as the second row and [0 0 1] as the third row, the numbers a, b and c are elements of the set [0, 1, 2]. If scalar addition and multiplication are modulo 3, show that G is a group under ordinary matrix multiplication. Determine the cycle index of G.
17. A regular polytope (a solid in which all faces are congruent polygons and each vertex is incident with the same number of faces) with 12 vertices, 20 faces (congruent equilateral triangles) and 30 edges is called a regular icosahedron. Identify the rotational symmetries of this solid, and obtain the cycle indices of the groups of (a) vertex permutations and (b) face permutations.

18. With respect to the β rotational symmetries of a cube, in how many ways can the faces be painted red, blue, or green, if each color must be used at least once ?

19. Find the number of ways, under the rotational group of coloring the vertices and faces of a regular octahedron so that 4 vertices are red, 2 vertices are blue 4 faces are green, and 4 faces are yellow.

20. Find the number of inequivalent ways of seating 2 men, 2 women, and 1 child at a round dining table.

21. If $G = <x>$ is a cyclic group of order 12, list the orders of x^k for $k = 0, 1, 2, ..., 11$.

22. With respect to the group of rotational symmetries of the cube, in how many ways can 6 edges be colored red and the remaining 6 blue ?

23. Prove that, for every integer r, $r^2(r^2 + 11) \equiv (\text{mod } 12)$.

24. Find the number of (rotationally) distinct ways of painting the faces of a regular dodecahedron in 3 or fewer colors.

25. How many distinct 7-horse merry-to-rounds are there with 2 red horses, 3 white horses, and 2 blue horses ?

26. Find the number of distinguishable ways of coloring the cells of a 3×3 chessboard so that 2 cells are red, 4 cells are white, and 3 cells are blue.

27. Show that the stabilizer G_x is a subgroup of G.

28. Prove that $r^8 + r^4 + 2r^2 + 4r$ is divisible by 8, for all positive integers r.

29. Express the permutation (1 2 3 4)(5 6 7)(1 6 7 2 9)(3 4) as a product of disjoint cycles.

30. Find the cycle index of the group of edge permutations indcued by the rotational symmetries of the cube.

31. A complex number θ is a primitive n^{th} root of unity if $\theta^n = 1$, but $\theta^k \neq 1$ for $k = 1, 2, ..., n - 1$. Count the primitive n^{th} roots of unity.

32. Use the Burnside-Frobenius theorem to find the number of distinguishable ways to colorings the sides of a square using 2 colors.

33. If $x = (1\ 3\ 5\ 7)(2\ 4\ 6)$ and $y = (1\ 2\ 3\ 4\ 5)$, find the order of xy.

34. Find the number of permutations of type [3 1 0 0 0].

35. Evaluate the cycle indices of the dihedral groups (*a*) H_{12} and (*b*) H_{14}.

36. If f is the permutation that maps 1, 2, 3, 4, 5, 6, 7, 8 and 9 into 9, 8, 5, 4, 1, 6, 3, 2 and 7. Write the disjoint-cycle representation of f.

37. Find the number of ways, under the rotational group, of coloring a regular tetrahedron so that 2 vertices are red, 2 vertices are blue, 2 faces are green, 2 faces are yellow, 3 edges are black, 3 edges are white.

38. Find the number of (rotationally) distinct ways of painting the faces of a regular icosahedron so that 4 faces are red and the other faces are blue.

39. If the vertices of a square are painted in 3 or fewer colors, in how many patterns will 2 vertices be of 1 color and 2 of another color.

40. Find the number of distinguishable necklaces with 10 stones of at most 2 colors.

41. If $x = (a\ b\ c\ d)$ and $y = (b\ d)$, express x^2, x^3, x^4, xy and x^2y as products of disjoint cycles.

42. Find the number of (rotationally) distinct ways of coloring the vertices of a cube using at most 3 colors.

43. If $x = (1\ 2\ 3\ 4\ 5\ 6\ 7\ 8)$ find the groups generated by x^2 and x^4.

44. Find the number of (rotationally) distinct ways of painting the faces of a cube using 6 colors so that each face is of a different color.

Answers 5.1

5. $k = \dfrac{n!}{n} = (n-1)!$

6. $k = 8$

9. $Z(H\,;x_1, x_2, x_3, x_4) = \dfrac{1}{4}(x_1^4 + 3x_2^2)$

12. $Z(G\,;x_1, x_2, \ldots\ldots x_6) = \dfrac{1}{24}(x_1^6 + 3x_1^2x_2^2 + 6x_2^3 + 6x_1^2x_4 + 8x_3^2)$

11. $Z(G\,;x_1, x_2, \ldots\ldots, x_8) = \dfrac{1}{24}(x_1^8 + 9x_2^4 + 8x_1^2x_3^2 + 6x_4^2)$

12. $\{e, f, f^2, \ldots\ldots, f^{2m}, h, hf, hf^2, \ldots\ldots, hf^{2m}\}$

13. $\{e, f, f^2, \ldots\ldots f^{2m-1}, g, gf, gf^2, gf^{2m-1}\}$

14. $Z(H_{2n}\,;x_1, x_2, \ldots\ldots, x_n\} = \begin{cases} \dfrac{1}{2n}(U+V) & n \text{ even} \\ \dfrac{1}{2n}(U+V') & n \text{ odd} \end{cases}$

15. $Z(G\,;x_1, x_2, x_3, x_4) = \dfrac{1}{12}(x_1^4 + 8x_1x_3 + 3x_2^2)$

16. $Z(G\,;x_1, x_2, \ldots\ldots x_{27}) = \dfrac{1}{27}(x_1^{27} + 26x_3^9)$

17. (*a*) $Z(G\,;x_1, x_2, \ldots\ldots x_{12}) = \dfrac{1}{60}(x_1^{12} + 15x_2^6 + 20x_3^4 + 24x_1^2x_5^2)$

(*b*) $Z(G\,;x_1, x_2, \ldots\ldots x_{20}) = \dfrac{1}{60}(x_1^{20} + 15x_2^{10} + 20x_1^2x_3^6 + 24x_5^4)$

18. 30

19. 7

20. 21

21. 1, 12, 6, 4, 3, 12, 2, 12, 3, 4, 6, 12

22. 48

24. 9099

25. 30

26. 174

29. (1 7 3)(2 9)(5 6)

30. $Z(G\,;x_1, x_2, \ldots\ldots x_{12}) = \dfrac{1}{24}(x_1^{12} + 6x_1^2x_2^5 + 3x_2^6 + 8x_3^4 + 6x_4^3)$

31. $\phi(n)$

32. 21

33. 12

34. 10

35. (*a*) $\dfrac{1}{12}(x_1^6 + 3x_1^2x_2^2 + 4x_2^3 + 2x_3^2 + 2x_6)$ (*b*) $\dfrac{1}{14}(x_1^7 + 7x_1x_2^3 + 6x_7)$

36. (1 9 7 3 5)(2 8)(4)(6)

37. 4

38. 96

39. 6

40. 78

41. (*a c*)(*b d*), (*a d c b*), (*a*)(*b*)(*c*)(*d*), (*ad*)(*bc*), (*ac*)(*b*)(*d*)

42. 333

43. $\{x^2, x^4, x^6, x^8\}, \{x^4, x^8\}$

44. 30.